Introduction to Matrix Analysis for Engineering and Science

Carl C. Cowen

West Pickle Press
707 Crestview Place
West Lafayette, Indiana 47906–2313
765–743–2782

Fifth Preliminary Edition
ISBN 0-9650717-6-6

Contents

Matrix analysis and linear algebra are second only to calculus in terms of mathematics of importance to engineering and scientific applications. In a typical application, the problem is formulated mathematically, it is then converted to a matrix analysis problem (possibly without your knowing it), the matrix problem is solved using a computer, and, finally, the results are interpreted. For example, many numerical routines for solving differential equations change the problem into a matrix analysis problem first.

This is a book describing applicable mathematics. While there are a few projects involving applications, the direct applications will not be the main focus of the book. Many of you already know or will soon know the application material and need to better understand the tools. The goal of the book is to provide the background you need to recognize the matrix analysis problems that occur in your area of expertise and to solve them or to be able to understand someone else's solution.

Throughout the book, we will remain conscious of the reliance on computers for real world computation. This is not a book on numerical linear algebra but there will be occasional mention of MATLAB approaches to problems and some exercises will be inconvenient to do without a machine of some kind. MATLAB is a scientific software package with a matrix analysis foundation developed, maintained, and marketed by THE MATHWORKS, Natick, MA (508-653-1415). Because it is widely used in the engineering community and increasing used in university engineering, science, and mathematics curricula, MATLAB is specifically described when machine computation issues come up.

However, most of the exercises are designed for paper and pencil computation. The exercises are designed to reinforce the learning of the material: this can only be effective when you reconsider the background for each problem as you work it. In many situations, a paper and pencil approach will be different from a machine computation approach and thinking about both approaches will strengthen your understanding. Even if you have found the theory to be an expendable part of your mathematical work in the past, it will be an essential part in matrix analysis because the problem set up and the understanding of the answer are both thoroughly grounded in the theory. Exercises that are especially suited to machine computation are marked with a $\Diamond$.

The importance of computer computation will be most apparent in the choice of topics for the book. In many situations in matrix analysis, the obvious method is not the one used in practice because it is too prone to error or too time consuming. I will always try to indicate the practical algorithms for solving linear algebra problems, and one of the goals of the book is to make it possible for you to read the documentation for the matrix analysis software packages you have access to and make appropriate choices from the available methods.

While concentrating on the applicable side of the subject, the book includes proofs that illuminate the subject and proofs are included in the exercises to help you master the theory. Linear transformations and abstract vector spaces are (barely) defined but are not covered in depth or very explicitly. An effort is made in the text to prepare for understanding these ideas later and mention is made of facts like the orthogonality underlying the theory of Fourier series that may be encountered later. The focus of the book is on $\mathbf{R}^n$ (and $\mathbf{C}^n$ to some extent) and its subspaces and on matrices. Although many students at this level are uncomfortable with the complex numbers, they are mentioned occasionally in the first few chapters and more often in Chapters 6 and 7 because of their importance in some engineering problems and in understanding the eigenvalues. Appendix A develops complex arithmetic for students needing review.

There are two equivalent ways to implement the Euclidean inner product on $\mathbf{C}^n$: conjugate linear in the first variable or conjugate linear in the second variable. Engineers tend to favor the first and mathematicians tend to favor the second. This book uses the engineers' convention! For all matrices, A' is used for the conjugate transpose, the adjoint, of the matrix (MATLAB represents the conjugate transpose by A'). Conjugate linear in the first variable makes natural the very useful identification of the number $\langle u, v \rangle$ with the 1×1 matrix $u'v$. Of course, mathematicians and engineers agree on the Euclidean inner product on $\mathbf{R}^n$, and in this case, the conjugate transpose is just the transpose.

Reduced row echelon form is developed but its use in the solution of systems is not encouraged, even though it is, of course, implicitly present in the Gaussian elimination presented for hand calculation. The function \ is the equation solver in MATLAB: to solve the system $AX = b$, type X=A\b. For $n \times n$ systems, MATLAB assumes the system will have a unique solution and calculates it with Gaussian elimination (with partial pivoting). For $m \times n$ systems, MATLAB calculates the least squares solution. In the book, the fact that MATLAB finds only one solution to a system of linear equations, and not always an actual solution, is used as a tool to emphasize the theory of solutions and ultimately an understanding of the least squares solution of systems. Chapters 2 and 3 are devoted to the solution of systems of equations and the theoretical foundation (subspace, basis, rank, etc.) for this study.

In contrast, the Euclidean inner product and QR–factorization (a matrix factorization form of the Gram–Schmidt algorithm) are more prominent than in many books for mathematics majors. These ideas are much relied on in engineering partly because the scalars are nearly always real or complex and partly because machine computations using orthogonality are more accurate than those based on not necessarily orthogonal bases. A substantial effort is devoted to understanding the inner product, the QR–factorization, and their applications to least squares, basis calculations, and rank determination. Although not presented in this book, the students will be prepared for the singular value decomposition (SVD) if they see it later. The study of the inner product and its applications comprise Chapters 4 and 5 of the book.

The final two chapters of the book are devoted to the study of eigenvectors and eigenvalues, their application to solution of systems of linear differential equations, and the diagonalization of Hermitian matrices. Section 6.6 presents the matrix exponential function and bases the study of differential equations on the matrix generalization of the easy equation $\dfrac{dy}{dt} = ay$ that most students study in calculus. It can replace much of Section 6.3 which has the more traditional *ad hoc* approach. Some instructors view the matrix exponential as too abstract while others view it as more natural.

Nearly every chapter begins with a motivational problem from science or engineering. Each of these problems is solved as an example later in the chapter it introduces. Sections 2.5, 2.7, 4.6, 4.7, 5.4, 5.5, 6.4, 6.6, and 6.7 can be omitted without losing the main thread and the material in Appendix B can be beneficially inserted any time after the finishing the first chapter if there is time for it.

The projects, meant for self study, are easy illustrations of the application of linear algebra to engineering. Section 2.7 includes applications to simple resistance circuits, of interest to electrical engineers, and an accounting application and a naive approach to linear programming, of interest to industrial engineers. Section 5.5, "Circles in Space", invites readers to decide whether ten points in space do or do not lie on a circle. This requires an integration of mathematical knowledge, both from linear algebra and from previous experience, to solve a real world problem. In addition to the problems included, there is software available on the World Wide Web (`http://www.math.purdue.edu/~cowen`) to produce individual data sets (and their solutions).

I would like to take this opportunity to thank Professor Martin Corless of the Purdue University School of Aeronautics and Astronautics, Professor Ray DeCarlo of the Purdue University School of Electrical Engineering, Professor Ron Rardin of the Purdue University School of Industrial Engineering, and Professor Michael Lachance of the University of Michigan, Dearborn, Department of Mathematics for their time and expertise in talking with me about the projects included in this book. I would also like to thank Frank Bria and my colleagues Professor Guershon Harel, Professor Joe Lipman, Professor Jim McClure, Professor Clarence Wilkerson, and Professor Bob Zink of the Purdue Mathematics Department for their helpful discussions about the issues involved in teaching and writing about linear algebra. Of course, if there are any errors in these presentations, they are due entirely to my lack of care and not to any advice I have been given by others.

I hope that you find this area of mathematics as stimulating and exciting as I have and that you will find a growing world of applications open to you.

Carl C. Cowen
West Lafayette, Indiana
July 1997

1

Introduction to Matrix Algebra

1.1   An Inventory Example

Applications frequently lead to creation of relevant mathematics and almost always, mathematical definitions and formalism come after the important discoveries of the mathematical facts, even though this is not the order things are presented in books and courses. In order to motivate our study, we will begin each chapter with an example that motivates the definitions and results that follow. These examples will help you appreciate the variety of applications that linear algebra has in engineering and science.

If Carol's Manufacturing Co. has one warehouse, containing 183 widgets and each widget is worth \$47 the total value of the inventory is found by multiplying

$$183 \cdot \$47 = \$8601$$

If she had 209 widgets at the end of last month, we might compute her average inventory as

$$.5(209 + 183) = 196$$

If Carol expands her product line with several versions of widgets and has warehouses across the country, it is important to organize these calculations. For example, if the company has 210 deluxe widgets, 488 regular widgets, and 255 economy widgets in the warehouse in Hartford, 346 deluxe widgets, 807 regular widgets, and 320 economy widgets in the warehouse in Indianapolis, 249 deluxe widgets, 561 regular widgets, and 382 economy widgets in Austin, and 478 deluxe widgets, 627 regular widgets, and 266 economy widgets in Oakland, and the deluxe widgets are worth \$56, the regular widgets \$47, and the economy widgets \$41, then the computation of the total value of the inventory in each warehouse is a more complicated version of the multiplication above. Moreover, writing the information down in an organized way can make the arithmetic easier to follow. Writing the information in a table will help us conceptualize it:

	Deluxe	Regular	Economy
Hartford	210	488	255
Indianapolis	346	807	320
Austin	249	561	382
Oakland	478	627	266

Remembering the labels in the table, but not writing them, we get the inventory *matrix*

$$\begin{pmatrix} 210 & 488 & 255 \\ 346 & 807 & 320 \\ 249 & 561 & 382 \\ 478 & 627 & 546 \end{pmatrix}$$

Similarly, we can write the *vector* of prices: $\begin{pmatrix} \$56 \\ \$47 \\ \$41 \end{pmatrix}$

We can calculate the total value of the inventory in Hartford as the "product" of the first row of the inventory matrix, which gives the number of widgets in Hartford, and the price vector

$$210 \cdot \$56 + 488 \cdot \$47 + 255 \cdot \$41 = \$45,151$$

To find the total values of the inventories in each location, calculate a "product" of the inventory matrix and the price vector as

$$\begin{pmatrix} 210 & 488 & 255 \\ 346 & 807 & 320 \\ 249 & 561 & 382 \\ 478 & 627 & 546 \end{pmatrix} \begin{pmatrix} \$56 \\ \$47 \\ \$41 \end{pmatrix} = \begin{pmatrix} 210 \cdot \$56 + 488 \cdot \$47 + 255 \cdot \$41 \\ 346 \cdot \$56 + 807 \cdot \$47 + 320 \cdot \$41 \\ 249 \cdot \$56 + 561 \cdot \$47 + 382 \cdot \$41 \\ 478 \cdot \$56 + 627 \cdot \$47 + 546 \cdot \$41 \end{pmatrix}$$

$$= \begin{pmatrix} \$45,151 \\ \$70,425 \\ \$55,973 \\ \$78,623 \end{pmatrix}$$

That is, remembering the labels on the table, the total inventory is \$45,151 at Hartford, \$70,425 at Indianapolis, \$55,973 at Austin, and \$78,623 at Oakland.

If last month's inventory matrix was

$$\begin{pmatrix} 218 & 591 & 328 \\ 345 & 721 & 105 \\ 274 & 607 & 186 \\ 470 & 446 & 502 \end{pmatrix}$$

then the matrix of average inventories is

$$.5 \left(\begin{pmatrix} 210 & 488 & 255 \\ 346 & 807 & 320 \\ 249 & 561 & 382 \\ 478 & 627 & 546 \end{pmatrix} + \begin{pmatrix} 218 & 591 & 328 \\ 345 & 721 & 105 \\ 274 & 607 & 186 \\ 470 & 446 & 502 \end{pmatrix} \right)$$

$$= \begin{pmatrix} .5(210+218) & .5(488+591) & .5(255+328) \\ .5(346+345) & .5(807+721) & .5(320+105) \\ .5(249+274) & .5(561+607) & .5(382+186) \\ .5(478+470) & .5(627+446) & .5(546+502) \end{pmatrix}$$

$$= \begin{pmatrix} 214 & 539.5 & 291.5 \\ 345.5 & 764 & 212.5 \\ 261.5 & 584 & 284 \\ 474 & 536.5 & 524 \end{pmatrix}$$

so, for example, the average inventory of economy widgets in Austin was 284.

In these calculations, the meaning of the desired answers determines the arithmetic needed. In the next section, we will define vector, matrix, and arithmetic operations for these objects — the arithmetic defined is exactly the arithmetic needed and used above.

1.2 Addition and Scalar Multiplication of Vectors and Matrices

We are ready to formalize the ideas of the previous section so that we can begin our study of linear algebra. The first objects of linear algebra we will meet are vectors and matrices. In the background, however, are the *scalars*, that is, the numbers, that we will use. For most of this book, the real numbers will be sufficient, but we will occasionally mention complex numbers; in particular, unless explicitly stated otherwise, all numbers in Chapters 1 through 5 can be assumed to be real. Complex scalars will be essential in understanding some phenomena related to eigenvalues and eigenvectors in the later chapters. In most cases, there is no difference between considering the scalars to be real or complex; if there is, it will be mentioned explicitly. Appendix A describes the arithmetic of complex numbers.

If you have studied vectors in physics or in calculus, you are familiar with visualizing vectors as arrows with "length" and "direction". This picture of vectors is also appropriate in linear algebra, except that we will not restrict our vectors to two or three dimensions, and, usually, our vectors will have their tails at the origin. Thus, we might consider the vector v as an arrow with tail at the origin and tip at the point $(-1, 4)$. However, if all our vectors have their tails at the origin, we can just as well view the point $(-1, 4)$ and the vector v as the same thing and write $v = (-1, 4)$. Similarly, as shown in Figure 1.1, we will consider the vectors u and w to be the same as the points where their tips are and write $u = (2, 3)$ and $w = (2, -.5)$.

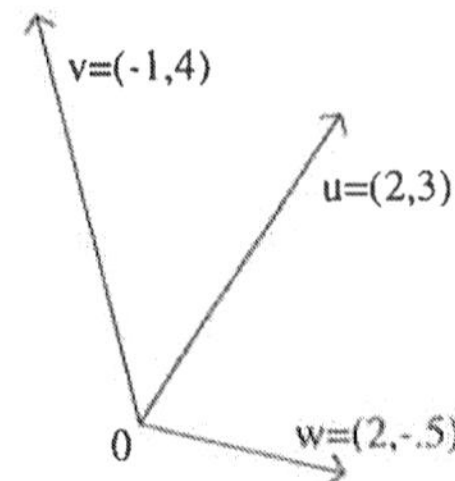

FIGURE 1.1

Vectors in the plane

Although there are more abstract definitions of vectors that are useful in science and engineering, we will begin with a narrow definition of vector that extends the ideas of vectors as arrows in the plane to higher dimensions.

DEFINITION For n a positive integer, $\mathbf{R}^n$ denotes the set of n–tuples of real numbers, that is, $(x_1, x_2, \ldots, x_n)$ is an element of $\mathbf{R}^n$ if x_1, x_2, $\ldots$, and x_n are real numbers. If $x = (x_1, x_2, \ldots, x_n)$, the numbers x_1, x_2, $\ldots$, x_n are called the components *of the vector x. The elements of $\mathbf{R}^n$ are called* real n–vectors *or, more often, just* vectors.

For example, the vectors u, v, and w in Figure 1.1 are vectors in $\mathbf{R}^2$; indeed, $\mathbf{R}^2$

is just another name for the familiar x, y–plane. Similarly, $\mathbf{R}^3$ is another name for x, y, z–space considered in multivariable calculus. The vectors $(1, -2.7, 3.16)$ and $(-3.1, 2.7, 6.7)$ are vectors in $\mathbf{R}^3$. But we will not limit our use of the word *vector* to situations in which we can imagine arrows in the plane or space. We will also consider vectors like $(-1, 0, 2, 3)$ in $\mathbf{R}^4$ or $(1, 1, -1, 2, 0, 5, 2)$ in $\mathbf{R}^7$. Moreover, these vectors are not any less useful because we have difficulty visualizing them! The three components of position and the three components of momentum are frequently combined to form a single vector in $\mathbf{R}^6$ that describes the position and momentum of a particle. An aeronautical engineer studying temperature variations in a jet engine might measure the temperature at fifty different points in the engine at several times; at each time, the temperatures in the different locations would be combined into a single data vector in $\mathbf{R}^{50}$.

In the same way, $\mathbf{C}^n$ is the set of n–tuples of complex numbers and the elements of $\mathbf{C}^n$ are called *complex n–vectors*. The vector $(2 - 3i, .6 + .2i, 5, -2.6 + 6.3i)$ is a vector in $\mathbf{C}^4$. Two vectors in $\mathbf{R}^n$ or $\mathbf{C}^n$ will be *equal* if they have the same first components, the same second components, and so on.

Sometimes it is important to distinguish between row vectors and column vectors:

$$u = \begin{pmatrix} -1 & 0 & 4 & -3 \end{pmatrix} \qquad v = \begin{pmatrix} 1 \\ -2.7 \\ 3.16 \end{pmatrix}$$

Here u is a *row vector* and v is a *column vector*. In fact, we will use column vectors almost exclusively throughout this book as is common practice in most applications. However, it is typographically inconvenient to use column vectors like $v = \begin{pmatrix} 1 \\ -2.7 \\ 3.16 \end{pmatrix}$ everywhere (it would make the book longer, more expensive, and harder to read). To avoid this inconvenience, we adopt the convention that column vectors will be written in row form with commas $v = (1, -2.7, 3.16)$ and row vectors will be written without commas $u = \begin{pmatrix} -1 & 0 & 4 & -3 \end{pmatrix}$.

If you have studied physics, you know that there are algebraic combinations of vectors that are important, for example, combining two forces applied to the same point corresponds to adding the vectors representing them. We will define the operations of multiplying a vector by a number and adding two vectors algebraically. More formally, these are called *scalar multiplication* and *vector addition*.

DEFINITION *If* $v = (v_1, v_2, \ldots, v_n)$ *and* $w = (w_1, w_2, \ldots, w_n)$ *are vectors in* $\mathbf{R}^n$ $v + w$ *is the vector* $(v_1 + w_1, v_2 + w_2, \ldots, v_n + w_n)$. *If* α *is a scalar, αv is the*

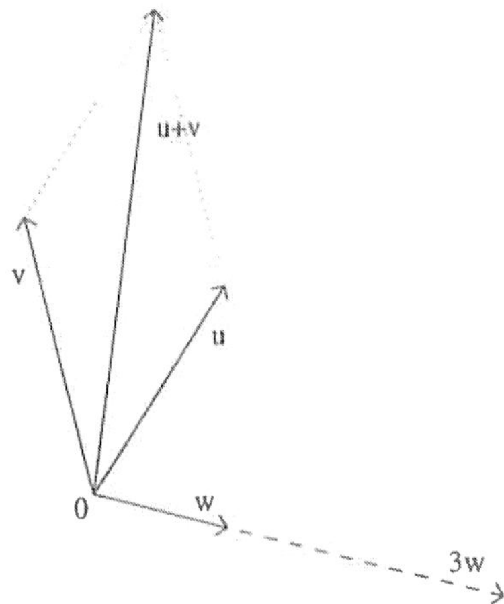

FIGURE 1.2

Addition and scalar multiplication of vectors

vector $(\alpha v_1, \alpha v_2, \ldots, \alpha v_n)$.

Geometrically, if we imagine vectors u and v as arrows, to find $u + v$, we use the copy of v (that is, the arrow is the same length and parallel to v) whose tail is at the tip of u, then $u + v$ is the arrow whose tail is at the tail of u and whose tip is at the tip of the copy of v (see Figure 1.2). The vector αw and the vector w have the same direction if α is positive and the opposite direction if α is negative and the length of αw is $|\alpha|$ times the length of w (see Figure 1.2).

EXAMPLE 1.1

The following computations illustrate the operations of vector addition and scalar multiplication.

$$(4, 1, -1, 3) + (1, 0, 5, -2) = (5, 1, 4, 1)$$

$$3(2, -1, 0, 4.1) = (6, -3, 0, 12.3)$$

$$(-2 + 5)\begin{pmatrix} 4 \\ -7 \\ 2 \end{pmatrix} = 3\begin{pmatrix} 4 \\ -7 \\ 2 \end{pmatrix} = \begin{pmatrix} 12 \\ -21 \\ 6 \end{pmatrix}$$

$$-2\begin{pmatrix} 4 \\ -7 \\ 2 \end{pmatrix} + 5\begin{pmatrix} 4 \\ -7 \\ 2 \end{pmatrix} = \begin{pmatrix} -8 \\ 14 \\ -4 \end{pmatrix} + \begin{pmatrix} 20 \\ -35 \\ 10 \end{pmatrix} = \begin{pmatrix} 12 \\ -21 \\ 6 \end{pmatrix}$$

$$4\left(\begin{pmatrix} -3 \\ 5 \end{pmatrix} + \begin{pmatrix} 2 \\ 4 \end{pmatrix}\right) = 4\begin{pmatrix} -1 \\ 9 \end{pmatrix} = \begin{pmatrix} -4 \\ 36 \end{pmatrix}$$

$$4\begin{pmatrix} -3 \\ 5 \end{pmatrix} + 4\begin{pmatrix} 2 \\ 4 \end{pmatrix} = \begin{pmatrix} -12 \\ 20 \end{pmatrix} + \begin{pmatrix} 8 \\ 16 \end{pmatrix} = \begin{pmatrix} -4 \\ 36 \end{pmatrix}$$

$$(3.1, 2.5, -1.7) + (0, 0, 0) = (3.1, 2.5, -1.7)$$

$$(3.1, 2.5, -1.7) + (-3.1, -2.5, +1.7) = (0, 0, 0)$$

These calculations and other similar calculations illustrate several properties of vector addition and scalar multiplication that are worth noting. It is not possible to add vectors of different sizes. The associative law and the commutative law hold for vector addition and scalar multiplication distributes over vector addition:

$$(\alpha + \beta)v = \alpha v + \beta v$$

and

$$\alpha(u + v) = \alpha u + \alpha v$$

Two obvious properties of scalar multiplication are: $1v = v$ for any vector v and $\alpha(\beta v) = (\alpha\beta)v$ for all scalars α and β and vectors v.

The vector $(0, 0, \ldots, 0)$, called the *zero vector*, plays a special role because adding it to any vector gives the same vector again, that is, it is the identity for vector addition. We will denote the zero vector by the same symbol as the zero scalar, 0, since the context will determine the correct interpretation. That is, we have $v + 0 = v$ and it is also easily seen that $0v = 0$.

If $v = (v_1, v_2, \ldots, v_n)$ is a vector, the *additive inverse* of v is the vector $-v = (-v_1, -v_2, \ldots, -v_n)$ so that $v + (-v) = 0 = (-v) + v$. Of course, $-v = (-1)v$ as with scalars.

All in all, the arithmetic and algebra of vectors works much as the arithmetic and algebra of numbers because the operations on vectors are done component by component. The solution of the following vector equation is further illustration of this principle.

EXAMPLE 1.2

Solve for v:

$$3v + 2 \begin{pmatrix} 3 \\ -1 \\ 1 \end{pmatrix} = \begin{pmatrix} 9 \\ 11 \\ -7 \end{pmatrix}$$

SOLUTION

$$3v + \begin{pmatrix} 6 \\ -2 \\ 2 \end{pmatrix} = \begin{pmatrix} 9 \\ 11 \\ -7 \end{pmatrix}$$

$$3v + \begin{pmatrix} 6 \\ -2 \\ 2 \end{pmatrix} - \begin{pmatrix} 6 \\ -2 \\ 2 \end{pmatrix} = \begin{pmatrix} 9 \\ 11 \\ -7 \end{pmatrix} - \begin{pmatrix} 6 \\ -2 \\ 2 \end{pmatrix}$$

$$3v = \begin{pmatrix} 3 \\ 13 \\ -9 \end{pmatrix}$$

$$\frac{1}{3}(3v) = \frac{1}{3} \begin{pmatrix} 3 \\ 13 \\ -9 \end{pmatrix}$$

$$v = \begin{pmatrix} 1 \\ \frac{13}{3} \\ -3 \end{pmatrix}$$

$\square$

Matrices

Matrices are at the heart of our study of linear algebra. As was noted in the first section, matrices can be useful in the organization of data: indeed, the principal importance of matrices is that they represent data in a structured way. Giving a single name to a collection of data makes it easier to think of the collection as a single object and manipulate it. As was illustrated in the first section, the arithmetic we use to combine matrices arises from our understanding of the relations between the numbers in the matrix, its entries, and how they combine with other collections of numbers.

DEFINITION *For m and n positive integers, an $m \times n$ matrix is a rectangular array of numbers with m rows and n columns.*

In symbols we might have the $m \times n$ matrix A

$$A = \begin{pmatrix} a_{11} & a_{12} & \cdots & a_{1n} \\ a_{21} & a_{22} & \cdots & a_{2n} \\ a_{31} & a_{32} & \cdots & a_{3n} \\ \vdots & \vdots & & \vdots \\ a_{m1} & a_{m2} & \cdots & a_{mn} \end{pmatrix}$$

where the entries a_{ij} are scalars. Note that the entry in the ith row and j^{th} column is denoted a_{ij}. Occasionally, we will write $A = (a_{ij})$ to indicate that A is a matrix whose entries are the numbers a_{ij}. Note that a $1 \times n$ matrix can be considered to be a row vector (u on page 5 is a 1×4 matrix) and an $m \times 1$ matrix can be considered to be a column vector (v on page 5 is a 3×1 matrix). The 1×1 matrix whose only entry is α, the 1–vector whose only component is α, and the scalar α will be considered to be the same. Two matrices will be *equal* when they have the same entries in the same positions.

EXAMPLE 1.3

$$A = \begin{pmatrix} -1 & 3 & 2 & 3 \\ 1 & 0 & -6 & 4 \\ -3 & 1 & -2 & 1 \end{pmatrix}$$

$$B = \begin{pmatrix} 2.1 & -1 & 1.2 & 0 \\ .7 & 0 & -1.6 & 4.6 \\ -1.3 & 3.5 & 0 & -2.9 \end{pmatrix}$$

$$C = \begin{pmatrix} -6 & 2 \\ 4 & 0 \\ 0 & 7 \\ -3 & -4 \\ 1 & 5 \end{pmatrix}$$

$$D = \begin{pmatrix} -8 & 2-3i & 1+2i \\ 4+i & -5i & 0 \end{pmatrix}$$

In this case, A is a 3×4 real matrix with entries $a_{11} = -1$, $a_{12} = 3$, $a_{13} = 2$, $a_{14} = 3$, $a_{21} = 1$, $a_{22} = 0$, $a_{23} = -6$, $a_{24} = 4$, $a_{31} = -3$, etc. The matrix B is a

3×4 real matrix with entries $b_{11} = 2.1$, $b_{12} = -1$, $b_{13} = 1.2$, $b_{14} = 0$, $b_{21} = .7$, $b_{22} = 0$, etc. The matrix C is a 5×2 real matrix with entries $c_{11} = -6$, $c_{12} = 2$, $c_{21} = 4$, $c_{22} = 0$, $c_{31} = 0$, $c_{32} = 7$, etc. The matrix D is a 2×3 complex matrix with entries $d_{11} = -8$, $d_{12} = 2 - 3i$, $d_{13} = 1 + 2i$, $d_{21} = 4 + i$, $d_{22} = -5i$, and $d_{23} = 0$. ☐

Addition and scalar multiplication of matrices are defined in the same way as these operations were defined for vectors.

DEFINITION *If $A = (a_{ij})$ and $B = (b_{ij})$ are $m \times n$ matrices, then $A + B$ is the $m \times n$ matrix whose entries are $a_{ij} + b_{ij}$. If α is a scalar, αA is the $m \times n$ matrix whose entries are αa_{ij}.*

EXAMPLE 1.4

$$5B = 5 \begin{pmatrix} 2.1 & -1 & 1.2 & 0 \\ .7 & 0 & -1.6 & 4.6 \\ -1.3 & 3.5 & 0 & -2.9 \end{pmatrix} = \begin{pmatrix} 10.5 & -5 & 6 & 0 \\ 3.5 & 0 & -8 & 23 \\ -6.5 & 17.5 & 0 & -14.5 \end{pmatrix}$$

$$A + B = \begin{pmatrix} -1 & 3 & 2 & 3 \\ 1 & 0 & -6 & 4 \\ -3 & 1 & -2 & 1 \end{pmatrix} + \begin{pmatrix} 2.1 & -1 & 1.2 & 0 \\ .7 & 0 & -1.6 & 4.6 \\ -1.3 & 3.5 & 0 & -2.9 \end{pmatrix}$$

$$= \begin{pmatrix} 1.1 & 2 & 3.2 & 3 \\ 1.7 & 0 & -7.6 & 8.6 \\ -4.3 & 4.5 & -2 & -1.9 \end{pmatrix}$$

$$2iD = 2i \begin{pmatrix} -8 & 2 - 3i & 1 + 2i \\ 4 + i & -5i & 0 \end{pmatrix} = \begin{pmatrix} -16i & 6 + 4i & -4 + 2i \\ -2 + 8i & 10 & 0 \end{pmatrix}$$

Here is how similar calculations look in MATLAB. Notice that in MATLAB, matrices are entered by listing the entries row by row, separating the rows by semi-colons, and surrounding the matrix by square brackets. Scalar multiplication and matrix multiplication are denoted by * in MATLAB.

```
> a = [1 -2; -3  2; 4 -1]

a =
```

```
        1        -2
       -3         2
        4        -1

> b = [3 1; 1 -1; 0 3]

b =

        3         1
        1        -1
        0         3

> 2*a-3*b

ans =

       -7        -7
       -9         7
        8       -11
```

□

It is not possible to add matrices of different sizes. The associative law and the commutative law hold for matrix addition and scalar multiplication distributes over matrix addition: $(\alpha + \beta)A = \alpha A + \beta A$ and $\alpha(A + B) = \alpha A + \alpha B$.

EXAMPLE 1.5

The following examples illustrate the properties of matrix arithmetic.
The commutative property:

$$\begin{pmatrix} 0 & 6 & -1 & 3 \\ 2 & 1 & -3 & 4 \\ -6 & 2 & -5 & -4 \end{pmatrix} + \begin{pmatrix} 2 & -1 & 1 & 0 \\ 7 & 0 & -2 & 5 \\ -3 & -4 & 0 & -3 \end{pmatrix} = \begin{pmatrix} 2 & 5 & 0 & 3 \\ 9 & 1 & -5 & 9 \\ -9 & -3 & -5 & 6 \end{pmatrix}$$

$$\begin{pmatrix} 2 & -1 & 1 & 0 \\ 7 & 0 & -2 & 5 \\ -3 & -4 & 0 & -3 \end{pmatrix} + \begin{pmatrix} 0 & 6 & -1 & 3 \\ 2 & 1 & -3 & 4 \\ -6 & 2 & -5 & -4 \end{pmatrix} = \begin{pmatrix} 2 & 5 & 0 & 3 \\ 9 & 1 & -5 & 9 \\ -9 & -3 & -5 & 6 \end{pmatrix}$$

The distributive property:

$$3\left(\begin{pmatrix} 1 & -1 \\ 2 & -1 \end{pmatrix} + \begin{pmatrix} 2 & 0 \\ 1 & 3 \end{pmatrix}\right) = 3\begin{pmatrix} 3 & -1 \\ 3 & 2 \end{pmatrix} = \begin{pmatrix} 9 & -3 \\ 9 & 6 \end{pmatrix}$$

$$3\begin{pmatrix} 1 & -1 \\ 2 & -1 \end{pmatrix} + 3\begin{pmatrix} 2 & 0 \\ 1 & 3 \end{pmatrix} = \begin{pmatrix} 3 & -3 \\ 6 & -3 \end{pmatrix} + \begin{pmatrix} 6 & 0 \\ 3 & 9 \end{pmatrix} = \begin{pmatrix} 9 & -3 \\ 9 & 6 \end{pmatrix}$$

Additive identity:

$$\begin{pmatrix} 3 & 1 \\ 7 & 0 \\ -1 & 1 \\ 0 & -2 \end{pmatrix} + \begin{pmatrix} 0 & 0 \\ 0 & 0 \\ 0 & 0 \\ 0 & 0 \end{pmatrix} = \begin{pmatrix} 3 & 1 \\ 7 & 0 \\ -1 & 1 \\ 0 & -2 \end{pmatrix}$$

Additive inverse:

$$\begin{pmatrix} 1 & -1 \\ 2 & -1 \end{pmatrix} + \begin{pmatrix} -1 & 1 \\ -2 & 1 \end{pmatrix} = \begin{pmatrix} 0 & 0 \\ 0 & 0 \end{pmatrix}$$

□

The $m \times n$ matrix all of whose entries are zero is called the $m \times n$ *zero matrix*.

$$\begin{pmatrix} 0 & 0 \\ 0 & 0 \end{pmatrix} \qquad \begin{pmatrix} 0 & 0 \\ 0 & 0 \\ 0 & 0 \\ 0 & 0 \end{pmatrix} \qquad \begin{pmatrix} 0 & 0 & 0 & 0 \\ 0 & 0 & 0 & 0 \\ 0 & 0 & 0 & 0 \end{pmatrix}$$

are the 2×2, the 4×2, and the 4×3 zero matrices. A zero matrix plays a special role because adding it to any matrix of the same size gives the same matrix again, that is, it is the identity for matrix addition. We will denote the zero matrices by the same symbol as the zero scalar, and the zero vectors, since the context will determine the correct interpretation. That is, we have $A + 0 = A$ and it is also easily seen that $0A = 0$. In MATLAB, the command `zeros(j,k)` produces the $j \times k$ zero matrix.

```
> Z=zeros(2,3)

Z =
        0        0        0
        0        0        0
```

If $A = (a_{ij})$ is an $m \times n$ matrix, the *additive inverse* of A is the $m \times n$ matrix $-A = (-a_{ij})$ so that $A + (-A) = 0 = (-A) + A$. Of course, $-A = (-1)A$ as with scalars and vectors.

As in the case of vectors, our expectations about the algebra of matrix addition and scalar multiplication are met. The solution of simple matrix equations is similar to the solutions of numerical equations.

EXAMPLE 1.6

Solve for P:

$$3P + 2\begin{pmatrix} 3 & 2 \\ 0 & -1 \\ -2 & 1 \end{pmatrix} = \begin{pmatrix} 9 & 0 \\ 6 & 11 \\ -1 & -7 \end{pmatrix}$$

SOLUTION

$$3P + \begin{pmatrix} 6 & 4 \\ 0 & -2 \\ -4 & 2 \end{pmatrix} = \begin{pmatrix} 9 & 0 \\ 6 & 11 \\ -1 & -7 \end{pmatrix}$$

$$3P + \begin{pmatrix} 6 & 4 \\ 0 & -2 \\ -4 & 2 \end{pmatrix} - \begin{pmatrix} 6 & 4 \\ 0 & -2 \\ -4 & 2 \end{pmatrix} = \begin{pmatrix} 9 & 0 \\ 6 & 11 \\ -1 & -7 \end{pmatrix} - \begin{pmatrix} 6 & 4 \\ 0 & -2 \\ -4 & 2 \end{pmatrix}$$

$$3P = \begin{pmatrix} 3 & -4 \\ 6 & 13 \\ 3 & -9 \end{pmatrix}$$

$$\frac{1}{3}(3P) = \frac{1}{3}\begin{pmatrix} 3 & -4 \\ 6 & 13 \\ 3 & -9 \end{pmatrix}$$

$$P = \begin{pmatrix} 1 & -\frac{4}{3} \\ 2 & \frac{13}{3} \\ 1 & -3 \end{pmatrix}$$

It should be clear from the examples of vector and matrix computations that vectors can be considered to be a special case of matrices. Indeed a row vector in $\mathbf{R}^n$ is just a $1 \times n$ matrix and a column vector in $\mathbf{R}^n$ is just an $n \times 1$ matrix. We will frequently take advantage of this point of view.

EXAMPLE 1.7

The average of two numbers a and b is $.5(a + b)$. In the inventory example of Section 1.1, we wrote the matrix

$$A = \begin{pmatrix} 210 & 488 & 255 \\ 346 & 807 & 320 \\ 249 & 561 & 382 \\ 478 & 627 & 546 \end{pmatrix}$$

to describe the inventories of various widgets at the four warehouses and the matrix

$$B = \begin{pmatrix} 218 & 591 & 328 \\ 345 & 721 & 105 \\ 274 & 607 & 186 \\ 470 & 446 & 502 \end{pmatrix}$$

describes the inventories the previous month. The definitions of matrix, scalar multiplication, and matrix addition have developed historically so that the calculation of all the average inventories using the matrix calculation: $.5(A + B)$ is

$$.5\left(\begin{pmatrix} 210 & 488 & 255 \\ 346 & 807 & 320 \\ 249 & 561 & 382 \\ 478 & 627 & 546 \end{pmatrix} + \begin{pmatrix} 218 & 591 & 328 \\ 345 & 721 & 105 \\ 274 & 607 & 186 \\ 470 & 446 & 502 \end{pmatrix}\right) = \begin{pmatrix} 214 & 539.5 & 291.5 \\ 345.5 & 764 & 212.5 \\ 261.5 & 584 & 284 \\ 474 & 536.5 & 524 \end{pmatrix}$$

gives the answers we expect, the answers we would have gotten if we had done the calculation separately for each type of widget and each warehouse. ☐

If A is an $m \times n$ matrix, the *transpose of A* is the $n \times m$ matrix A^t whose ij–entry is a_{ji}. An $m \times n$ matrix is called *square* if $m = n$. A square matrix is called *symmetric* if $A = A^t$. For complex matrices, the transpose is not very important, rather it is the complex conjugate transpose, or adjoint[1], of the matrix that is important. Recall that the complex conjugate of a complex number is obtained by changing the sign of the imaginary part, for example, $\overline{3 + 5i} = 3 - 5i$ or $\overline{2.6 - 3.7i} = 2.6 + 3.7i$. This means that the complex conjugate of a real number is the same as the number; for example, $\overline{4} = \overline{4 + 0i} = 4 - 0i = 4$. It follows that the adjoint of a real matrix is the same as the transpose of the matrix. We will not use the notation for the transpose, except for rare situations, rather we will use the adjoint notation to emphasize the unity of the real and complex cases.

DEFINITION *If A is an $m \times n$ matrix, the* adjoint *of A is the $n \times m$ matrix A' whose ij–entry is $\overline{a_{ji}}$, that is, A' is the conjugate transpose of A.*

The definition says that in taking the adjoint of a matrix, the rows become columns, the columns become rows, and every entry is conjugated. This amounts

---

[1] Unfortunately, the terminology is not completely standardized: some authors use the word adjoint to mean a certain matrix associated with the inverse of A; we will call that matrix the *classical adjoint* or *adjugate* of A.

to flipping the matrix across its main diagonal and replacing the entries by their conjugates.

A square matrix A is called *self adjoint* or *Hermitian* if $A = A'$. A real matrix is Hermitian precisely when it is symmetric. Hermitian matrices arise quite frequently in applications to engineering and science because of symmetries in the problem or as a consequence of physical laws. We will study Hermitian matrices extensively in Chapter 7.

EXAMPLE 1.8

The following computations illustrate the definition of adjoint. The adjoint is also denoted by ' in MATLAB.

$$\begin{pmatrix} 1 & 2 \\ 3 & 4 \end{pmatrix}' = \begin{pmatrix} 1 & 3 \\ 2 & 4 \end{pmatrix}$$

$$\begin{pmatrix} 3 & 1 & 0 & -2 \\ -1 & 2 & -3 & 0 \\ -1 & 1 & 0 & 2 \end{pmatrix}' = \begin{pmatrix} 3 & -1 & -1 \\ 1 & 2 & 1 \\ 0 & -3 & 0 \\ -2 & 0 & 2 \end{pmatrix}$$

Here is how similar calculations look in MATLAB.

```
> a = [1 -2; -3  2; 4 -1]

a =
     1    -2
    -3     2
     4    -1

> a'

ans =
     1    -3     4
    -2     2    -1

> c = [2-3*i 1+2*i; 3-4*i 2+i]

c =
   2.0000 - 3.0000i   1.0000 + 2.0000i
   3.0000 - 4.0000i   2.0000 + 1.0000i
```

```
> c'

ans =
   2.0000 + 3.0000i   3.0000 + 4.0000i
   1.0000 - 2.0000i   2.0000 - 1.0000i
```

Notice that MATLAB uses i for $\sqrt{-1}$ and it knows that the adjoint is the conjugate transpose of a complex matrix. Finally, we see that

$$\begin{pmatrix} 2 & 3 & -2 \\ 3 & -1 & 0 \\ -2 & 0 & 4 \end{pmatrix}' = \begin{pmatrix} 2 & 3 & -2 \\ 3 & -1 & 0 \\ -2 & 0 & 4 \end{pmatrix}$$

so this matrix is Hermitian. ☐

The following result relates matrix arithmetic and the adjoint.

THEOREM 1.9

If A and B are matrices of the same size, then

(i) $(A')' = A.$

(ii) $(A + B)' = A' + B'.$

(iii) *For α a scalar, $(\alpha A)' = \overline{\alpha} A'.$*

(iv) *For scalars α and β, $(\alpha A + \beta B)' = \overline{\alpha} A' + \overline{\beta} B'.$*

PROOF We show how the property in *(iv)* follows from properties *(ii)* and *(iii)*. The proofs of the first three properties are not difficult, but are somewhat tedious, so they will be omitted.

$$(\alpha A + \beta B)' = (\alpha A)' + (\beta B)'$$

by property *(ii)*. Using property *(iii)* gives

$$(\alpha A)' + (\beta B)' = \overline{\alpha} A' + \overline{\beta} B'$$

and putting the two equations together gives *(iv)*. ∎

In analyzing a problem symbolically, it is usually better to use the results of Theorem 1.9 than working with the matrix entries.

Exercises 1.2

1. Let $u = (1, -1, 3)$, $v = (0, 2, -1)$, and $w = (3, 1, 1)$. Evaluate the following expressions:

   (a) $4u$ (b) $-3v$ (c) $u + w$ (d) $4u - 3v$ (e) $2u - 4v + 3w$

2. Let $u = (2, 1, 0, -3)$, $v = (1, 0, 3, -1)$, $w = (2, 0, 6, -2)$, and $x = (1, -2, 1)$. Evaluate the following expressions when possible; say *Undefined* when the arithmetic in the expression cannot be carried out.

   (a) $3u - 2v$ (b) $2u + v - 3w$ (c) $3x + w$ (d) $\alpha u + \beta v + \gamma w$

3. Let u, v, and w be vectors as in the previous problem.
   (a) Find α and β, if possible, so that $\alpha u + \beta v = (1, 2, -9, -3)$.
   (b) Find α and β, if possible, so that $\alpha u + \beta w = (3, -1, 2, 0)$.
   (c) Find α and β, if possible, so that $\alpha u + \beta v = (0, -2.5, 15, 2.5)$.
   (d) Find α and β, if possible, so that $\alpha v + \beta w = (-1, 0, -3, 1)$.

4. Let $M = \begin{pmatrix} 0 & 2 & -1 \\ 3 & 1 & 1 \end{pmatrix}$ and $N = \begin{pmatrix} -1 & 0 & 4 \\ 1 & 1 & -2 \end{pmatrix}$. Evaluate the following expressions.

   (a) $3M$ (b) $-2N$ (c) $M + N$ (d) $3M - 2N$
   (e) M' (f) N' (g) $3M' - 2N'$ (h) $(3M - 2N)'$

5. Show that if E is square, then $E + E'$ is Hermitian.

1.3 Multiplication of Matrices

From the experience of the last section, at this point it might seem natural to define matrix multiplication analogously to matrix addition: as entrywise multiplication. Even though this notion of multiplication is sometimes useful, it is not nearly as applicable as the notion of matrix multiplication that arose in the inventory example of Section 1.1. In this book, as in all other elementary treatments, we will not use the entrywise notion of multiplication but will use the following, seemingly complicated, but more useful notion of multiplication.

DEFINITION If $A = (a_{ij})$ is an $m \times n$ matrix and $B = (b_{jk})$ is a $n \times p$ matrix, then the product AB is defined and $C = AB$ is the $m \times p$ matrix such that

$$c_{ij} = a_{i1}b_{1j} + a_{i2}b_{2j} + \cdots + a_{in}b_{nj}$$

That is, in order for AB to be defined, the number of columns of A must be the same as the number of rows of B and the product has the same number of rows as A and the same number of columns as B. Each entry in the product matrix is obtained by combining a single row of A with a single column of B, and the whole product matrix is the result of doing this in all possible ways. In the inventory example, the inventory matrix was 4×3 and the price vector was 3×1 so the product was the 4×1 value vector.

EXAMPLE 1.10

Find the product $C = AB$ where

$$A = \begin{pmatrix} 3 & -1 \\ 5 & -2 \end{pmatrix} \text{ and } B = \begin{pmatrix} 1 & -2 \\ 3 & 4 \end{pmatrix}$$

SOLUTION According to the definition, since A has 2 columns and B has 2 rows, the product $C = AB$ is defined. Since A has 2 rows, C has 2 rows and since B has 2 columns, C has 2 columns, that is, C is a 2×2 matrix.

Now, the entry in the first row, first column of C is obtained by combining the first row of A and the first column of B:

$$c_{11} = a_{11}b_{11} + a_{12}b_{21} = (3)(1) + (-1)(3) = 0$$

Similarly, the entry in the first row, second column of C is obtained by combining the first row of A with the second column of B:

$$c_{12} = a_{11}b_{12} + a_{12}b_{22} = (3)(-2) + (-1)(4) = -10$$

Continuing, we find

$$c_{21} = a_{21}b_{11} + a_{22}b_{21} = (5)(1) + (-2)(3) = -1$$

$$c_{22} = a_{21}b_{12} + a_{22}b_{22} = (5)(-2) + (-2)(4) = -18$$

so

$$C = AB = \begin{pmatrix} 0 & -10 \\ -1 & -18 \end{pmatrix}$$

▯

We give another, slightly larger example.

EXAMPLE 1.11

Let $D = \begin{pmatrix} 3 & -1 & 2 & 0 \\ -1 & 0 & -6 & 4 \\ -3 & 5 & 0 & -2 \end{pmatrix}$ and $E = \begin{pmatrix} 1 & -1 \\ 3 & 1 \\ 0 & 2 \\ 1 & 4 \end{pmatrix}$

Then DE is the 3×2 matrix

$$\begin{pmatrix} 3 \cdot 1 + (-1) \cdot 3 + 2 \cdot 0 + 0 \cdot 1 & 3 \cdot (-1) + (-1) \cdot 1 + 2 \cdot 2 + 0 \cdot 4 \\ (-1) \cdot 1 + 0 \cdot 3 + (-6) \cdot 0 + 4 \cdot 1 & (-1) \cdot (-1) + 0 \cdot 1 + (-6) \cdot 2 + 4 \cdot 4 \\ (-3) \cdot 1 + 5 \cdot 3 + 0 \cdot 0 + (-2) \cdot 1 & (-3) \cdot (-1) + 5 \cdot 1 + 0 \cdot 2 + (-2) \cdot 4 \end{pmatrix}$$

$$= \begin{pmatrix} 0 & 0 \\ 3 & 5 \\ 10 & 0 \end{pmatrix}$$

Notice that in this case, the product ED is not defined because the sizes aren't compatible: E has 2 columns and D has 3 rows. Even in cases for which both AB and BA are defined and are the same size, they are not necessarily equal, for example, for the matrices A and B of Example 1.10,

$$BA = \begin{pmatrix} -7 & 3 \\ 29 & -11 \end{pmatrix} \neq \begin{pmatrix} 0 & -10 \\ -1 & -18 \end{pmatrix} = AB$$

Thus, the commutative law does *not* hold for matrix multiplication. We will not prove these results here, but matrix multiplication is associative, $A(BC) = (AB)C$, if either product is defined, and the distributive laws hold, $A(B + C) = AB + AC$ and $(B + C)A = BA + CA$. Even though the commutative law does not hold for matrices, multiplication by numbers can be done in any order: if α is a number and A and B are matrices, $A(\alpha B) = \alpha AB$. Thus, most of the arithmetic calculations that you would expect to work do work in matrix arithmetic also, as long as you keep track of the order of the matrix products.

Recall that an $m \times n$ matrix is called *square* if $m = n$. If A is a square matrix, then AA is defined and we write A^2 for it, and $A(AA) = (AA)A$ is defined and is denoted by A^3, etc.

For example, if

$$A = \begin{pmatrix} 3 & 5 & 4 \\ 2 & 4 & 3 \\ -5 & -9 & -7 \end{pmatrix}$$

then

$$A^2 = \begin{pmatrix} -1 & -1 & -1 \\ -1 & -1 & -1 \\ 2 & 2 & 2 \end{pmatrix}$$

and

$$A^3 = \begin{pmatrix} 0 & 0 & 0 \\ 0 & 0 & 0 \\ 0 & 0 & 0 \end{pmatrix}$$

This illustrates another difference between usual arithmetic and matrix arithmetic: the product of a zero matrix and another matrix is a zero matrix, but, as above, it is possible for the product of non–zero matrices to be zero also.

EXAMPLE 1.12

If each widget is worth \$47 and there are 488 widgets, the total value of the widgets is $488 \cdot \$47 = \$22936$. In the example of Section 1.1, we described the inventories of the various widgets at the four warehouses with the matrix

$$\begin{pmatrix} 210 & 488 & 255 \\ 346 & 807 & 320 \\ 249 & 561 & 382 \\ 478 & 627 & 546 \end{pmatrix}$$

and the vector of prices as $\begin{pmatrix} \$56 \\ \$47 \\ \$41 \end{pmatrix}$. The definition of matrix multiplication has developed historically so that the product of the inventory matrix and the price vector as

$$\begin{pmatrix} 210 & 488 & 255 \\ 346 & 807 & 320 \\ 249 & 561 & 382 \\ 478 & 627 & 546 \end{pmatrix} \begin{pmatrix} \$56 \\ \$47 \\ \$41 \end{pmatrix} = \begin{pmatrix} \$45,151 \\ \$70,425 \\ \$55,973 \\ \$78,623 \end{pmatrix}$$

gives the same answers we would get if we had done the total inventory calculation product by product and city by city; indeed, the arithmetic is the same.   ⬚

DEFINITION *An $n \times n$ matrix A is called* diagonal *if $a_{ij} = 0$ for $i \neq j$.*

Diagonal matrices are particularly simple to understand. For example, multiplying a matrix on the left multiplies each row by the corresponding diagonal entry, and multiplying on the right by a diagonal matrix multiplies the columns in the same way.

$$\begin{pmatrix} -1 & 0 & 0 \\ 0 & 2 & 0 \\ 0 & 0 & 3 \end{pmatrix} \begin{pmatrix} 3 & 1 & 0 & -2 \\ -1 & 2 & -3 & 0 \\ -1 & 1 & 0 & 2 \end{pmatrix} = \begin{pmatrix} -3 & -1 & 0 & 2 \\ -2 & 4 & -6 & 0 \\ -3 & 3 & 0 & 6 \end{pmatrix}$$

$$\begin{pmatrix} 2 & 1 & -1 \\ -4 & 2 & 0 \\ 2 & 1 & -2 \\ 3 & 1 & -1 \end{pmatrix} \begin{pmatrix} -1 & 0 & 0 \\ 0 & 2 & 0 \\ 0 & 0 & 3 \end{pmatrix} = \begin{pmatrix} -2 & 2 & -3 \\ 4 & 4 & 0 \\ -2 & 2 & -6 \\ -3 & 2 & -3 \end{pmatrix}$$

DEFINITION *The $n \times n$ diagonal matrix whose diagonal entries are all ones is called the $n \times n$ identity matrix. The $n \times n$ identity matrix will be denoted by I_n or just I if it is not important to emphasize the size.*

For example,

$$\begin{pmatrix} 1 & 0 \\ 0 & 1 \end{pmatrix} \qquad \begin{pmatrix} 1 & 0 & 0 \\ 0 & 1 & 0 \\ 0 & 0 & 1 \end{pmatrix} \qquad \begin{pmatrix} 1 & 0 & 0 & 0 \\ 0 & 1 & 0 & 0 \\ 0 & 0 & 1 & 0 \\ 0 & 0 & 0 & 1 \end{pmatrix}$$

are the 2×2, the 3×3, and the 4×4 identity matrices.

The identity matrices are the multiplicative identities: if A is an $m \times n$ matrix and B is an $n \times p$ matrix, $AI_n = A$ and $I_nB = B$. For example,

$$\begin{pmatrix} 1 & 0 \\ 0 & 1 \end{pmatrix} \begin{pmatrix} 3 & -1 & 7 \\ 2 & 0 & -5 \end{pmatrix} = \begin{pmatrix} 3 & -1 & 7 \\ 2 & 0 & -5 \end{pmatrix}$$

and

$$\begin{pmatrix} 3 & -1 & 7 \\ 2 & 0 & -5 \end{pmatrix} \begin{pmatrix} 1 & 0 & 0 \\ 0 & 1 & 0 \\ 0 & 0 & 1 \end{pmatrix} = \begin{pmatrix} 3 & -1 & 7 \\ 2 & 0 & -5 \end{pmatrix}$$

In MATLAB, the command eye(k) produces the $k \times k$ identity matrix.

```
> I=eye(3)

I =
     1     0     0
     0     1     0
     0     0     1
```

Adjoints relate to products of matrices in a perhaps unexpected way: the adjoint of a product is the product of the adjoints, *but the order of the product is reversed!* A consideration of sizes provides one reason something like this must happen: if A is an $m \times n$ matrix and B is an $n \times p$ matrix so that AB is $m \times p$, then A' is $n \times m$, B' is $p \times n$ and $(AB)'$ is $p \times m$: the product $A'B'$ need not even be defined. Again, the proof of this theorem is not difficult, but is a bit tedious, so we will omit it.

THEOREM 1.13

If A and B are matrices, then $(AB)' = B'A'$

EXAMPLE 1.14

Let us calculate both sides of the equation in the conclusion of the theorem separately to see what the result says.

$$\left(\begin{pmatrix} 3 & 1 \\ -1 & 2 \\ -3 & 0 \end{pmatrix} \begin{pmatrix} -1 & 4 & 1 \\ 1 & -2 & 3 \end{pmatrix} \right)' = \begin{pmatrix} -2 & 10 & 6 \\ 3 & -8 & 5 \\ 3 & -12 & -3 \end{pmatrix}'$$

$$= \begin{pmatrix} -2 & 3 & 3 \\ 10 & -8 & -12 \\ 6 & 5 & -3 \end{pmatrix}$$

and, as in the conclusion,

$$\begin{pmatrix} -1 & 4 & 1 \\ 1 & -2 & 3 \end{pmatrix}' \begin{pmatrix} 3 & 1 \\ -1 & 2 \\ -3 & 0 \end{pmatrix}' = \begin{pmatrix} -1 & 1 \\ 4 & -2 \\ 1 & 3 \end{pmatrix} \begin{pmatrix} 3 & -1 & -3 \\ 1 & 2 & 0 \end{pmatrix}$$

$$= \begin{pmatrix} -2 & 3 & 3 \\ 10 & -8 & -12 \\ 6 & 5 & -3 \end{pmatrix}$$

□

We finish the section by defining the matrix analog of division.

DEFINITION *If A is an $n \times n$ matrix, we say the $n \times n$ matrix B is an* inverse *of A if $AB = I_n = BA$. If A has an inverse, we say A is* invertible.

Notice that only square matrices can be invertible, by definition, and that both AB and BA are required to be the identity. We will see later that, for finite matrices, only one of these is necessary as the other follows from it. Some writers, and the developers of MATLAB, use the word *singular* for matrices that are not invertible and call matrices that are invertible *non-singular*.

$$\begin{pmatrix} 1 & -2 & 2 \\ -2 & 6 & -5 \\ 1 & -4 & 4 \end{pmatrix} \begin{pmatrix} 2 & 0 & -1 \\ \frac{3}{2} & 1 & \frac{1}{2} \\ 1 & 1 & 1 \end{pmatrix} = \begin{pmatrix} 1 & 0 & 0 \\ 0 & 1 & 0 \\ 0 & 0 & 1 \end{pmatrix}$$

and

$$\begin{pmatrix} 2 & 0 & -1 \\ \frac{3}{2} & 1 & \frac{1}{2} \\ 1 & 1 & 1 \end{pmatrix} \begin{pmatrix} 1 & -2 & 2 \\ -2 & 6 & -5 \\ 1 & -4 & 4 \end{pmatrix} = \begin{pmatrix} 1 & 0 & 0 \\ 0 & 1 & 0 \\ 0 & 0 & 1 \end{pmatrix}$$

so the two matrices are inverses of each other.

Not all matrices are invertible, for example,

$$\begin{pmatrix} 1 & -2 \\ 0 & 0 \end{pmatrix} \begin{pmatrix} a & b \\ c & d \end{pmatrix} = \begin{pmatrix} a - 2c & b - 2d \\ 0 & 0 \end{pmatrix} \neq \begin{pmatrix} 1 & 0 \\ 0 & 1 \end{pmatrix}$$

no matter what values a, b, c, and d have, so the matrix $\begin{pmatrix} 1 & -2 \\ 0 & 0 \end{pmatrix}$ does not have an inverse.

How many inverses can a matrix have? If A is a square matrix and both B and C are inverses, that is, $AB = BA = I$ and $AC = CA = I$, then

$$C = IC = (BA)C = B(AC) = BI = B$$

(where the middle equality follows from the associativity of matrix multiplication), so $C = B$. That is, there is only one inverse for an invertible matrix. This justifies the terminology that for an invertible matrix A, *THE* inverse of A is denoted A^{-1}.

We will find ways to find the inverse in the next chapter, but we can, if we are given two matrices, check if they are inverses by multiplying.

EXAMPLE 1.15

Sarah claims that

$$\begin{pmatrix} 1 & -1 & 2 & 1 \\ 2 & -3 & 5 & 3 \\ -3 & 2 & -4 & 0 \\ -2 & 2 & -5 & -3 \end{pmatrix}^{-1} = \begin{pmatrix} 12 & -2 & 1 & 2 \\ 0 & -1 & 0 & -1 \\ -9 & 1 & -1 & -2 \\ 7 & -1 & 1 & 1 \end{pmatrix}$$

George, a classmate in her matrix analysis class, is not sure she is correct. Is Sarah correct?

SOLUTION According to the definition, we just need to multiply the matrices in both orders:

$$\begin{pmatrix} 1 & -1 & 2 & 1 \\ 2 & -3 & 5 & 3 \\ -3 & 2 & -4 & 0 \\ -2 & 2 & -5 & -3 \end{pmatrix} \begin{pmatrix} 12 & -2 & 1 & 2 \\ 0 & -1 & 0 & -1 \\ -9 & 1 & -1 & -2 \\ 7 & -1 & 1 & 1 \end{pmatrix} = \begin{pmatrix} 1 & 0 & 0 & 0 \\ 0 & 1 & 0 & 0 \\ 0 & 0 & 1 & 0 \\ 0 & 0 & 0 & 1 \end{pmatrix}$$

and

$$\begin{pmatrix} 12 & -2 & 1 & 2 \\ 0 & -1 & 0 & -1 \\ -9 & 1 & -1 & -2 \\ 7 & -1 & 1 & 1 \end{pmatrix} \begin{pmatrix} 1 & -1 & 2 & 1 \\ 2 & -3 & 5 & 3 \\ -3 & 2 & -4 & 0 \\ -2 & 2 & -5 & -3 \end{pmatrix} = \begin{pmatrix} 1 & 0 & 0 & 0 \\ 0 & 1 & 0 & 0 \\ 0 & 0 & 1 & 0 \\ 0 & 0 & 0 & 1 \end{pmatrix}$$

so Sarah is correct. $\square$

EXAMPLE 1.16

In doing a problem for their physics course, Lydia and Alex need to find the inverse of the matrix

$$R = \begin{pmatrix} 2 & \alpha \\ 0 & 3 \end{pmatrix}$$

for each value of the parameter α. How can they do this?

SOLUTION Lydia and Alex need to find a matrix S so that $RS = I$ and $SR = I$. Since R is a 2×2 matrix, S will also be a 2×2 matrix. We don't know what the entries of S will be, but we do know that they will be some numbers: let us call the entries w, x, y, and z and try to find their values. We need S to satisfy

$$RS = \begin{pmatrix} 2 & \alpha \\ 0 & 3 \end{pmatrix} \begin{pmatrix} w & x \\ y & z \end{pmatrix} = \begin{pmatrix} 1 & 0 \\ 0 & 1 \end{pmatrix}$$

Since the product on the left is

$$\begin{pmatrix} 2w + \alpha y & 2x + \alpha z \\ 3y & 3z \end{pmatrix}$$

and we know that matrices that are equal must have the same entries, we see that

$$2w + \alpha y = 1$$
$$2x + \alpha z = 0$$
$$3y = 0$$
$$3z = 1$$

From the latter two equations, we see that $y = 0$ and $z = 1/3$. Using these results in the first two equations, we see that $w = 1/2$ and $x = -\alpha/6$. In other words, if R has any inverse, it must be

$$S = \begin{pmatrix} \frac{1}{2} & -\frac{\alpha}{6} \\ 0 & \frac{1}{3} \end{pmatrix}$$

We can check that this works by multiplying:

$$RS = \begin{pmatrix} 2 & \alpha \\ 0 & 3 \end{pmatrix} \begin{pmatrix} \frac{1}{2} & -\frac{\alpha}{6} \\ 0 & \frac{1}{3} \end{pmatrix} = \begin{pmatrix} 1 & 0 \\ 0 & 1 \end{pmatrix}$$

and

$$SR = \begin{pmatrix} \frac{1}{2} & -\frac{\alpha}{6} \\ 0 & \frac{1}{3} \end{pmatrix} \begin{pmatrix} 2 & \alpha \\ 0 & 3 \end{pmatrix} = \begin{pmatrix} 1 & 0 \\ 0 & 1 \end{pmatrix}$$

$\square$

The inverse shares with adjoints the property of reversing the order of products!

THEOREM 1.17

If A and B are invertible $n \times n$ matrices, then AB is invertible and

$$(AB)^{-1} = B^{-1}A^{-1}$$

PROOF   We need to check the multiplication in both orders:

$$(AB)(B^{-1}A^{-1}) = A(BB^{-1})A^{-1} = AIA^{-1} = AA^{-1} = I$$

and

$$(B^{-1}A^{-1})(AB) = B^{-1}(A^{-1}A)B = BIB^{-1} = BB^{-1} = I$$

This means that both $(B^{-1}A^{-1})(AB) = I$ and $(AB)(B^{-1}A^{-1}) = I$ so $B^{-1}A^{-1}$ is the inverse of AB, i.e., $(AB)^{-1} = B^{-1}A^{-1}$. ∎

Matrix Multiplication as a Linear Transformation

One of the ways in which multiplication by a matrix A can be understood is to consider the multiplication process as "transforming" the vector v into the vector Av. We can think of the the transformation in geometric way, in which we visualize the transformation as moving the vector v onto the vector Av. We can also think of the transformation in a more physical way as linear filter or a linear amplifier. In this view, the vector v (representing a signal, say) is input into a device that multiplies the vector by A and the transformed vector Av is the output of the device. (See Figure 1.3.)

FIGURE 1.3

The device that multiplies by the matrix A

A filter that removes from all parts of a signal whose frequencies are higher than 20 kilohertz or an amplifier that doubles the amplitude of all signals are examples of transformations that are of interest in signal processing.

Figure 1.4 shows a vector v and (on a different set of axes) the transformed vector Av. We can imagine the vector v being moved and stretched to become the vector Av. In general, multiplication by a matrix moves and stretches different vectors in different ways, for example, see Figure 1.5 in which eight vectors are shown and their images after multiplying by A are shown. In fact, the vectors whose tips lie on the unit circle are moved and stretched until their images lie on the ellipse shown (see Exercise 1.3.17.).

v   Av

FIGURE 1.4

Transformation by multiplying by a matrix

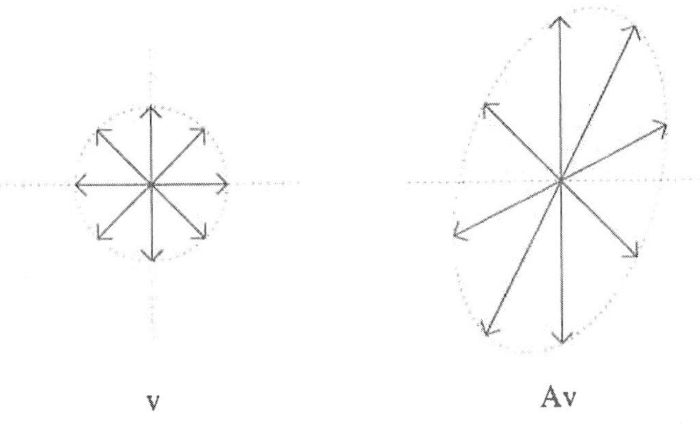

v   Av

FIGURE 1.5

Several vectors transformed by multiplying by a matrix

Stretches, rotations, and shears are transformations that are easy to visualize. For example, multiplication by the matrix

$$\begin{pmatrix} 2 & 0 \\ 0 & 2 \end{pmatrix}$$

is a stretch that doubles the length of every vector, while leaving directions of vectors the same (see Figure 1.6).

Multiplication by the matrix

$$\begin{pmatrix} \frac{1}{2} & \frac{\sqrt{3}}{2} \\ -\frac{\sqrt{3}}{2} & \frac{1}{2} \end{pmatrix}$$

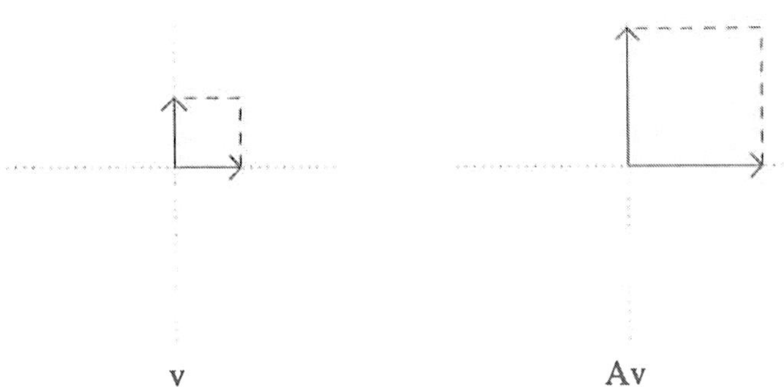

v   Av

FIGURE 1.6

Stretch

is a rotation. This transformation keeps the lengths of all vectors the same but rotates them all 60 degrees clockwise (see Figure 1.7).

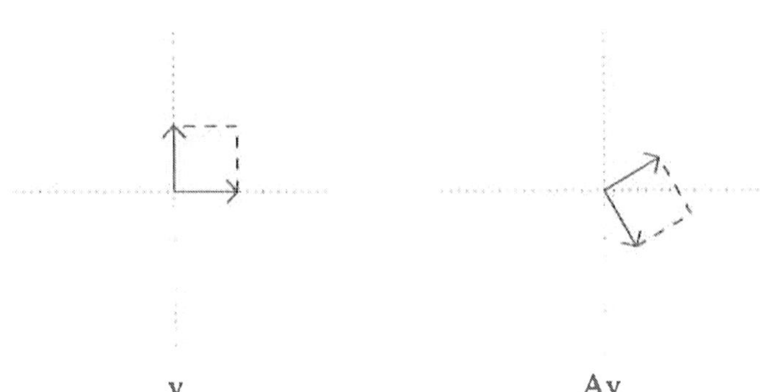

v   Av

FIGURE 1.7

Rotation

Multiplication by the matrix

$$\begin{pmatrix} 1 & .5 \\ 0 & 1 \end{pmatrix}$$

is a shear. It fixes vectors on the x–axis but pushes other vectors, that is, it shears parallel to the x–axis (see Figure 1.8).

FIGURE 1.8

Shear

Multiplication by a matrix is a *linear transformation*. A thorough study of linear transformations is beyond the scope of this book, but we will make the formal definition and show that multiplication by a matrix is an example. Less obvious is that, after choosing coordinates, every linear transformation can be considered to be multiplication by a matrix.

DEFINITION A function T defined on $\mathbf{R}^n$ with values in $\mathbf{R}^m$ is called a linear transformation if

$$T(u + v) = T(u) + T(v)$$

for all vectors u and v in $\mathbf{R}^n$ and

$$T(\alpha u) = \alpha T(u)$$

for all numbers α and all vectors u.

Most functions we studied in calculus are not linear transformations, for example, $T(x) = x^2$ is not linear because $T(2) = 4$, $T(3) = 9$ but

$$T(2 + 3) = T(5) = 25 \neq 13 = T(2) + T(3)$$

However, multiplication by a matrix is a linear transformation.

If A is an $m \times n$ matrix, then let $T(v) = Av$. For v in $\mathbf{R}^n$, Av makes sense and Av is in $\mathbf{R}^m$. Moreover, the distributive law says that

$$T(u + v) = A(u + v) = Au + Av = T(u) + T(v)$$

Also, because a matrices can be multiplied by a number in any order with the same result,

$$T(\alpha u) = A(\alpha u) = \alpha Au = \alpha T(u)$$

Linear transformations are called "linear" because, except in degenerate cases, they transform lines into other lines, as in the example below or in Exercise 1.3.16.

EXAMPLE 1.18

Multiplication by the matrix

$$A = \begin{pmatrix} 1 & 2 \\ -1 & 1 \end{pmatrix}$$

moves the line through $(-1, 3)$ and $(2, 1)$ onto the line through $(5, 4)$ and $(4, -1)$.

To see this, we will parametrize the line through $(-1, 3)$ and $(2, 1)$ as $\gamma(t) = t(-1, 3) + (1 - t)(2, 1) = (2 - 3t, 1 + 2t)$. In this parametrization, $\gamma(0) = (2, 1)$ and $\gamma(1) = (-1, 3)$ and each value of t corresponds to a different point on the line.

To see what happens to the line under the transformation A, we multiply each vector (representing a point) by the matrix A. Then

$$A\gamma(t) = \begin{pmatrix} 1 & 2 \\ -1 & 1 \end{pmatrix} \begin{pmatrix} 2 - 3t \\ 1 + 2t \end{pmatrix} = \begin{pmatrix} 4 + t \\ -1 + 5t \end{pmatrix} = t \begin{pmatrix} 5 \\ 4 \end{pmatrix} + (1 - t) \begin{pmatrix} 4 \\ -1 \end{pmatrix}$$

The parametrization shows that the curve $A\gamma$ is the straight line through $(4, -1)$ (corresponding to $t = 0$) and $(5, 4)$ (corresponding to $t = 1$). ⬜

Exercises 1.3

1. Let $P = \begin{pmatrix} 1 & 2 & 0 \\ -1 & 0 & -2 \end{pmatrix}$, $Q = \begin{pmatrix} 1 & -3 & 1 \\ 0 & 1 & -1 \end{pmatrix}$, and $R = \begin{pmatrix} -2 & 1 & 3 \\ -1 & 0 & 3 \\ 1 & 1 & -2 \end{pmatrix}$

   Evaluate the following expressions.

   (a) $P + 2Q$ (b) PR (c) QR (d) $PR + 2QR$ (e) $(P + 2Q)R$
   (f) PQ' (g) $Q'P$ (h) PP' (i) RR' (j) $R'R$

2. Let $M = \begin{pmatrix} 0 & 2 & -1 \\ 3 & 1 & 1 \end{pmatrix}$ and $N = \begin{pmatrix} -1 & 0 & 4 \\ 1 & 1 & -2 \end{pmatrix}$.

   Evaluate the following expressions.

   (a) MM' (b) $M'M$ (c) MN' (d) $N'M$

3. Let $A = \begin{pmatrix} 1 & 1 & 0 \\ 0 & 1 & -1 \end{pmatrix}$, $B = \begin{pmatrix} -1 & 1 & 1 \\ -1 & 0 & 3 \end{pmatrix}$, $C = \begin{pmatrix} 1 & -3 \\ -1 & 3 \\ -1 & 3 \end{pmatrix}$,

$D = \begin{pmatrix} 1 & -1 \\ 1 & 1 \end{pmatrix}$, and $E = \begin{pmatrix} -2 \\ 3 \\ -1 \end{pmatrix}$ Evaluate the following expressions when

possible; say *Undefined* when the arithmetic in the expression cannot be carried out.

(a) $3A - 2B$	(b) AE	(c) AB	(d) AC	(e) CA
(f) EA	(g) $E'A$	(h) $AB' + D$	(i) A^2	(j) D^2

4. Let $A = \begin{pmatrix} 5 & -4 & 1 \\ 12 & -11 & 6 \\ 10 & -10 & 8 \end{pmatrix}$

and $u = (1, -1, 2)$, $v = (1, 1, 0)$, $w = (1, 2, 1)$, $e_1 = (1, 0, 0)$, and $e_2 = (0, 1, 0)$.
(a) Find Au.
(b) Find Av.
(c) Find Aw.
(d) Find Ae_1 and Ae_2. Let $e_3 = (0, 0, 1)$; guess what Ae_3 is, then compute it.

5. Let $S = \begin{pmatrix} 0 & 2 & -1 \\ 2 & 1 & 3 \\ -1 & 3 & -2 \end{pmatrix}$

(a) Find S'.
(b) What special property does S have?
(c) What is $S + I$. How do you know which identity matrix to add to S?
(d) Find S^2 and S^3.
(e) Show that if T is any Hermitian matrix, then T^2 is Hermitian also.

6. Let $P = \begin{pmatrix} 1 & 2 & 3 \\ 4 & 5 & 6 \\ 7 & 8 & 9 \end{pmatrix}$, $Q = \begin{pmatrix} 2 & -1 & 1 \\ -3 & 4 & -2 \\ 5 & 3 & -5 \end{pmatrix}$, and let D be the diagonal matrix

$D = \begin{pmatrix} -1 & 0 & 0 \\ 0 & 2 & 0 \\ 0 & 0 & 3 \end{pmatrix}$

(a) Find DP and DQ.
(b) If E is the diagonal matrix with diagonal entries α, β, and γ, and R is a matrix, describe ER.
(c) Find PD and QD.
(d) If E is the diagonal matrix with diagonal entries α, β, and γ, and R is a matrix, describe RE.

7. Let $S = \begin{pmatrix} 1 & 1 \\ -1 & 2 \end{pmatrix}$ and let $T = \begin{pmatrix} 2 & 1 & 3 \\ 1 & -1 & -2 \end{pmatrix}$

   Let $C_1 = \begin{pmatrix} 2 \\ 1 \end{pmatrix}$, let $C_2 = \begin{pmatrix} 1 \\ -1 \end{pmatrix}$, and let $C_3 = \begin{pmatrix} 3 \\ -2 \end{pmatrix}$

   (a) Find SC_1, SC_2, and SC_3.

   (b) Find ST and compare your answer with the results of part a).

8.(a) Let $A = \begin{pmatrix} 1 & -1 & 0 \\ -1 & 2 & 2 \\ 2 & -1 & 1 \end{pmatrix}$ and let $B = \begin{pmatrix} -4 & -1 & 2 \\ -5 & -1 & 2 \\ 3 & 1 & -1 \end{pmatrix}$

   Explain why $A = B^{-1}$.

   (b) Is $B = A^{-1}$? Explain!

   (c) Let $C = \begin{pmatrix} 1 & -1 & 0 \\ -1 & 1 & 1 \end{pmatrix}$, and $D = \begin{pmatrix} -2 & 1 \\ -3 & 1 \\ 1 & 1 \end{pmatrix}$ Is $D = C^{-1}$? Explain!

9. Let $E = \begin{pmatrix} 1 & 1 \\ 2 & -1 \end{pmatrix}$ Find a matrix $F = \begin{pmatrix} a & b \\ c & d \end{pmatrix}$ so that $F = E^{-1}$.

10.(a) Show that if G is an invertible matrix, then G' is also invertible and
   $(G')^{-1} = (G^{-1})'$.

   (b) Use your answer to part (a) and problem 1.3.8. above to find the inverse of $\begin{pmatrix} 1 & -1 & 2 \\ -1 & 2 & -1 \\ 0 & 2 & 1 \end{pmatrix}$

   (c) Show that if H is an invertible Hermitian matrix, then H^{-1} is also Hermitian.

11. Verify that if N is a matrix such that $N^4 = 0$, then

$$(I - N)^{-1} = I + N + N^2 + N^3.$$

   WARNING! Such matrices are called *nilpotent* and are **not** necessarily 0.
   For example, the matrix $M = \begin{pmatrix} 1 & 1 \\ -1 & -1 \end{pmatrix}$ satisfies $M^2 = 0$.

12. Let E be an $m \times n$ matrix.
   (a) Show that EE' and $E'E$ are both Hermitian.
   (b) Give an example to show that these are not always the same.

◇ 13.   Redo Exercises 1.3.3., 1.3.8., and 1.3.9. using a suitable machine. How does your machine react to undefined matrix operations?

◇ 14.

$$F = \begin{pmatrix} 1 & 2 & -1 & 1 \\ 0 & -1 & 4 & 3 \\ 4 & 2.6 & 0 & 3 \\ 3 & -.3 & 8 & 1.5 \end{pmatrix}$$

   (a) Use a suitable machine to find $G = F^{-1}$
   (b) Find the computed values of GF and $GF - I$. Explain the output of your machine.

15. For each of the following matrices, make sketches with separate axes for v and Av (as in Figures 1.6, 1.7, and 1.8) showing $v = (1, 0)$, $v = (0, 1)$, $v = (3, 0)$, and $v = (3, 1)$ on one set of axes, and the corresponding images Av on the other set.

$$\text{(a) } A = \begin{pmatrix} 3 & 0 \\ 0 & -1 \end{pmatrix} \quad \text{(b) } A = \begin{pmatrix} .6 & -.8 \\ .8 & .6 \end{pmatrix} \quad \text{(c) } A = \begin{pmatrix} 3 & 1 \\ 1 & -1 \end{pmatrix}$$

16. Let B be the matrix

$$B = \begin{pmatrix} 1 & 1 \\ 2 & -1 \end{pmatrix}$$

and let v be a point that lies on the line described by $v(t) = (3, -1) + t(1, 1) = (3 + t, -1 + t)$ for $-\infty < t < \infty$. (That is, the tip of the vector v lies on the line through $v(0) = (3, -1)$ and $v(1) = (4, 0)$.) Show that Bv lies on a line also, by finding the parametric form of the line.

17. Let A and B be the matrices

$$A = \begin{pmatrix} 2 & 0 \\ 0 & 3 \end{pmatrix} \quad B = \begin{pmatrix} 1 & 1 \\ 2 & -1 \end{pmatrix}$$

   (a) Show that multiplication by A takes the unit circle to an ellipse, that is, show that there is an ellipse so that if the tip of the vector $v = (r, s)$ lies on the circle $x^2 + y^2 = 1$, then the tip of the vector Av lies on the ellipse. (Can you identify the ellipse?)
   (b) Show that multiplication by B takes the unit circle to an ellipse, that is, show that there is an ellipse so that if the tip of the vector $v = (r, s)$ lies on the circle $x^2 + y^2 = 1$, then the tip of the vector Bv lies on the ellipse. (Can you identify the ellipse?)

2

Systems of Linear Equations

2.1   A Circuit Analysis Problem

The most important problem in linear algebra is the solution of systems of linear equations. Such systems arise in many application areas, but as an example we will consider the problem of determining the currents in a simple resistance circuit. For clarity, we will approach the problem naively from Ohm's and Kirchhoff's Laws without using any of the simplifications that result from considering consequences of these laws. Ohm's Law says that the voltage drop across a resistor is proportional to the current. The proportionality constant is called resistance; symbolically we have $V = iR$, where V is the voltage drop, i is the current, and R is the resistance. Kirchhoff's Laws are that the algebraic sum of the currents leaving any node of a circuit must be zero and that the algebraic sum of the voltage drops around any loop in a circuit must be zero.

Consider the circuit in Figure 2.1 consisting of six resistors and a voltage source, connected as shown. We want to determine the currents in each of the resistors and we assign the positive directions of the currents as indicated by the arrows. Then we get the following system of eight equations for the currents.

$$
\begin{aligned}
i_{AB} + i_{AD} + i_{AC} &= 0 & \text{(Node } A\text{)} \\
-i_{AB} + i_{BC} + i_{BD} &= 0 & \text{(Node } B\text{)} \\
-i_{AC} - i_{BC} + i_{CD} &= 0 & \text{(Node } C\text{)} \\
-i_{AD} - i_{BD} - i_{CD} &= 0 & \text{(Node } D\text{)} \\
-10 + 10i_{AB} + 4i_{BC} - i_{AC} &= 0 & \text{(Loop } ABC\text{)} \\
-10 + 10i_{AD} + 5i_{BD} - 2i_{AD} &= 0 & \text{(Loop } ABD\text{)}
\end{aligned}
$$

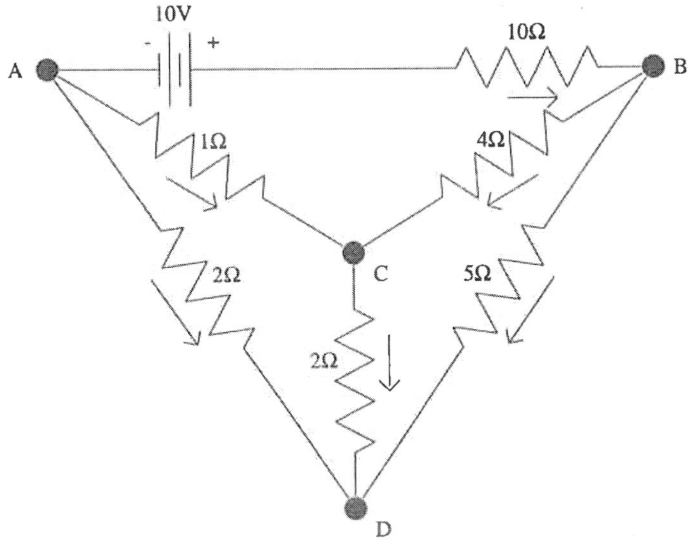

FIGURE 2.1

A linear resistance circuit

$$i_{AC} + 2i_{CD} - 2i_{AD} = 0 \qquad \text{(Loop } ACD)$$

$$-4i_{BC} + 5i_{BD} - 2i_{CD} = 0 \qquad \text{(Loop } BDC)$$

Since each of the unknowns has exponent 1 and none of the unknowns is multiplied by another unknown, this is called a *system of linear equations*. It is not difficult to check that $i_{AB} = 255/329$, $i_{AD} = -95/329$, $i_{AC} = -160/329$, $i_{BC} = 145/329$, $i_{BD} = 110/329$, and $i_{CD} = -15/329$ is a solution to each of the equations in the system. Are there other solutions to this system? For physical reasons, we would expect this is the only solution. Does every such system have a solution, that is, if we change the values of the resistances, or relocate the voltage source, or add new resistors and voltage sources, does the resulting system always have a solution? As we study systems of linear equations, we will answer the questions "How can we find solutions to systems of linear equations?" and "How many solutions does a system have?" as well as others that are raised in the process of looking at these questions. In this chapter, we will learn a systematic way, *Gaussian elimination*, to solve systems of linear equations and discuss some related problems such as finding inverses of matrices and calculating determinants.

2.2   Linear Systems

Solution of linear equations is the single most important problem in the application of mathematics beyond arithmetic. Typical problems may involve hundreds of equations in hundreds of unknowns, while very large systems may involve millions of equations. In approaching solution of linear equations by machine, one must be conscious of issues of efficiency, accuracy, and occasionally, machine storage conservation. The techniques introduced in this section are typical of efficient methods for the solution of linear equations but may not represent the most efficient method in each case. We will not discuss techniques used to minimize errors due to rounding numbers off in the course of computation.

DEFINITION *If a_1, a_2, $\cdots$, a_n and b are constants,*

$$a_1 x_1 + a_2 x_2 + \cdots + a_n x_n = b$$

is called a linear equation in the variables x_1, x_2, $\cdots$, x_n; *the numbers a_1, a_2, $\cdots$, a_n are called the* coefficients *of the equation.*

For example, the equation $3x - 4y + 6z = 8$ is a linear equation in the variables x, y, z; the coefficient of x is 3, the coefficient of y is -4, and the coefficient of z is 6. We say $(0, 1, 2)$ is a solution of the equation because substituting $x = 0$, $y = 1$, and $z = 2$ gives $3(0) - 4(1) + 6(2) = 8$ which is true.

DEFINITION *A system of linear equations in the variables x_1, x_2, $\cdots$, x_n or, more simply, a* linear system *is a set S of linear equations in the variables x_1, x_2, $\cdots$, x_n. If S is a linear system, we say $(c_1, c_2, \ldots, c_n)$ is a* solution *of the system if the assignments of the variables $x_1 = c_1$, $x_2 = c_2$, $\cdots$, $x_n = c_n$ make all the equations true.*

For example,

$$\left\{ \begin{array}{rcrcrcr} 2x & + & y & + & z & = & 3 \\ x & & & - & z & = & -1 \\ x & + & y & + & 2z & = & 4 \\ & & y & + & 3z & = & 5 \end{array} \right.$$

is a system of 4 linear equations in the three variables x, y, and z. We can see that substitution of $x = 1$, $y = -1$, and $z = 2$ into each of the four equations makes them true, that is, $(1, -1, 2)$ is a solution of this system. Indeed, checking, we see

$$\begin{cases} 2(1) & + & (-1) & + & 2 & = & 3 \\ 1 & & & - & 2 & = & -1 \\ 1 & + & (-1) & + & 2(2) & = & 4 \\ & & -1 & + & 3(2) & = & 5 \end{cases}$$

so that all four equations are true. In addition, we can check that $x = 0$, $y = 2$, and $z = 1$ is a solution of the system, and $(-1/3, 3, 2/3)$ is another solution. Apparently, some systems have many solutions.

$$\begin{cases} 2x & & = & 3 \\ x & - & y & = & 1 \end{cases}$$

is a system of 2 linear equations in 2 unknowns. We can see that $x = 3/2$ and $y = 1/2$ is a solution, and that it is the only solution. Thus, some systems have exactly one solution.

On the other hand, it is easy to see that the system

$$\begin{cases} x & - & y & = & 4 \\ 2x & - & 2y & = & -1 \end{cases}$$

has no solutions because if $x - y = 4$, then $2x - 2y = 8 \neq -1$.

It is a simple matter to check whether a given potential solution of a system is really a solution, but it is clear guessing solutions of a system is not a practical way to find them. We will analyze the problem of finding solutions and construct an algorithm for finding all solutions of a given system if it has solutions, and telling us it has none if that is the case.

The first observation is that some systems are easier to solve than others. For example, the system

$$\begin{cases} 2x & + & y & & & = & -1 \\ x & & & & & = & 2 \\ -x & + & 2y & + & z & = & 0 \end{cases} \tag{2.2.1}$$

is very easy to solve because the second equation says $x = 2$ which we may use in the first equation to get $y = -5$ and we may use both values in the third equation to get $z = 12$. Similarly the system

$$\begin{cases} x & + & 2y & - & z & = & 3 \\ & & y & + & 3z & = & -2 \\ & & & & z & = & 1 \end{cases}$$

is very easy to solve: from the last equation, $z = 1$, from the second equation, then, $y = -5$, so from the first equation, $x = 14$ and $(14, -5, 1)$ is the solution.

These two examples show that if a system has a triangular structure in its coefficients, the variables can be solved for one by one until the whole system has been solved. We will exploit this observation by making every system into an easy "triangular" system.

In all of the methods used in practice, the system of equations is viewed in matrix form and the corresponding matrix problem is solved.

Suppose we want to solve the system

$$\begin{cases} 2x_1 + x_2 & = 4 \\ x_1 + x_2 - x_3 & = 4 \\ 2x_1 + x_3 & = 1 \end{cases}$$

This system can be written in the form $AX = b$ where

$$A = \begin{pmatrix} 2 & 1 & 0 \\ 1 & 1 & -1 \\ 2 & 0 & 1 \end{pmatrix}, \quad X = \begin{pmatrix} x_1 \\ x_2 \\ x_3 \end{pmatrix}, \text{ and } b = \begin{pmatrix} 4 \\ 4 \\ 1 \end{pmatrix} \qquad (2.2.2)$$

indeed, multiplying the matrix A by the vector X gives

$$\begin{pmatrix} 2 & 1 & 0 \\ 1 & 1 & -1 \\ 2 & 0 & 1 \end{pmatrix} \begin{pmatrix} x_1 \\ x_2 \\ x_3 \end{pmatrix} = \begin{pmatrix} 2x_1 + x_2 \\ x_1 + x_2 - x_3 \\ 2x_1 + x_3 \end{pmatrix}$$

and equating this to the vector b gives the original system. Similarly, every system of m linear equations in n unknowns can be written as $AX = b$ where A is the $m \times n$ matrix of coefficients, X is the vector of n unknowns, and b is the vector of righthand sides of the m equations.

DEFINITION *If S is a system of m linear equations in n unknowns that has been written as $AX = b$, we say S is a homogeneous system if $b = 0$ and we say S is a non–homogeneous system if $b \neq 0$.*

EXAMPLE 2.1

Write the following systems in the form $AX = b$ and decide if the systems are homogeneous or non–homogeneous.

$$\begin{cases} x_1 - 3x_2 + x_3 = 1 \\ -x_1 + 2x_2 - 3x_3 = -2 \\ 2x_1 - 5x_2 + 4x_3 = 3 \end{cases} \qquad \begin{cases} x_1 - 3x_2 + x_3 = 0 \\ -x_1 + 2x_2 - 3x_3 = 0 \\ 2x_1 - 5x_2 + 4x_3 = 0 \end{cases}$$

SOLUTION The first equation is $AX = b$ where

$$A = \begin{pmatrix} 1 & -3 & 1 \\ -1 & 2 & -3 \\ 2 & -5 & 4 \end{pmatrix} \qquad X = \begin{pmatrix} x_1 \\ x_2 \\ x_3 \end{pmatrix} \quad \text{and} \quad b = \begin{pmatrix} 1 \\ -2 \\ 3 \end{pmatrix}$$

Since $b = (1, -2, 3) \neq (0, 0, 0)$, the first system is a non–homogeneous system.

The second system is $AX = b'$ where

$$A = \begin{pmatrix} 1 & -3 & 1 \\ -1 & 2 & -3 \\ 2 & -5 & 4 \end{pmatrix} \qquad X = \begin{pmatrix} x_1 \\ x_2 \\ x_3 \end{pmatrix} \quad \text{and} \quad b' = \begin{pmatrix} 0 \\ 0 \\ 0 \end{pmatrix}$$

Since $b' = (0, 0, 0)$, the second system is a homogeneous system.

Moreover, since the coefficient matrix for both systems is the same, we will call the second system *the homogeneous system associated with the first system*. In the next few sections, we will investigate the relation between a system and its associated homogeneous system. ▯

An immediate payoff in writing systems of equations as a single matrix equation is the following theorem. The theorem expresses the relation between the solutions of the homogeneous and non–homogeneous systems and the "principle of super-position" that applies to linear problems in engineering. If you are familiar with the solution of second order linear differential equations, you should compare this with the techniques presented in that context of solving the homogeneous equation completely, finding one solution of the non–homogeneous equation, then getting the general solution of the non–homogeneous equation as a sum of these solutions. These are the same because the differential equation can be written in the form $Ly = b$ where y is the unknown function, b is the "forcing function", and L is a linear differential operator.

THEOREM 2.2

Let S be a system of m linear equations in n unknowns that has been written as $AX = b$, where A is the $m \times n$ matrix of coefficients, X is the vector of unknowns, and b is the vector of righthand sides.

(a) *If X_1 and X_2 are both solutions of the homogeneous system $AX = 0$ and α is a number, then αX_1 and $X_1 + X_2$ are also solutions of the homogeneous system.*

(b)   *If X_3 and X_4 are solutions of the system $AX = b$, then $X_3 - X_4$ is a solution of the homogeneous system $AX = 0$.*

(c)   *If X_p is a solution of the non-homogeneous system $AX = b$, then every solution of the non-homogeneous system can be written as $X = X_p + X_0$, where X_0 is a solution of the homogeneous system $AX = 0$. Conversely, if X_p satisfies $AX_p = b$ and $AX_0 = 0$, then $X_p + X_0$ satisfies $AX = b$.*

Theorem 2.2 discusses the relation of the solutions of homogeneous and non–homogeneous systems but does not talk about solving either system. Part *(a)* says that the sum of two solutions of a *homogeneous system* is again a solution and that a multiple of a solution is again a solution; this tells us ways to create new solutions from old solutions of homogeneous systems. Part *(b)* gives a method for finding solutions of the homogeneous system from solutions of the non–homogeneous system. Part *(c)*, a kind of converse to part *(b)*, gives a method for creating new solutions of a non–homogeneous system from a particular solution of the non–homogeneous system and solutions of the homogeneous system.

EXAMPLE 2.3

$X = (2, 1, 1)$ is a solution of the system

$$
\begin{cases}
x_1 & - & 3x_2 & + & x_3 & = & 0 \\
-2x_1 & + & x_2 & + & 3x_3 & = & 0 \\
x_1 & + & 2x_2 & - & 4x_3 & = & 0
\end{cases}
$$

Find three other solutions of this system.

SOLUTION This is a homogeneous system, so part *(a)* of Theorem 2.2 says that any multiple of a solution is another solution. Thus, $2X = (4, 2, 2)$, $3.6X = (7.2, 3.6, 3.6)$, and $-3X = (-6, -3, -3)$ are also solutions of the given system.
☐

EXAMPLE 2.4

$X = (-1, 1, -2)$ is a solution of the system

$$
\begin{cases}
x_1 & + & x_2 & - & 3x_3 & = & 6 \\
-2x_1 & + & x_2 & + & x_3 & = & 1 \\
-x_1 & + & 2x_2 & - & 2x_3 & = & 7
\end{cases}
$$

and $Z = (4, 5, 3)$ is a solution of the associated homogeneous system. Find three other solutions of the given system.

SOLUTION This associated homogeneous system is

$$
\begin{cases}
x_1 & + & x_2 & - & 3x_3 & = & 0 \\
-2x_1 & + & x_2 & + & x_3 & = & 0 \\
-x_1 & + & 2x_2 & - & 2x_3 & = & 0
\end{cases}
$$

and part *(a)* of Theorem 2.2 says that any multiple of Z is another solution of the homogeneous system. Thus, $2Z = (8, 10, 6)$ and $-Z = (-4, -5, -3)$ are also solutions of the homogeneous system. Now, part *(c)* says that any solution of the homogeneous system added to the particular solution $X = (-1, 1, -2)$ will be another solution of the given system. Thus $X + Z = (3, 6, 1)$ is a solution of the given system; indeed, substituting into the system gives

$$
\begin{cases}
3 & + & 6 & - & 3(1) & = & 6 \\
-2(3) & + & 6 & + & 1 & = & 1 \\
-(3) & + & 2(6) & - & 2(1) & = & 7
\end{cases}
$$

as promised. Also, $X + 2Z = (7, 11, 4)$ and $X - Z = (-5, -4, -5)$ are solutions of the given system. $\quad\square$

EXAMPLE 2.5

$X = (-1, 1, 1, 1)$ and $Y = (0, -2, 3, 1)$ are solutions of the system

$$
\begin{cases}
x_1 & - & x_2 & - & 2x_3 & + & 2x_4 & = & -2 \\
x_1 & + & x_2 & + & x_3 & + & x_4 & = & 2 \\
3x_1 & + & x_2 & & & + & 4x_4 & = & 2
\end{cases}
$$

Find three other solutions of the given system.

SOLUTION According to part *(b)* of Theorem 2.2, the difference of two solutions of the given system will be a solution of the associated homogeneous system. That is, $X - Y = X_0 = (-1, 3, -2, 0)$ is a solution of the associated homogeneous system. Now, part *(a)* says that any multiple of X_0 is another solution of the homogeneous system. For example, $-X_0 = (1, -3, 2, 0)$ and $2X_0 = (-2, 6, -4, 0)$ are solutions of the homogeneous system. Finally, $X + 2X_0 = (-3, 7, -3, 1)$, $Y + 2X_0 = (-2, 4, -1, 1)$, $X - X_0 = (0, -2, 3, 1)$, and $Y - X_0 = (1, -5, 5, 1)$ are all solutions of the given system. (Note, however, that the third solution listed is actually Y again, so we only have three new solutions, not four.) $\quad\square$

The proof of Theorem 2.2 is not difficult and it illustrates principles that are important in much of what follows. Readers will find that any effort needed to understand the ideas of the proof will be well worth it.

PROOF   (of Theorem 2.2)

(a) Let X_1 and X_2 be solutions of the homogeneous system $AX = 0$ and let α be a number. Then $A(X_1 + X_2) = AX_1 + AX_2 = 0 + 0 = 0$, so $X_1 + X_2$ is also a solution of the homogeneous system. Similarly, $A(\alpha X_1) = \alpha AX_1 = \alpha \cdot 0 = 0$, so αX_1 is also a solution of the homogeneous system.

(b) We are given that $AX_3 = b$ and $AX_4 = b$, so $A(X_3 - X_4) = AX_3 - AX_4 = b - b = 0$. In other words, $X_3 - X_4$ is a solution of the homogeneous system $AX = 0$.

(c) Let X_p the given solution of the non-homogeneous system $AX = b$ and suppose X_q is another solution of $AX = b$. Let $X_0 = X_q - X_p$. Then $X_q = X_p + X_0$ and $AX_0 = A(X_q - X_p) = AX_q - AX_p = b - b = 0$, so X_0 is a solution of the homogeneous system $AX = 0$.

Now suppose $AX_p = b$ and $AX_0 = 0$, then $A(X_p + X_0) = AX_p + AX_0 = b + 0 = b.$ ∎

One consequence of Theorem 2.2 is that no system of linear equations has exactly three solutions or exactly seven solutions: the number of solutions of a system is either 0, 1, or infinity. A system of linear equations that has no solutions is said to be *inconsistent* while a system that has solutions is said to be *consistent*. If a linear system has exactly one solution, we say it has a *unique solution*.

COROLLARY 2.6

Every system of linear equations has, either no solution, exactly one solution, or infinitely many solutions. A homogeneous system either has exactly one solution, 0, or it has infinitely many solutions.

PROOF   Notice that $X = 0$ is a solution of the homogeneous system $AX = 0$, so homogeneous systems always have a solution. Moreover, if $AX = 0$ has a non–zero solution, X_1, then for all numbers α, part *(a)* says that αX_1 is also a solution, and for non–zero X_1, the set αX_1 as α runs over all numbers is an infinite set. That is, the system $AX = 0$ has either exactly one solution, namely $X = 0$, or infinitely many solutions and the conclusion is true in this case.

For $b \neq 0$, if $AX = b$ has no solutions, then the conclusion of the corollary is true. On the other hand, if X_p is a solution of $AX = b$, then either the corresponding homogeneous system $AX = 0$ has exactly one solution, $X_0 = 0$ and by *(b)*, $AX = b$ has only one solution $X_p = X_p + 0 = X_p + X_0$, or $AX = 0$ has infinitely many

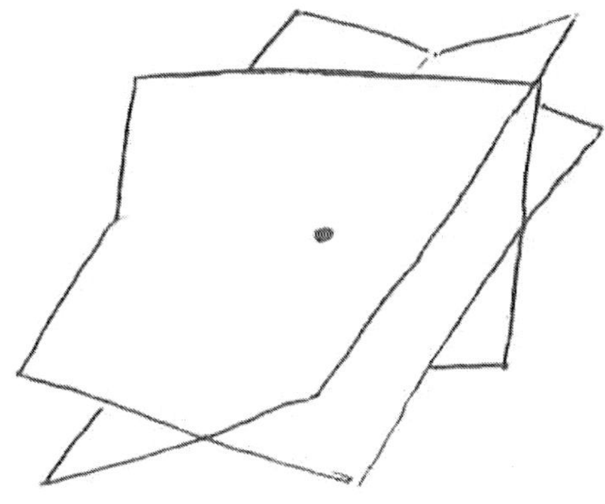

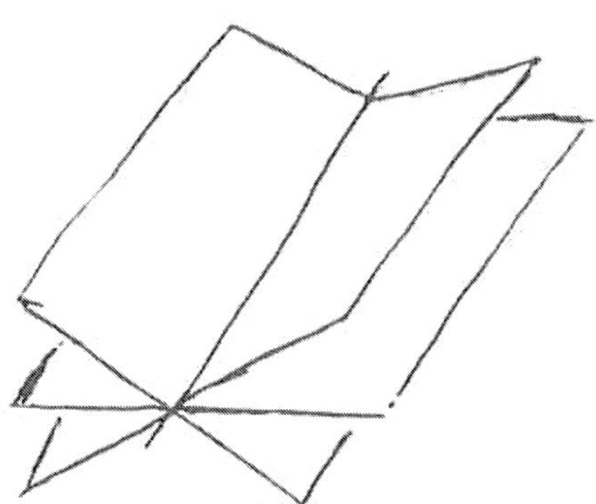

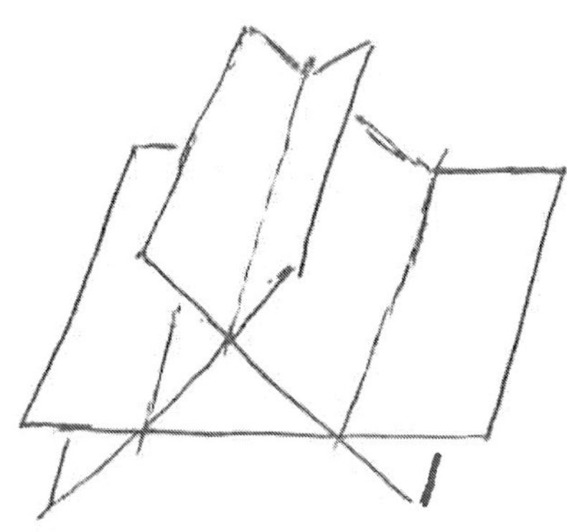

FIGURE 2.2

Geometry of systems of equations

solutions X_0 and $AX = b$ has infinitely many solutions also, $X_p + X_0$ for each solution X_0 of $AX = 0$.   ∎

This Theorem 2.2 and Corollary 2.6 show that there are three types of systems, classified by the number of solutions they have, and tell the relation of the solutions of the system and associated the homogeneous system. The proof of the corollary shows why we will concentrate on answering two questions:

When does $AX = b$ have a solution?

What sort of solutions does the homogeneous system $AX = 0$ have?

Since $AX = 0$ always has the solution $X = 0$, this is called the *trivial solution*; non–zero solutions of $AX = 0$ are called *non–trivial solutions*.

At least in systems with three unknowns, we can see geometrically how these conditions come about (see Figure 2.2). Recall that when there are three unknowns, each equation in the system represents a plane in $\mathbf{R}^3$. A system has a unique solution when the planes representing the different equations intersect in a single point. The system has infinitely many solutions in case the planes intersect in a line. There no solutions when the planes have parallel interesections. Figure 2.2 shows how these conditions can be achieved.

The plan for solving the system $AX = b$ is to successively replace the system by easier systems with the same set of solutions until, finally, the last system is called *the solution*. This process, called *Gaussian elimination*, is accomplished by *row operations* which correspond to operations on equations that do not change the solution set. We will explore this important algorithm in the next section.

The MATLAB command A\b gives one solution to the system $AX = b$ whenever the sizes of A and b are compatible. We will learn more about which solution MATLAB gives to the system and what the "solution" means when the system has no solutions in Section 5.2. It is safest to check the "solution" obtained by hand or by MATLAB by substituting into the original system or, what is the same thing, by computing AX and comparing it with b.

Exercises 2.2

Solve the following systems.

1.
$$\begin{cases} w & = & 5 \\ 2w + x & = & 2 \\ w + x + y & = & -1 \\ w - x + 2y + z & = & 4 \end{cases}$$

2.
$$\begin{cases} a + 2b + c & = & 2 \\ -a + 3b - c + d & = & 3 \\ a & = & 4 \\ 2a + b & = & -1 \end{cases}$$

3.
$$\begin{cases} 2w - x + 2y + z & = & 1 \\ x + y - z & = & -2 \\ 3y + z & = & 0 \\ 2z & = & 6 \end{cases}$$

4.
$$\begin{cases} 2w - x + 2y + z & = & 0 \\ x + y - z & = & 0 \\ y - z & = & 0 \end{cases}$$

(Hint: solve for w, x, and y in terms of z. There will be infinitely many solutions, one for each value of z.)

5. Write each system in Problems 2.2.6.–2.2.9. as a matrix equation.

Use your software to solve the following systems. Be sure to check your answers!

$\diamond$ 6.
$$\begin{cases} x + 2y & = & 3 \\ 3x + 4y & = & -2 \end{cases}$$

$\diamond$ 7.
$$\begin{cases} x - y + z & = & 1 \\ -x + 3y + 3z & = & 5 \\ 2x + 3z & = & 4 \end{cases}$$

$\diamond$ 8.
$$\begin{cases} w - y + 2z & = & 0 \\ -w + x + 3y - z & = & 5 \\ 2w + 5z & = & 3 \\ w + x + y + 2z & = & 4 \end{cases}$$

$\diamond$ 9.
$$\begin{cases} 2w + 3x + y - z & = & 1 \\ -w + 2x + 3y + z & = & -1 \\ 2w + x - 2y + 3z & = & 0 \\ w - x + y + 2z & = & 2 \end{cases}$$

10. Consider the system:

$$\begin{cases} u + 2v + w - x - 2y & = & 3 \\ -2u + v + w + x + 2y & = & 5 \\ u + v - w + 2x + 4y & = & -2 \\ u - v + 3x + y & = & -7 \\ -u + 3v + w + x + 3y & = & 7 \end{cases}$$

(a) Choose A and b so that the system can be written in matrix form as $AX = b$ where $X = (u, v, w, x, y)$.

(b) Check that $X_p = (-1, 1, 2, -2, 1)$ is a solution of the system and check that $X_0 = (-1, 1, -2, 1, -1)$ is a solution of the associated homogeneous system $AX = 0$.

(c) Without using Gaussian elimination or a machine, find two other non-trivial solutions of $AX = 0$.

(d) Without using Gaussian elimination or a machine, find two other solutions of $AX = b$.

11. The five-tuples $(2, 2, 1, -1, 1)$ and $(1, 1, 2, -1, -1)$ are both solutions of the system:

$$\begin{cases} a + b + 4c + d + e &= 8 \\ a - b + 2c + 2d + e &= 1 \\ 2a + b - c - d - 2e &= 4 \\ b + 3c + d + e &= 5 \\ 2a - b + c + 3d &= 0 \end{cases}$$

(a) Without using Gaussian elimination or a machine, write down two non-trivial solutions of the associated homogeneous system.

(b) Write down two other solutions of the given system.

12. Let A be the matrix

$$\begin{pmatrix} 1 & -1 & 2 & 1 \\ 2 & 1 & -3 & -1 \\ 1 & 1 & 3 & -2 \\ -1 & 2 & -2 & 3 \end{pmatrix}$$

and let $b = (3, -1, 3, 2)$ and let $c = (0, 4, -4, 4)$.

(a) Check that $Y = (1, 1, 1, 1)$ solves the system $AX = b$ and that $Z = (1, 0, -1, 1)$ solves the system $AX = c$.

(b) Without using Gaussian Elimination or a machine, find a solution of the system $AX = (6, -2, 6, 4) = 2b$.

(c) Without using Gaussian Elimination or a machine, find a solution of the system $AX = (3, 3, -1, 6) = b + c$.

(d) Without using Gaussian Elimination or a machine, find a solution of the system $AX = (9, 5, 1, 14) = 3b + 2c$.

13. Let $A = \begin{pmatrix} 2 & -1 & 1 & 0 \\ 1 & 1 & -2 & -1 \\ 0 & 1 & -2 & 1 \\ 1 & -1 & 0 & 1 \end{pmatrix}$

(a) Find, by guessing, a vector X so that $AX = \begin{pmatrix} 2 \\ 1 \\ 0 \\ 1 \end{pmatrix}$. Notice that $(2, 1, 0, 1)$ is the first column of A.

(b) Guess a vector X so that $AX = (-1, 1, 1, -1)$, the second column of A.

(c) Guess a vector X so that $AX = (1, -2, -2, 0)$, the third column of A.

(d) Guess a vector X so that $AX = (0, -1, 1, 1)$, the fourth column of A.

(e) Guess a vector X so that $AX = (-1, 4, 5, 0)$. Notice that

$$\begin{pmatrix} -1 \\ 4 \\ 5 \\ 0 \end{pmatrix} = \begin{pmatrix} 2 \\ 1 \\ 0 \\ 1 \end{pmatrix} + 2\begin{pmatrix} -1 \\ 1 \\ 1 \\ -1 \end{pmatrix} - \begin{pmatrix} 1 \\ -2 \\ -2 \\ 0 \end{pmatrix} + \begin{pmatrix} 0 \\ -1 \\ 1 \\ 1 \end{pmatrix}$$

14. Let $A = \begin{pmatrix} 1 & -1 & 1 & 2 \\ 2 & -3 & 1 & 0 \\ 1 & 3 & -1 & 1 \\ 2 & 1 & 0 & 1 \end{pmatrix}$

(a) Guess a vector X so that $AX = (1, 2, 1, 2)$. Notice that $(1, 2, 1, 2)$ is the first column of A.

(b) Guess a vector X so that $AX = (-1, -3, 3, 1)$, the second column of A.

(c) Guess a vector X so that $AX = (1, 1, -1, 0)$, the third column of A.

(d) Guess a vector X so that $AX = (2, 0, 1, 1)$, the fourth column of A.

(e) Guess a vector X so that $AX = (2, 10, -6, 1)$. Notice that

$$\begin{pmatrix} 2 \\ 10 \\ -6 \\ 1 \end{pmatrix} = 2\,(\ 1, 2, 1, 2\ ) - (\ -1, -3, 3, 1\ ) + 3\,(\ 1, 1, -1, 0\ ) - 2\,(\ 2, 0, 1, 1\ )$$

2.3 Gaussian Elimination

Two systems of equations are called *equivalent* if they have the same set of solutions. In Gaussian elimination, the original system is replaced by a succession of easier, but equivalent systems until the final equivalent system represents an easy system to solve or perhaps even the solution to the system. For example, consider the system

$$\begin{cases} x_1 & - & 2x_2 & + & x_3 & = & 3 \\ -2x_1 & + & 3x_2 & - & x_3 & = & -4 \\ x_1 & - & 5x_2 & + & 5x_3 & = & 12 \end{cases}$$

If x_1, x_2, and x_3 satisfy these three equations, then multiplying the first equation by 2, we see that they satisfy $2x_1 - 4x_2 + 2x_3 = 6$ and they must satisfy the sum of this and the second equation: $-x_2 + x_3 = 2$. If we replace the second equation in the original system by this one, we see that every solution of the original system satisfies the system

$$\begin{cases} x_1 & - & 2x_2 & + & x_3 & = & 3 \\ & - & x_2 & + & x_3 & = & 2 \\ x_1 & - & 5x_2 & + & 5x_3 & = & 12 \end{cases}$$

On the other hand, if x_1, x_2, and x_3 satisfy these latter equations, then they satisfy the sum of -2 times the first equation of this system and the second equation of this system. Replacing the second equation of the second system by this gives the original system, so every solution of the second system is also a solution of the first system. Thus, the two systems are equivalent. Now, the variable x_1 does not occur in the second equation of the second system, that is, we have eliminated x_1 from this equation. The goal is to eliminate x_1 and x_2 from one of the equations and thereby have an equation that only contains x_3 which can be easily solved. In fact, we end up with a triangular system which, as we have seen, is easily solved. To continue, we may replace the third equation by the third equation minus the first equation to get the system

$$\begin{cases} x_1 & - & 2x_2 & + & x_3 & = & 3 \\ & - & x_2 & + & x_3 & = & 2 \\ & - & 3x_2 & + & 4x_3 & = & 9 \end{cases}$$

in which the variable x_1 has been eliminated from the third equation as well. In the same way, we may replace the third equation by the third equation minus 3 times the second equation to get the system

$$\begin{cases} x_1 & - & 2x_2 & + & x_3 & = & 3 \\ & - & x_2 & + & x_3 & = & 2 \\ & & & + & x_3 & = & 3 \end{cases}$$

We easily find the solution of the triangular system is $x_3 = 3$, $x_2 = 1$, and $x_1 = 2$. But we were careful to make sure this triangular system is equivalent to the original system, that is, the solution of the original system is also $x_3 = 3$, $x_2 = 1$, and $x_1 = 2$.

Now, we can see that the important features of this approach deal only with the coefficients and the right hand sides of the system, that the variables x_1, etc., only serve as markers for the manipulations with the coefficients. With this in mind, we can replace the system by the *augmented matrix*

$$\left(\begin{array}{ccc|c} 1 & -2 & 1 & 3 \\ -2 & 3 & -1 & -4 \\ 1 & -5 & 5 & 12 \end{array} \right)$$

which we remember represents the system above. Replacing the system by the second equivalent system can be represented by

$$\left(\begin{array}{ccc|c} 1 & -2 & 1 & 3 \\ -2 & 3 & -1 & -4 \\ 1 & -5 & 5 & 12 \end{array} \right) \longrightarrow \left(\begin{array}{ccc|c} 1 & -2 & 1 & 3 \\ 0 & -1 & 1 & 2 \\ 1 & -5 & 5 & 12 \end{array} \right)$$

In fact the whole solution process can be represented by

$$\begin{pmatrix} 1 & -2 & 1 & \Big| & 3 \\ -2 & 3 & -1 & \Big| & -4 \\ 1 & -5 & 5 & \Big| & 12 \end{pmatrix} \longrightarrow \begin{pmatrix} 1 & -2 & 1 & \Big| & 3 \\ 0 & -1 & 1 & \Big| & 2 \\ 1 & -5 & 5 & \Big| & 12 \end{pmatrix}$$

$$\longrightarrow \begin{pmatrix} 1 & -2 & 1 & \Big| & 3 \\ 0 & -1 & 1 & \Big| & 2 \\ 0 & -3 & 4 & \Big| & 9 \end{pmatrix} \longrightarrow \begin{pmatrix} 1 & -2 & 1 & \Big| & 3 \\ 0 & -1 & 1 & \Big| & 2 \\ 0 & 0 & 1 & \Big| & 3 \end{pmatrix}$$

in which the matrices are the augmented matrices for the systems and the arrows signify the replacement of a system by an equivalent system.

Concentrating on the augmented matrices, each is obtained from the previous one by a *row operation* and the process by which we replace an augmented matrix with the augmented matrix of an equivalent system is called *row reduction*. From the point of view of systems of equations, we are interested in row operations that do not change the set of solutions, that is, that result in a matrix that represents an equivalent system. There are three convenient row operations that have this property, that is, if an augmented matrix represents a system and one of the following row operations is done on it, the resulting system is equivalent to original system.

The row operations of Gaussian Elimination

1. Replace a row by the sum of that row and a multiple of another row.

2. Replace a row by a non–zero multiple of that row.

3. Interchange two rows.

The goal of using these operations is to replace the original system with an equivalent upper triangular system. As we have seen, a triangular system is easy to solve, and because it is equivalent to the original system, the two systems have the same set of solutions.

EXAMPLE 2.7

As a further example, decide whether the following system has no solutions, exactly one solution, or infinitely many solutions.

$$\begin{cases} x_1 & - & 2x_2 & + & x_3 & - & x_4 & = & 1 \\ 2x_1 & - & 4x_2 & + & 3x_3 & - & x_4 & = & 4 \\ -x_1 & + & 3x_2 & - & 2x_3 & + & 4x_4 & = & 7 \\ x_1 & - & 4x_2 & + & 4x_3 & - & 4x_4 & = & -7 \end{cases}$$

If the system has no solutions, say so; if the system has a unique solution, find it and say it is unique; if the system has infinitely many solutions, say that it has infinitely many solutions and find three solutions explicitly.

SOLUTION The augmented matrix for this system is

$$\left(\begin{array}{cccc|c} 1 & -2 & 1 & -1 & 1 \\ 2 & -4 & 3 & -1 & 4 \\ -1 & 3 & -2 & 4 & 7 \\ 1 & -4 & 4 & -4 & -7 \end{array} \right)$$

and replacing the second row by the second minus twice the first, we obtain

$$\left(\begin{array}{cccc|c} 1 & -2 & 1 & -1 & 1 \\ 2 & -4 & 3 & -1 & 4 \\ -1 & 3 & -2 & 4 & 7 \\ 1 & -4 & 4 & -4 & -7 \end{array} \right) \longrightarrow \left(\begin{array}{cccc|c} 1 & -2 & 1 & -1 & 1 \\ 0 & 0 & 1 & 1 & 2 \\ -1 & 3 & -2 & 4 & 7 \\ 1 & -4 & 4 & -4 & -7 \end{array} \right)$$

Now replacing the third by the third plus the first, we obtain

$$\left(\begin{array}{cccc|c} 1 & -2 & 1 & -1 & 1 \\ 0 & 0 & 1 & 1 & 2 \\ -1 & 3 & -2 & 4 & 7 \\ 1 & -4 & 4 & -4 & -7 \end{array} \right) \longrightarrow \left(\begin{array}{cccc|c} 1 & -2 & 1 & -1 & 1 \\ 0 & 0 & 1 & 1 & 2 \\ 0 & 1 & -1 & 3 & 8 \\ 1 & -4 & 4 & -4 & -7 \end{array} \right)$$

Next, we will use the first row again to eliminate the 1 in the fourth row, first column. In fact, we have used the first row in each of these row operations, so it is permissible, and shorter to do all in one step:

$$\left(\begin{array}{cccc|c} 1 & -2 & 1 & -1 & 1 \\ 2 & -4 & 3 & -1 & 4 \\ -1 & 3 & -2 & 4 & 7 \\ 1 & -4 & 4 & -4 & -7 \end{array} \right) \longrightarrow \left(\begin{array}{cccc|c} 1 & -2 & 1 & -1 & 1 \\ 0 & 0 & 1 & 1 & 2 \\ 0 & 1 & -1 & 3 & 8 \\ 0 & -2 & 3 & -3 & -8 \end{array} \right)$$

(If we did not use the same row in each of the row operations, it would be dangerous to combine the steps in this way because it is possible to throw away information and not have an equivalent system.)

Since the entry in the second row, second column is 0, we cannot use it to eliminate x_2 from the remaining equations: we will therefore interchange the second and third rows before proceeding further:

$$\left(\begin{array}{cccc|c} 1 & -2 & 1 & -1 & 1 \\ 0 & 0 & 1 & 1 & 2 \\ 0 & 1 & -1 & 3 & 8 \\ 0 & -2 & 3 & -3 & -8 \end{array}\right) \longrightarrow \left(\begin{array}{cccc|c} 1 & -2 & 1 & -1 & 1 \\ 0 & 1 & -1 & 3 & 8 \\ 0 & 0 & 1 & 1 & 2 \\ 0 & -2 & 3 & -3 & -8 \end{array}\right)$$

Then using the new second row, we get

$$\left(\begin{array}{cccc|c} 1 & -2 & 1 & -1 & 1 \\ 0 & 1 & -1 & 3 & 8 \\ 0 & 0 & 1 & 1 & 2 \\ 0 & -2 & 3 & -3 & -8 \end{array}\right) \longrightarrow \left(\begin{array}{cccc|c} 1 & -2 & 1 & -1 & 1 \\ 0 & 1 & -1 & 3 & 8 \\ 0 & 0 & 1 & 1 & 2 \\ 0 & 0 & 1 & 3 & 8 \end{array}\right)$$

Replacing the fourth row by it minus the third row, we get

$$\left(\begin{array}{cccc|c} 1 & -2 & 1 & -1 & 1 \\ 0 & 1 & -1 & 3 & 8 \\ 0 & 0 & 1 & 1 & 2 \\ 0 & 0 & 1 & 3 & 8 \end{array}\right) \longrightarrow \left(\begin{array}{cccc|c} 1 & -2 & 1 & -1 & 1 \\ 0 & 1 & -1 & 3 & 8 \\ 0 & 0 & 1 & 1 & 2 \\ 0 & 0 & 0 & 2 & 6 \end{array}\right)$$

This triangular system is equivalent to the original system, but it is easily solved to obtain $x_4 = 3$, $x_3 = -1$, $x_2 = -2$, and $x_1 = 1$, so this must be the solution of the original system. We would probably simply write

$$\left(\begin{array}{cccc|c} 1 & -2 & 1 & -1 & 1 \\ 2 & -4 & 3 & -1 & 4 \\ -1 & 3 & -2 & 4 & 7 \\ 1 & -4 & 4 & -4 & -7 \end{array}\right) \longrightarrow \left(\begin{array}{cccc|c} 1 & -2 & 1 & -1 & 1 \\ 0 & 0 & 1 & 1 & 2 \\ 0 & 1 & -1 & 3 & 8 \\ 0 & -2 & 3 & -3 & -8 \end{array}\right)$$

$$\longrightarrow \left(\begin{array}{cccc|c} 1 & -2 & 1 & -1 & 1 \\ 0 & 1 & -1 & 3 & 8 \\ 0 & 0 & 1 & 1 & 2 \\ 0 & -2 & 3 & -3 & -8 \end{array}\right) \longrightarrow \left(\begin{array}{cccc|c} 1 & -2 & 1 & -1 & 1 \\ 0 & 1 & -1 & 3 & 8 \\ 0 & 0 & 1 & 1 & 2 \\ 0 & 0 & 1 & 3 & 8 \end{array}\right)$$

$$\longrightarrow \left(\begin{array}{cccc|c} 1 & -2 & 1 & -1 & 1 \\ 0 & 1 & -1 & 3 & 8 \\ 0 & 0 & 1 & 1 & 2 \\ 0 & 0 & 0 & 2 & 6 \end{array}\right)$$

Since this augmented matrix is upper triangular, the system it represents

$$\begin{cases} x_1 & - & 2x_2 & + & x_3 & - & x_4 & = & 1 \\ & & x_2 & - & x_3 & + & 3x_4 & = & 8 \\ & & & & x_3 & + & x_4 & = & 2 \\ & & & & & & 2x_4 & = & 6 \end{cases}$$

is easy to solve by solving the last equation for x_4, replacing that value in the third equation, and so on. We can do essentially the same arithmetic and finish the problem in the augmented matrix form by doing row operations to eliminate the non-zero entries above the diagonal insofar as is possible.

Dividing the last row of the augmented matrix by 2, the original system is equivalent to

$$\left(\begin{array}{cccc|c} 1 & -2 & 1 & -1 & 1 \\ 0 & 1 & -1 & 3 & 8 \\ 0 & 0 & 1 & 1 & 2 \\ 0 & 0 & 0 & 1 & 3 \end{array} \right)$$

We will use the 1 in the $(4,4)$ entry to eliminate the other non-zero entries in the fourth column; this is equivalent to putting the newly found value of x_4, namely, $x_4 = 3$, into the other equations.

$$\longrightarrow \left(\begin{array}{cccc|c} 1 & -2 & 1 & 0 & 4 \\ 0 & 1 & -1 & 0 & -1 \\ 0 & 0 & 1 & 0 & -1 \\ 0 & 0 & 0 & 1 & 3 \end{array} \right)$$

Now, we can continue using the 1's in the $(3,3)$ entry and the $(2,2)$ entry to eliminate the non-zero entries above the diagonal.

$$\longrightarrow \left(\begin{array}{cccc|c} 1 & -2 & 0 & 0 & 5 \\ 0 & 1 & 0 & 0 & -2 \\ 0 & 0 & 1 & 0 & -1 \\ 0 & 0 & 0 & 1 & 3 \end{array} \right) \longrightarrow \left(\begin{array}{cccc|c} 1 & 0 & 0 & 0 & 1 \\ 0 & 1 & 0 & 0 & -2 \\ 0 & 0 & 1 & 0 & -1 \\ 0 & 0 & 0 & 1 & 3 \end{array} \right)$$

Our calculations show that the original system has the same set of solutions as the system

$$\begin{cases} x_1 & & & & = & 1 \\ & x_2 & & & = & -2 \\ & & x_3 & & = & -1 \\ & & & x_4 & = & 3 \end{cases}$$

This system obviously has exactly one solution: $x_1 = 1$, $x_2 = -2$, $x_3 = -1$, and $x_4 = 3$, so we conclude that the original system has a unique solution and this is the solution. We will usually write this as $X = (1, -2, -1, 3)$. Of course, we can check to see that this is a solution by putting $x_1 = 1$, $x_2 = -2$, $x_3 = -1$, and $x_4 = 3$ into the original system; indeed, each equation is true, so this is a solution of the system. $\Box$

The example above is typical of the computation, by hand, of a system that has a unique solution. As we will soon see, if the system has no solutions or infinitely many solutions, the computation does not work out in exactly this way, but the basic strategy is the same.

Notice that in each step, one row was kept and used to create zeros in other rows. The entries that were eliminated in other rows were the entries in the same column of the first non-zero entry of the row being kept: this entry is called the "pivot". The pivot cannot be zero because there is nothing to multiply zero by to get a non-zero number for the elimination. It is for this reason that we interchanged the second and third rows before doing any work in the second column: the zero in the $(2, 2)$ entry before the switch could not be the pivot. (It is not really necessary to interchange the rows — one could simply use the $(3, 2)$ entry as the pivot. This is commonly done in machines, but when computing by hand, the rows will usually be exchanged.)   Notice also that in doing the first half of the problem, we were careful not to reuse the first row after creating the 0's in the first column: had we used the first row again, we would have reintroduced non-zero entries in the first column. Similarly, we did not use the second row again on the rows below it: had we done so, we would have reintroduced non-zero entries in the second column.

The strategy suggested below is convenient for hand calculation because it proceeds methodically from top to bottom, left to right in the first phase, and bottom to top, right to left in the second phase, thus preventing circularity. In most cases, the machine computation of solutions of systems of linear equations uses this sort of procedure with some modifications to ensure that accuracy is not lost in the process of calculation due to round-off errors: usually, the pivots are chosen in a different way.

Typical Strategy for Solving Systems of Linear Equations

1. Express the system as an augmented matrix.

2. Interchange rows if necessary, then keep the first row and use the $(1,1)$ entry as a pivot to create zeros in the first column below the $(1,1)$ entry.

3. Interchange rows below the first if necessary, then keep the second row and use the $(2,2)$ entry as a pivot to create zeros in the second column below the $(2,2)$ entry.

4. Continue using rows to eliminate entries below the pivot until there are no more non-zero rows. If, at any stage, the desired pivot entry and all entries below it are zero, use the entry next to the right as a pivot.

5. Use the first non-zero entry of the last non-zero row as a pivot to create zeros in the column above the pivot.

6. Use the first non-zero entry of the next-to-last non-zero row as a pivot to create zeros in the column above the pivot.

7. Continue from the bottom to the top until each pivot is the only non-zero entry in its column: this augmented matrix represents the solution of the system.

8. Interpret the final augmented matrix to express the solution of the system in a convenient form.

EXAMPLE 2.8

Let us consider the following system.

$$\begin{cases} x_1 + 2x_2 - x_3 + x_4 = 2 \\ -x_1 - x_2 \qquad\quad - 2x_4 = -1 \\ 2x_1 + 5x_2 - 4x_3 + 3x_4 = 7 \\ -x_1 - x_2 + x_3 - 4x_4 = -3 \end{cases}$$

The augmented matrix for this system is

$$\left(\begin{array}{cccc|c} 1 & 2 & -1 & 1 & 2 \\ -1 & -1 & 0 & -2 & -1 \\ 2 & 5 & -4 & 3 & 7 \\ -1 & -1 & 1 & -4 & -3 \end{array} \right)$$

The $(1, 1)$ entry being non-zero, we may use it as the pivot and save the first row. To emphasize the role of the pivot, we will circle each pivot entry. Replacing the second row by the second plus the first, replacing the third row by the third minus twice the first, and replacing the fourth by the fourth plus the first, we obtain

$$
\left(\begin{array}{cccc|c}
① & 2 & -1 & 1 & 2 \\
-1 & -1 & 0 & -2 & -1 \\
2 & 5 & -4 & 3 & 7 \\
-1 & -1 & 1 & -4 & -3
\end{array}\right)
\longrightarrow
\left(\begin{array}{cccc|c}
① & 2 & -1 & 1 & 2 \\
0 & 1 & -1 & -1 & 1 \\
0 & 1 & -2 & 1 & 3 \\
0 & 1 & 0 & -3 & -1
\end{array}\right)
$$

Now we can use the second row for elimination, since the $(2, 2)$ entry can be used as the pivot, to put zeros in the $(3, 2)$ and $(4, 2)$ places. Replacing the third row by the third row minus the second, and replacing the fourth row by the fourth row minus the second, we get

$$
\left(\begin{array}{cccc|c}
① & 2 & -1 & 1 & 2 \\
0 & ① & -1 & -1 & 1 \\
0 & 1 & -2 & 1 & 3 \\
0 & 1 & 0 & -3 & -1
\end{array}\right)
\longrightarrow
\left(\begin{array}{cccc|c}
① & 2 & -1 & 1 & 2 \\
0 & ① & -1 & -1 & 1 \\
0 & 0 & -1 & 2 & 2 \\
0 & 0 & 1 & -2 & -2
\end{array}\right)
$$

Finally, replacing the last row by the last row plus the third row, we get

$$
\left(\begin{array}{cccc|c}
① & 2 & -1 & 1 & 2 \\
0 & ① & -1 & -1 & 1 \\
0 & 0 & ⊝ & 2 & 2 \\
0 & 0 & 1 & -2 & -2
\end{array}\right)
\longrightarrow
\left(\begin{array}{cccc|c}
① & 2 & -1 & 1 & 2 \\
0 & ① & -1 & -1 & 1 \\
0 & 0 & ⊝ & 2 & 2 \\
0 & 0 & 0 & 0 & 0
\end{array}\right)
$$

The system represented by this augmented matrix

$$
\left\{
\begin{array}{rcrcrcrcr}
x_1 & + & 2x_2 & - & x_3 & + & x_4 & = & 2 \\
 & & x_2 & - & x_3 & - & x_4 & = & 1 \\
 & & & & -x_3 & + & 2x_4 & = & 2 \\
0x_1 & + & 0x_2 & + & 0x_3 & + & 0x_4 & = & 0
\end{array}
\right.
$$

is equivalent to the original system, so the systems have the same set of solutions. On the other hand, the last equation of this system

$$
0x_1 + 0x_2 + 0x_3 + 0x_4 = 0
$$

is true for *all* choices of values of the variables! In other words, this equation does not contribute to the solution of the system. There are no longer enough conditions to determine each variable uniquely: we will do the best we can by solving for three of the variables in terms of the other one, say, x_1, x_2, and x_3 in terms of x_4. The third equation is $-x_3 + 2x_4 = 2$ which gives a relationship between x_3 and x_4 but does not determine either of them. Solving for x_3, we get

$$x_3 = -2 + 2x_4$$

Now, we can substitute for x_3 in the first two equations to get

$$\begin{cases} x_1 + 2x_2 - (-2 + 2x_4) + x_4 = 2 \\ x_2 - (-2 + 2x_4) - x_4 = 1 \end{cases}$$

which is the same as

$$\begin{cases} x_1 + 2x_2 - x_4 = 0 \\ x_2 - 3x_4 = -1 \end{cases}$$

Solving the second equation for x_2 gives $x_2 = -1 + 3x_4$ so we have x_2 in terms of x_4 also. Finally, using this in the first equation implies $x_1 + 2(-1 + 3x_4) - x_4 = 0$, so we get $x_1 = 2 - 5x_4$. Altogether, this is

$$\begin{cases} x_1 = 2 - 5x_4 \\ x_2 = -1 + 3x_4 \\ x_3 = -2 + 2x_4 \\ x_4 \quad \text{arbitrary} \end{cases}$$

In other words, for each value of x_4, there is a solution of the system. Since x_4 can take any value, there are infinitely many solutions. For example, $x_1 = 2$, $x_2 = -1$, $x_3 = -2$, and $x_4 = 0$ is a solution and $x_1 = 12$, $x_2 = -7$, $x_3 = -6$, and $x_4 = -2$ is another solution of the system. Since x_4 can take any value, it is frequently called a "free variable". ▯

In fact, the backsolving necessary to complete the solution of the system from the upper triangular system can be accomplished by row operations in essentially the same way we did in Example 2.7.

EXAMPLE 2.8, continued. We had found that the original system is equivalent to the system represented by the following augmented matrix.

$$\left(\begin{array}{cccc|c} ① & 2 & -1 & 1 & 2 \\ 0 & ① & -1 & -1 & 1 \\ 0 & 0 & ⊝ & 2 & 2 \\ 0 & 0 & 0 & 0 & 0 \end{array}\right)$$

We are half done because we have reached the bottom of the matrix and we have eliminated all the non-zero entries below the pivots. Our goal now will be to use row operations to replace the non-zero entries above each of the pivots without changing the zeros we have created below the pivots.

Using the -1 in the third row as the pivot will allow us to eliminate the non-zero entries above it and, because there are only zeros in the third row to the left of the pivot, we will not ruin any of the work we had previously done in creating zeros below the pivots.

Replacing the first row by the first minus the third and replacing the second row by the second minus the third, gives an augmented matrix in which the pivot -1 has no other non-zero entries in its column.

$$\left(\begin{array}{cccc|c} ① & 2 & 0 & -1 & 0 \\ 0 & ① & 0 & -3 & -1 \\ 0 & 0 & ⊝ & 2 & 2 \\ 0 & 0 & 0 & 0 & 0 \end{array}\right)$$

Finally, replacing the first row by the first row plus -2 times the second row gives an equivalent system in which x_2 has been eliminated from all but one equation.

$$\left(\begin{array}{cccc|c} ① & 0 & 0 & 5 & 2 \\ 0 & ① & 0 & -3 & -1 \\ 0 & 0 & ⊝ & 2 & 2 \\ 0 & 0 & 0 & 0 & 0 \end{array}\right)$$

This system is

$$\left\{\begin{array}{rcrcrcrcr} x_1 & & & & & + & 5x_4 & = & 2 \\ & & x_2 & & & - & 3x_4 & = & -1 \\ & & & - & x_3 & + & 2x_4 & = & 2 \end{array}\right.$$

or

$$\begin{cases} x_1 &= 2 - 5x_4 \\ x_2 &= -1 + 3x_4 \\ x_3 &= -2 + 2x_4 \\ x_4 & \quad \text{arbitrary} \end{cases}$$

as we obtained above.

We would probably more often write this answer in the vector form

$$X = \begin{pmatrix} 2 - 5x_4 \\ -1 + 3x_4 \\ -2 + 2x_4 \\ x_4 \end{pmatrix} = \begin{pmatrix} 2 \\ -1 \\ -2 \\ 0 \end{pmatrix} + \begin{pmatrix} -5x_4 \\ 3x_4 \\ 2x_4 \\ x_4 \end{pmatrix} = \begin{pmatrix} 2 \\ -1 \\ -2 \\ 0 \end{pmatrix} + x_4 \begin{pmatrix} -5 \\ 3 \\ 2 \\ 1 \end{pmatrix}$$

where there is one solution for every choice of x_4. $\square$

To understand more clearly the form of the answer we have just found, let's consider the homogeneous system associated with the system of Example 2.8.

EXAMPLE 2.9

The homogeneous system associated with the system of Example 2.8 is

$$\begin{cases} x_1 &+ 2x_2 &- &x_3 &- &x_4 &= 0 \\ -x_1 &- &x_2 & & &- 2x_4 &= 0 \\ 2x_1 &+ 5x_2 &- &4x_3 &+ &3x_4 &= 0 \\ -x_1 &- &x_2 &+ &x_3 &- 4x_4 &= 0 \end{cases}$$

and the augmented matrix for the system is

$$\left(\begin{array}{cccc|c} 1 & 2 & -1 & 1 & 0 \\ -1 & -1 & 0 & -2 & 0 \\ 2 & 5 & -4 & 3 & 0 \\ -1 & -1 & 1 & -4 & 0 \end{array} \right)$$

Since the coefficient matrix is the same as in the previous example, the same row operations will work: replacing the second row by the second plus the first, replacing the third row by the third minus twice the first, and replacing the fourth by the fourth plus the first, we obtain

$$\left(\begin{array}{cccc|c} \textcircled{1} & 2 & -1 & 1 & 0 \\ -1 & -1 & 0 & -2 & 0 \\ 2 & 5 & -4 & 3 & 0 \\ -1 & -1 & 1 & -4 & 0 \end{array} \right) \longrightarrow \left(\begin{array}{cccc|c} \textcircled{1} & 2 & -1 & 1 & 0 \\ 0 & 1 & -1 & -1 & 0 \\ 0 & 1 & -2 & 1 & 0 \\ 0 & 1 & 0 & -3 & 0 \end{array} \right)$$

Using the second row for elimination and replacing the third row by the third row minus the second and replacing the fourth row by the fourth row minus the second, we get

$$
\begin{pmatrix}
\textcircled{1} & 2 & -1 & 1 & | & 0 \\
0 & \textcircled{1} & -1 & -1 & | & 0 \\
0 & 1 & -2 & 1 & | & 0 \\
0 & 1 & 0 & -3 & | & 0
\end{pmatrix}
\longrightarrow
\begin{pmatrix}
\textcircled{1} & 2 & -1 & 1 & | & 0 \\
0 & \textcircled{1} & -1 & -1 & | & 0 \\
0 & 0 & -1 & 2 & | & 0 \\
0 & 0 & 1 & -2 & | & 0
\end{pmatrix}
$$

Finally, replacing the last row by the last row plus the third row, we get

$$
\begin{pmatrix}
\textcircled{1} & 2 & -1 & 1 & | & 0 \\
0 & \textcircled{1} & -1 & -1 & | & 0 \\
0 & 0 & \boxed{-1} & 2 & | & 0 \\
0 & 0 & 1 & -2 & | & 0
\end{pmatrix}
\longrightarrow
\begin{pmatrix}
\textcircled{1} & 2 & -1 & 1 & | & 0 \\
0 & \textcircled{1} & -1 & -1 & | & 0 \\
0 & 0 & \boxed{-1} & 2 & | & 0 \\
0 & 0 & 0 & 0 & | & 0
\end{pmatrix}
$$

For the backsolving stage, replacing the first row by the first minus the third and replacing the second row by the second minus the third, gives

$$
\begin{pmatrix}
\textcircled{1} & 2 & 0 & -1 & | & 0 \\
0 & \textcircled{1} & 0 & -3 & | & 0 \\
0 & 0 & \boxed{-1} & 2 & | & 0 \\
0 & 0 & 0 & 0 & | & 0
\end{pmatrix}
$$

and finally, replacing the first row by the first row plus -2 times the second row gives

$$
\begin{pmatrix}
\textcircled{1} & 0 & 0 & 5 & | & 0 \\
0 & \textcircled{1} & 0 & -3 & | & 0 \\
0 & 0 & \boxed{-1} & 2 & | & 0 \\
0 & 0 & 0 & 0 & | & 0
\end{pmatrix}
$$

This system is $x_1 + 5x_4 = 0$, $x_2 - 3x_4 = 0$, and $-x_3 + 2x_4 = 0$, or $x_1 = -5x_4$, $x_2 = 3x_4$, and $x_3 = 2x_4$ which we can write in the vector form $X = (-5x_4, 3x_4, 2x_4, x_4) = x_4(-5, 3, 2, 1)$. Thus, $X_0 = (-5, 3, 2, 1)$ is a solution of the homogeneous system above and there are infinitely many solutions, namely, all of the multiples of X_0.

We can interpret the solution of the system of Example 2.8 with reference to Theorem 2.2. We saw that $X_p = (2, -1, -2, 0)$ is the solution of the non-homogeneous system of Example 2.8 corresponding to $x_4 = 0$. We have just found the solutions of the associated homogeneous system are the multiples of X_0 and we wrote the set of solutions of Example 2.8 as a solution X_p, of the non-homogeneous system, plus solutions $x_4 X_0$ of the homogeneous system.   ⬜

EXAMPLE 2.10

Another example, closely related to the system we just considered, is the system

$$\begin{cases} x_1 + 2x_2 - x_3 + x_4 = 1 \\ -x_1 - x_2 - 2x_4 = -1 \\ 2x_1 + 5x_2 - 4x_3 + 3x_4 = 7 \\ -x_1 - x_2 + x_3 - 4x_4 = -3 \end{cases}$$

which has augmented matrix

$$\left(\begin{array}{cccc|c} 1 & 2 & -1 & 1 & 1 \\ -1 & -1 & 0 & -2 & -1 \\ 2 & 5 & -4 & 3 & 7 \\ -1 & -1 & 1 & -4 & -3 \end{array} \right)$$

Proceeding exactly as before, we obtain

$$\left(\begin{array}{cccc|c} \textcircled{1} & 2 & -1 & 1 & 1 \\ -1 & -1 & 0 & -2 & -1 \\ 2 & 5 & -4 & 3 & 7 \\ -1 & -1 & 1 & -4 & -3 \end{array} \right) \longrightarrow \left(\begin{array}{cccc|c} \textcircled{1} & 2 & -1 & 1 & 1 \\ 0 & \textcircled{1} & -1 & -1 & 0 \\ 0 & 1 & -2 & 1 & 5 \\ 0 & 1 & 0 & -3 & -2 \end{array} \right)$$

$$\longrightarrow \left(\begin{array}{cccc|c} \textcircled{1} & 2 & -1 & 1 & 1 \\ 0 & \textcircled{1} & -1 & -1 & 0 \\ 0 & 0 & -1 & 2 & 5 \\ 0 & 0 & 1 & -2 & -2 \end{array} \right)$$

Finally, replacing the last row by the last row plus the third row, we get

$$\left(\begin{array}{cccc|c} ① & 2 & -1 & 1 & 1 \\ 0 & ① & -1 & -1 & 0 \\ 0 & 0 & ⊖1 & 2 & 5 \\ 0 & 0 & 1 & -2 & -2 \end{array}\right) \longrightarrow \left(\begin{array}{cccc|c} ① & 2 & -1 & 1 & 1 \\ 0 & ① & -1 & -1 & 0 \\ 0 & 0 & ⊖1 & 2 & 5 \\ 0 & 0 & 0 & 0 & 3 \end{array}\right)$$

The system represented by this augmented matrix is equivalent to the original system, so the systems have the same set of solutions. On the other hand, the last equation of this system is

$$0x_1 + 0x_2 + 0x_3 + 0x_4 = 3$$

which is *not* true for *any* choice of the variables! In other words, this equation has no solutions which, in turn, means the original system has no solutions!   □

EXAMPLE 2.11

As a final example, consider the system

$$\begin{cases} x_1 + x_2 - x_3 & + x_5 = 1 \\ -x_1 - x_2 + 2x_3 - x_4 + x_5 = 2 \\ 2x_1 + 2x_2 - 3x_3 + 3x_4 + x_5 = 0 \\ x_1 + x_2 & - 3x_4 + 2x_5 = 3 \end{cases}$$

which has augmented matrix

$$\left(\begin{array}{ccccc|c} 1 & 1 & -1 & 0 & 1 & 1 \\ -1 & -1 & 2 & -1 & 1 & 2 \\ 2 & 2 & -3 & 3 & 1 & 0 \\ 1 & 1 & 0 & -3 & 2 & 3 \end{array}\right)$$

Proceeding exactly as before, using the $(1, 1)$ entry as the pivot and replacing the second row by itself plus the first, replacing the third row by itself minus twice the first, and replacing the fourth row by the fourth row minus the first, we obtain

$$\left(\begin{array}{ccccc|c} ① & 1 & -1 & 0 & 1 & 1 \\ -1 & -1 & 2 & -1 & 1 & 2 \\ 2 & 2 & -3 & 3 & 1 & 0 \\ 1 & 1 & 0 & -3 & 2 & 3 \end{array}\right) \longrightarrow \left(\begin{array}{ccccc|c} ① & 1 & -1 & 0 & 1 & 1 \\ 0 & 0 & 1 & -1 & 2 & 3 \\ 0 & 0 & -1 & 3 & -1 & -2 \\ 0 & 0 & 1 & -3 & 1 & 2 \end{array}\right)$$

We would expect to use the $(2, 2)$ entry as pivot, but it is zero. Moreover, rearranging the rows does not help because all the entries in the second column below the first row are zero. In this situation, we will use the $(3, 2)$ entry as pivot, replacing the third row by itself plus the second and replacing the fourth row by the fourth row minus the second.

$$\longrightarrow \left(\begin{array}{ccccc|c} ① & 1 & -1 & 0 & 1 & 1 \\ 0 & 0 & ① & -1 & 2 & 3 \\ 0 & 0 & -1 & 3 & -1 & -2 \\ 0 & 0 & 1 & -3 & 1 & 2 \end{array} \right) \longrightarrow \left(\begin{array}{ccccc|c} ① & 1 & -1 & 0 & 1 & 1 \\ 0 & 0 & ① & -1 & 2 & 3 \\ 0 & 0 & 0 & ② & 1 & 1 \\ 0 & 0 & 0 & -2 & -1 & -1 \end{array} \right)$$

$$\longrightarrow \left(\begin{array}{ccccc|c} ① & 1 & -1 & 0 & 1 & 1 \\ 0 & 0 & ① & -1 & 2 & 3 \\ 0 & 0 & 0 & ② & 1 & 1 \\ 0 & 0 & 0 & 0 & 0 & 0 \end{array} \right) \longrightarrow \left(\begin{array}{ccccc|c} ① & 1 & -1 & 0 & 1 & 1 \\ 0 & 0 & ① & -1 & 2 & 3 \\ 0 & 0 & 0 & ① & \frac{1}{2} & \frac{1}{2} \\ 0 & 0 & 0 & 0 & 0 & 0 \end{array} \right)$$

where, in the last step, we divided the third row by 2 to get a 1 in the pivot position. We are now half way through the elimination, having eliminated the non-zero entries below the pivots, and we are ready for the backsolving phase. We next use the rows from the bottom to the top to eliminate the non-zero entries above the pivots.

$$\left(\begin{array}{ccccc|c} ① & 1 & -1 & 0 & 1 & 1 \\ 0 & 0 & ① & -1 & 2 & 3 \\ 0 & 0 & 0 & ① & \frac{1}{2} & \frac{1}{2} \\ 0 & 0 & 0 & 0 & 0 & 0 \end{array} \right) \longrightarrow \left(\begin{array}{ccccc|c} ① & 1 & -1 & 0 & 1 & 1 \\ 0 & 0 & ① & 0 & \frac{5}{2} & \frac{7}{2} \\ 0 & 0 & 0 & ① & \frac{1}{2} & \frac{1}{2} \\ 0 & 0 & 0 & 0 & 0 & 0 \end{array} \right)$$

$$\longrightarrow \left(\begin{array}{ccccc|c} ① & 1 & 0 & 0 & \frac{7}{2} & \frac{9}{2} \\ 0 & 0 & ① & 0 & \frac{5}{2} & \frac{7}{2} \\ 0 & 0 & 0 & ① & \frac{1}{2} & \frac{1}{2} \\ 0 & 0 & 0 & 0 & 0 & 0 \end{array} \right)$$

This is the final equivalent system because we have eliminated all the non-zero entries from the pivot columns. This augmented matrix represents the system

$$\left\{ \begin{array}{rcl} x_1 \;+\; x_2 \;+\; \frac{7}{2}x_5 &=& \frac{9}{2} \\ x_3 \;+\; \frac{5}{2}x_5 &=& \frac{7}{2} \\ x_4 \;+\; \frac{1}{2}x_5 &=& \frac{1}{2} \end{array} \right.$$

or

$$\left\{ \begin{array}{rcl} x_1 &=& \frac{9}{2} \;-\; x_2 \;-\; \frac{7}{2}x_5 \\ x_3 &=& \frac{7}{2} \;-\; \frac{5}{2}x_5 \\ x_4 &=& \frac{1}{2} \;-\; \frac{1}{2}x_5 \end{array} \right.$$

In other words, there are infinitely many solutions, one for each choice of values for x_2 and x_5. For example, choosing 0 for both x_2 and x_5 gives the solution $X = (\frac{9}{2}, 0, \frac{7}{2}, \frac{1}{2}, 0)$; choosing $x_2 = 1$ and $x_5 = 0$ gives the solution $X = (\frac{7}{2}, 1, \frac{7}{2}, \frac{1}{2}, 0)$; and choosing $x_2 = 3$ and $x_5 = -1$ gives the solution $X = (5, 3, 6, 1, -1)$. We can write our general answer in vector form as

$$X = \begin{pmatrix} \frac{9}{2} - x_2 - \frac{7}{2}x_5 \\ x_2 \\ \frac{7}{2} - \frac{5}{2}x_5 \\ \frac{1}{2} - \frac{1}{2}x_5 \\ x_5 \end{pmatrix} = \begin{pmatrix} \frac{9}{2} \\ 0 \\ \frac{7}{2} \\ \frac{1}{2} \\ 0 \end{pmatrix} + x_2 \begin{pmatrix} -1 \\ 1 \\ 0 \\ 0 \\ 0 \end{pmatrix} + x_5 \begin{pmatrix} -\frac{7}{2} \\ 0 \\ -\frac{5}{2} \\ -\frac{1}{2} \\ 1 \end{pmatrix}$$

Usually, we will try to express our answers in this form. (As we might expect from Theorem 2.2, the vectors $(-1, 1, 0, 0, 0)$ and $(-7/2, 0, -5/2, -1/2, 1)$ are solutions of the associated homogeneous system.) $\Box$

All systems of linear equations can be solved in this manner and the nature of the set of solutions is revealed in the process: if the resulting equivalent system includes an equation of the form $0 = b$ where b is non-zero, there are no solutions; if the resulting equivalent system has as many non-zero equations as unknowns, there is exactly one solution; and if the resulting equivalent system has fewer non-zero equations than unknowns, there are infinitely many solutions. When the system has infinitely many solutions, we solve the equations for some of the variables in terms of other of the variables that can take arbitrary values. We obtain specific solutions by choosing values for the variables that can have arbitrary values, and obtain one solution for each choice of values for the variables.

We have solved systems of equations by using row operations to replace the system by an equivalent system with a simpler form. There is terminology for the matrices and this process that will help us formalize some of these ideas.

DEFINITION *Two $m \times n$ matrices are said to be* row equivalent *if one can be obtained from the other by row operations.*

In Example 2.7, we represented the system by the augmented matrix

$$\left(\begin{array}{cccc|c} 1 & -2 & 1 & -1 & 1 \\ 2 & -4 & 3 & -1 & 4 \\ -1 & 3 & -2 & 4 & 7 \\ 1 & -4 & 4 & -4 & -7 \end{array} \right)$$

and, by doing row operations, found that the matrix

$$\left(\begin{array}{cccc|c} 1 & 0 & 0 & 0 & 1 \\ 0 & 1 & 0 & 0 & -2 \\ 0 & 0 & 1 & 0 & -1 \\ 0 & 0 & 0 & 1 & 3 \end{array} \right)$$

is row equivalent to the original matrix. Since row equivalent augmented matrices represent equivalent systems, the simple final matrix represents the solution to the system.

Notice that if a matrix can be obtained from another by row operations, the first can be obtained from the second by row operations also because each row operation can be undone by another row operation. Thus, the row equivalence of two matrices is a symmetric relationship.

DEFINITION *Let E be an $m \times n$ matrix. The first non-zero entry in each row is called the* leading entry *of that row. The matrix E is said to be a* row echelon matrix *if the leading entry of each row of E is to the right of the leading entry of the row above it and all zero rows are below all non-zero rows.*

Symbolically, e_{ij_0} is the leading entry of row i if $e_{ij_0} \neq 0$ but $e_{ij} = 0$ for $j < j_0$. The matrix E is in row echelon form if e_{ij} is the leading entry of row i and $e_{i+1,k}$ is the leading entry of row $i+1$ implies $j < k$. In the process of Gaussian elimination, the pivots were chosen to be the leading entries of the appropriate row.

EXAMPLE 2.12

In the matrix

$$
E = \begin{pmatrix}
\textcircled{1} & -1 & 0 & 2 & 4 & -2 \\
0 & 0 & \textcircled{3} & -1 & 2 & -2 \\
0 & 0 & 0 & \textcircled{5} & -1 & 4 \\
0 & 0 & 0 & 0 & \textcircled{6} & -3 \\
0 & 0 & 0 & 0 & 0 & 0
\end{pmatrix}
$$

the leading entry of the first row is 1 (which is circled), the leading entry of the second row is 3, the leading entry of the third row is 5, the leading entry of the fourth row is 6, and the fifth row is zero, so it has no leading entry. The matrix E is a row echelon matrix because the 3 in the second row is to the right of the 1 in the first row, the 5 in the third row is to the right of the 3 in the second row, the 6 in the fourth row is to the right of the 5 in the third row, and the row of zeros is at the bottom.

On the other hand, the matrix

$$
F = \begin{pmatrix}
\textcircled{1} & -1 & 0 & 2 \\
0 & 0 & \textcircled{3} & -1 \\
0 & 0 & \textcircled{5} & -2 \\
0 & 0 & 0 & \textcircled{6}
\end{pmatrix}
$$

is *not* in row echelon form because the leading entry, 5, of the third row is in the same column as the leading entry, 3, of the second row, not to its right.   □

It should be noted that in the first stage of Gaussian elimination, we used row operations to find a row echelon matrix that is row equivalent to the given matrix. In the second stage, which is the same as backsolving, we further simplified the coefficient matrix.

DEFINITION *Let E be an $m \times n$ matrix. The matrix E is said to be a* reduced row echelon matrix *if E is a row echelon matrix in which the leading entry of each non-zero row of E is 1 and in any column containing a leading entry, the leading entry is the only non-zero entry in that column.*

EXAMPLE 2.13

The matrix

$$
E = \begin{pmatrix}
\textcircled{1} & 0 & 4 & 0 & 0 & -3 \\
0 & \textcircled{1} & -3 & 0 & 0 & 5 \\
0 & 0 & 0 & \textcircled{1} & 0 & 2 \\
0 & 0 & 0 & 0 & \textcircled{1} & -4 \\
0 & 0 & 0 & 0 & 0 & 0
\end{pmatrix}
$$

is a reduced row echelon matrix since the leading entry of each non-zero row is 1 and in each of the first, second, fourth, and fifth columns, the only non-zero entry is the leading entry.

On the other hand, the matrix

$$
F = \begin{pmatrix}
\textcircled{1} & -1 & 2 & 0 \\
0 & 0 & \textcircled{1} & 0 \\
0 & 0 & 0 & \textcircled{1} \\
0 & 0 & 0 & 0
\end{pmatrix}
$$

is *not* in reduced row echelon form because the leading entry, 1, of the second row is in the third column and the entry in the first row, third column is 2, not 0. $\square$

For a matrix, B, MATLAB calculates the reduced row echelon matrix E that is row equivalent to B with the command `rref(B)`. Although good for hand calculation, finding the reduced row echelon matrix E is not usually the best way to solve systems using a machine. The MATLAB command `rref` is good for checking your hand calculations, though, and thinking about the solution of a system of equations as the conversion, through row operations, of the augmented matrix into an equivalent reduced row echelon matrix that represents the solution of the system is helpful in understanding the sorts of results one gets in solving systems. In Chapter 3.2, we will learn more about the solution of systems that will enable us to use the MATLAB command \ effectively for more general systems.

EXAMPLE 2.14

Use the MATLAB command `rref` to solve the system

$$
\begin{cases}
x_1 & - & x_2 & - & 3x_3 & + & 3x_4 & + & x_5 & - & 2x_6 & = & -4 \\
    &   & x_2 & + & 2x_3 & - & x_4 & + & x_5 & + & 2x_6 & = & 3 \\
x_1 &   &   & - & x_3 & + & 2x_4 & - & x_5 & - & 3x_6 & = & -7 \\
2x_1 & + & x_2 &   &   & - & x_4 &   &   & + & 5x_6 & = & -5 \\
x_1 & + & x_2 & + & x_3 & - & x_4 & + & x_5 & + & 4x_6 & = & 0
\end{cases}
$$

If there is no solution, say so, if the solution is unique, indicate that it is unique, and if there are infinitely many solutions, find three solutions and find the general solution.

SOLUTION We enter the augmented matrix, $B = (A \mid b)$, and find `rref(B)`

```
> B=[ 1 -1 -3  3  1 -2 -4; 0  1  2 -1  1  2  3;...
      1  0 -1  2 -1 -3 -7; 2  1  0 -1  0  5 -5;...
      1  1  1 -1  1  4  0]

B =
     1    -1    -3     3     1    -2    -4
     0     1     2    -1     1     2     3
     1     0    -1     2    -1    -3    -7
     2     1     0    -1     0     5    -5
     1     1     1    -1     1     4     0

> rref(B)

ans =
     1     0    -1     0     0     2    -3
     0     1     2     0     0    -1     0
     0     0     0     1     0    -2    -1
     0     0     0     0     1     1     2
     0     0     0     0     0     0     0
```

We notice that there are not leading 1's in the third, sixth, and seventh columns. Since there is not a leading 1 in the last column, the system is consistent. Since there are not leading 1's in the third and sixth columns, we see that x_3 and x_6 are free variables. In fact, the equations say $x_1 - x_3 + 2x_6 = -3$, $x_2 + 2x_3 - x_6 = 0$, $x_4 - 2x_6 = -1$, and $x_5 + x_6 = 2$. Solving for the pivot variables in terms of the free variables, we get the solution of the system:

$$\begin{cases} x_1 &= x_3 - 2x_6 - 3 \\ x_2 &= -2x_3 + x_6 \\ x_4 &= 2x_6 - 1 \\ x_5 &= -x_6 + 2 \end{cases}$$

and x_3 and x_6 are arbitrary. Thus, there are infinitely many solutions for the original system, one for each choice of x_3 and x_6. This is one way of writing the *general solution*; it tells how to find a solution from any choices of the free variables. For example, taking $x_3 = x_6 = 0$, we get the solution $x_1 = -3$, $x_2 = 0$, $x_4 = -1$, and $x_5 = 2$, which we would usually write as $X = (-3, 0, 0, -1, 2, 0)$. Choosing $x_3 = 1$ and $x_6 = 0$, we get $X = (-2, -2, 1, -1, 2, 0)$ is another solution, and choosing $x_3 = 0$ and $x_6 = 1$, we get $X = (-5, 1, 0, 1, 1, 1)$ is a third solution. Another way of writing the general solution is in a vector form; the solutions are

$$X = \begin{pmatrix} x_3 - 2x_6 - 3 \\ -2x_3 + x_6 \\ x_3 \\ 2x_6 - 1 \\ -x_6 + 2 \\ x_6 \end{pmatrix} = x_3 \begin{pmatrix} 1 \\ -2 \\ 1 \\ 0 \\ 0 \\ 0 \end{pmatrix} + x_6 \begin{pmatrix} -2 \\ 1 \\ 0 \\ 2 \\ -1 \\ 1 \end{pmatrix} + \begin{pmatrix} -3 \\ 0 \\ 0 \\ -1 \\ 2 \\ 0 \end{pmatrix}$$

where x_3 and x_6 are arbitrary.

In the MATLAB input above, the ellipsis $\ldots$ at the end of the line is the continuation command: it allows the input to be continued on the next line without the current line being interpreted when the return key is pressed. $\Box$

EXAMPLE 2.15

We began the chapter with the system of linear equations on page 34 that arose in the analysis of a circuit. Rewriting the system slightly, we had

$$\begin{cases} i_{AB} + i_{AC} + i_{AD} & & & = 0 \\ -i_{AB} & + i_{BC} + i_{BD} & & = 0 \\ - i_{AC} & - i_{BC} & + i_{CD} & = 0 \\ - i_{AD} & - i_{BD} & - i_{CD} & = 0 \\ 10i_{AB} - i_{AC} & + 4i_{BC} & & = 10 \\ 10i_{AB} & - 2i_{AD} & + 5i_{BD} & & = 10 \\ i_{AC} - 2i_{AD} & & + 2i_{CD} & = 0 \\ & - 4i_{BC} + 5i_{BD} & - 2i_{CD} & = 0 \end{cases}$$

which is a system of eight linear equations in six unknowns. We can solve this system with the techiques of this section. It is easiest to use the equation solver from MATLAB.

```
> A=[1 1 1 0 0 0;-1 0 0 1 1 0; 0 -1 0 -1 0 1;0 0 -1 0 -1 -1;...
    10 -1 0 4 0 0;10 0 -2 0 5 0;0 1 -2 0 0 2; 0 0 0 -4 5 -2]

A =
      1      1      1      0      0      0
     -1      0      0      1      1      0
      0     -1      0     -1      0      1
      0      0     -1      0     -1     -1
     10     -1      0      4      0      0
     10      0     -2      0      5      0
      0      1     -2      0      0      2
      0      0      0     -4      5     -2

> b=[0;0;0;0;10;10;0;0]

b =
      0
      0
      0
      0
     10
     10
      0
      0

> X=A\b

X =
    0.7751
   -0.4863
   -0.2888
    0.4407
    0.3343
   -0.0456
```

We can check that this is indeed a solution of the system by comparing AX with b.

```
> A*X-b

ans =
    1.0e-14 *
        -0.2554
         0.1499
         0.0631
         0.0423
              0
        -0.1776
        -0.0014
         0.0347
```

If X is a solution of $AX = b$, then $AX - b = 0$. In fact, because of round-off errors, the computed value of $AX - b$ is not zero. However, it is nearly zero! The line `1.0e-14*` means that each number in the vector following is multiplied by 10^{-14}, so that the entry in the first component of $AX - b$ was computed to be $-.000000000000002554$. Since our errors in the measurement of the resistances and voltages is undoubtedly far greater than this, the calculated answer is zero for all practical purposes. In other words, this says that X is (nearly) a solution.

This agrees with the solution that was asserted on page 34 because, i_{AB} was given as $255/329$ which has the decimal approximation $.7750759878\ldots$, i_{AC} was given as $-160/329$ which has decimal approximation $-.4863221884\ldots$, and so on.

We can also solve the system by finding a matrix in reduced row echelon form that is row equivalent to the augmented matrix for the system.

```
> B=[A b]

B =
     1     1     1     0     0     0     0
    -1     0     0     1     1     0     0
     0    -1     0    -1     0     1     0
     0     0    -1     0    -1    -1     0
    10    -1     0     4     0     0    10
    10     0    -2     0     5     0    10
     0     1    -2     0     0     2     0
     0     0     0    -4     5    -2     0

> rref(B)
```

```
ans =
    1.00        0        0        0        0        0     0.7751
       0     1.00        0        0        0        0    -0.4863
       0        0     1.00        0        0        0    -0.2888
       0        0        0     1.00        0        0     0.4407
       0        0        0        0     1.00        0     0.3343
       0        0        0        0        0     1.00    -0.0456
       0        0        0        0        0        0        0
       0        0        0        0        0        0        0
```

Since there is no leading 1 in the last column, the system has a solution and since there is a leading 1 in each of the other columns, the solution is unique. We again get $i_{AB} = .7751$, etc. $\square$

We conclude this section with an observation about solutions of homogeneous systems that will be used later. A system that has more unknowns than equations is called an *underdetermined system.*

THEOREM 2.16

A homogeneous system with more unknowns than equations has infinitely many solutions.

PROOF First note that 0 is always a solution of a homogeneous system so the system cannot be inconsistent. If a system has more unknowns than equations, the augmented matrix for the system has more columns than rows. Because of this, after replacing the augmented matrix by a row equivalent matrix in reduced row echelon form, there must be columns of the matrix that do not contain a leading entry in any row. Since we solve each equation corresponding to a non-zero row for the variable corresponding to the leading entry, this means that there will be variables that do not occur on the left side of the solution to the system, that is, they may take on any value and there are therefore an infinite number of solutions to the system. $\blacksquare$

Of course, the converse of this result is not true, it is also possible for systems with more equations than unknowns to have infinitely many solutions also.

Summary of Gaussian Elimination

Our experience suggests that starting from any matrix one can use row operations to reach a row echelon matrix: this is true and is the content of Theorem 2.19 that will be explained in Section 2.5.

To summarize, in the matrix formulation of Gaussian elimination, we use row operations to replace the augmented matrix for a system by a row equivalent matrix in reduced row echelon form. We then use this convenient form to write down the solution of the system.

Explicitly, the reduced row echelon form of the augmented matrix represents an equivalent system for which the nature of the solutions is easy to understand. The system has no solution if the last column contains a leading 1. In Example 2.10, the original augmented matrix is row equivalent to the reduced row echelon matrix

$$
\left(\begin{array}{cccc|c}
1 & 2 & -1 & 1 & 1 \\
-1 & -1 & 0 & -2 & -1 \\
2 & 5 & -4 & 3 & 7 \\
-1 & -1 & 1 & -4 & -3
\end{array}\right)
\longrightarrow \cdots \longrightarrow
\left(\begin{array}{cccc|c}
① & 0 & 0 & 5 & 6 \\
0 & ① & 0 & -3 & -5 \\
0 & 0 & ① & -2 & -5 \\
0 & 0 & 0 & 0 & ①
\end{array}\right)
$$

Since the last row says that $0x_1 + 0x_2 + 0x_3 + 0x_4 = 1$, the system represented by the reduced row echelon matrix clearly has no solutions which means the original system has no solutions.

If each column except the last contains a leading 1, the reduced row echelon matrix represents a system that has a unique solution, so the original system has the same unique solution. For example, in Example 2.7, we did row operations to convert the original augmented matrix into the equivalent reduced row echelon matrix

$$
\left(\begin{array}{cccc|c}
1 & -2 & 1 & -1 & 1 \\
2 & -4 & 3 & -1 & 4 \\
-1 & 3 & -2 & 4 & 7 \\
1 & -4 & 4 & -4 & -7
\end{array}\right)
\longrightarrow \cdots \longrightarrow
\left(\begin{array}{cccc|c}
① & 0 & 0 & 0 & 1 \\
0 & ① & 0 & 0 & -2 \\
0 & 0 & ① & 0 & -1 \\
0 & 0 & 0 & ① & 3
\end{array}\right)
$$

Since the latter matrix clearly represents the system $x_1 = 1$, $x_2 = -2$, $x_3 = -1$, and $x_4 = 3$, this is also the unique solution of the original system.

Finally, if the last column does not contain a leading 1 and there are other columns that do not contain a leading 1, then the reduced row echelon matrix represents a system with free variables and infinitely many solutions, one for each choice of the free variables. We use the final matrix to solve for each of the pivot variables in terms of the non-pivot variables, that is, the free variables, as we did in Example 2.14.

Exercises 2.3

Solve the following systems by Gaussian elimination. If there is only one solution, say that the solution is unique; if there is no solution, say so; if there are infinitely many, say so **and** find the general solution **and** find three explicit solutions. (You should check your answers by substitution or using MATLAB!)

1. $\begin{cases} 2a + b = 3 \\ 3a + 4b = 2 \end{cases}$

2. $\begin{cases} 3u - 2v = 5 \\ u + 5v = -9 \end{cases}$

3. $\begin{cases} x - 2y = 2 \\ -2x + 4y = -3 \end{cases}$

4. $\begin{cases} 2x - y = 3 \\ -4x + 2y = -6 \end{cases}$

5. $\begin{cases} 3x - 2y = 7 \end{cases}$

6. $\begin{cases} 2x + y = 4 \\ 3x - y = 11 \\ x - 2y = 7 \end{cases}$

7. $\begin{cases} 2x + y = 3 \\ 3x - y = -1 \\ x - 2y = 2 \end{cases}$

8. $\begin{cases} x + y - z = 4 \\ -x + y - 2z = -3 \\ 2x - 4y + 8z = 6 \end{cases}$

9. $\begin{cases} -y + z = -4 \\ -x + 3y + 3z = 4 \\ 2x + 3z = 1 \end{cases}$

10. $\begin{cases} x + 2y - z = -3 \\ 2x - y + z = 8 \\ -2x - 2y + z = 1 \end{cases}$

11. $\begin{cases} w - x + 2y = -3 \\ 2w + x + y = 3 \\ 3w + 4y = -2 \end{cases}$

12. $\begin{cases} x + y - z = 1 \\ -x + 4z = 1 \\ 2x + 3y + z = 3 \end{cases}$

13. $\begin{cases} x + y - z = -1 \\ -x + 4z = 2 \\ 2x + 3y + z = -1 \end{cases}$

14. $\begin{cases} x - y + z = 0 \\ -x + 3y + 3z = 0 \\ 2x + 3z = 0 \end{cases}$

15. $\begin{cases} 3x + 2y + z = 4 \\ 2x + y + 2z = -2 \end{cases}$

16. $\begin{cases} x - y - 2z &= 3 \\ 2x + y + z &= 3 \\ -5x + y + 3z &= -11 \\ x + 2y - z &= 0 \end{cases}$

17. $\begin{cases} x - y - 2z &= 2 \\ 2x + y + z &= 1 \\ -5x + y + 3z &= 7 \\ x + 2y - z &= 0 \end{cases}$

18. $\{\ x - 3y + 2z = 6$

19. $\begin{cases} w - x + y + 2z &= -1 \\ -2w + 3x + y - 3z &= 2 \\ w + 2x + 9y + 7z &= -3 \\ 2w - x + 6y + 4z &= -1 \\ w - 2x \quad\ - 2z &= 2 \end{cases}$

20. $\begin{cases} w - x - y + 2z &= -5 \\ 2w - 2x \quad\ + 5z &= -1 \\ -w \quad\ + 3y - z &= 15 \\ w + x + y + 2z &= 3 \end{cases}$

21. $\begin{cases} w - x + y + z &= 0 \\ -2x + 6y + 7z &= 3 \\ -w + x + y + 2z &= -1 \\ -x + 4y + 5z &= 2 \end{cases}$

22. $\begin{cases} v - w + 2x + y + z &= 0 \\ -2w + 2x + 6y + 7z &= 9 \\ -v + w - 2x + y + 2z &= 5 \\ -w + x + 4y + 5z &= 7 \end{cases}$

23. $\begin{cases} v + 3w - x + y + 2z &= 0 \\ 2v + 5w - 3x - 2y + 4z &= 0 \\ v + 5w + 2x + 2y + z &= 0 \end{cases}$

24. $\begin{cases} w \quad\quad\ + y + z &= 2 \\ x + y + 2z &= 1 \\ -w + x - y + 2z &= 1 \\ w - x \quad\quad\ - z &= 1 \end{cases}$

25. $\{\ 2a + b - 3c - d + 4e = 11$

26. Letting $X = (a, b, c, d, e)$,

$$X = \begin{pmatrix} -2 \\ 1 \\ -1 \\ -1 \\ 1 \end{pmatrix} + s \begin{pmatrix} 1 \\ -1 \\ -1 \\ 1 \\ 0 \end{pmatrix} + t \begin{pmatrix} -1 \\ -2 \\ 1 \\ 0 \\ 1 \end{pmatrix}$$

is a solution of the system

$$\begin{cases} a + b - c - d + 4e &= 5 \\ 2a + b + c \quad\quad\ + 3e &= -1 \\ -a + 3b + c + 5d + 4e &= 3 \end{cases}$$

for every s and t and every solution can be written in this form for some s and t.

(a) Show that $X = (1, 4, -4, 0, -1)$ is a solution of the system and find values of s and t that correspond to this solution.

(b) Without using Gaussian elimination or a machine, find two non-trivial solutions of the system $AX = 0$.

(c) $X = (2, -1, 0, 1, 3)$ is a solution of the system $AX = (12, 12, 12)$. Without using Gaussian elimination or a machine, find all solutions of the system $AX = (12, 12, 12)$.

◇ 27. Use an appropriate machine to find the solution of the following system

$$\begin{cases} 5x_1 + x_2 + 3x_3 + 2x_4 + 4x_5 - x_6 &=& 0 \\ -2x_1 + 5x_2 + x_3 - 3x_4 - x_5 &=& 1 \\ -x_1 - 2x_2 + 5x_3 + x_4 + 3x_5 - 2x_6 &=& 6 \\ 3x_2 - 3x_3 + 2x_4 - 4x_5 - x_6 &=& -9 \\ 3x_1 + 4x_2 - x_4 + 5x_5 &=& 2 \\ -4x_1 - x_2 + 6x_3 + 2x_4 - 3x_5 + 2x_6 &=& 1 \end{cases}$$

◇ 28. In Example 2.11 and other places in this section, after getting a general solution for a system with infinitely many solutions, to find explicit solutions to the system, we had to choose values for some of the variables. Since the following homogeneous system has six equations in eight unknowns, we would expect to have to choose values for (at least) two of the variables to get explicit solutions. Perhaps we could choose values for variables at the beginning rather than at the end! For software that finds only one solution for a system, such as the \ command in MATLAB, this can give an *ad hoc* approach to finding many solutions for a system. We will see later ways to decide how many variables to choose and how this approach might fail. Coax your machine into finding all solutions of the following system (without using a command like `rref`) by first putting $x_7 = 1$ and $x_8 = 0$ and solving the resulting non-homogeneous system, then reversing their roles. (Note the resemblance to Problem 2.3.27.)

$$\begin{cases} 5x_1 + x_2 + 3x_3 + 2x_4 + 4x_5 - x_6 + 2x_7 - 3x_8 &=& 0 \\ -2x_1 + 5x_2 + x_3 - 3x_4 - x_5 + x_7 + 3x_8 &=& 0 \\ -x_1 - 2x_2 + 5x_3 + x_4 + 3x_5 - 2x_6 - 2x_7 - x_8 &=& 0 \\ 3x_2 - 3x_3 + 2x_4 - 4x_5 - x_6 + x_7 + x_8 &=& 0 \\ 3x_1 + 4x_2 - x_4 + 5x_5 + 4x_7 + 3x_8 &=& 0 \\ -4x_1 - x_2 + 6x_3 + 2x_4 - 3x_5 + 2x_6 + 5x_7 + 4x_8 &=& 0 \end{cases}$$

◇ 29. Use the results of Problems 2.3.27. and 2.3.28. to find the general solution (that is, a description of all solutions) of the system

$$\begin{cases} 5x_1 + x_2 + 3x_3 + 2x_4 + 4x_5 - x_6 + 2x_7 - 3x_8 &=& 0 \\ -2x_1 + 5x_2 + x_3 - 3x_4 - x_5 + x_7 + 3x_8 &=& 1 \\ -x_1 - 2x_2 + 5x_3 + x_4 + 3x_5 - 2x_6 - 2x_7 - x_8 &=& 6 \\ 3x_2 - 3x_3 + 2x_4 - 4x_5 - x_6 + x_7 + x_8 &=& -9 \\ 3x_1 + 4x_2 - x_4 + 5x_5 + +4x_7 + 3x_8 &=& 2 \\ -4x_1 - x_2 + 6x_3 + 2x_4 - 3x_5 + 2x_6 + 5x_7 + 4x_8 &=& 1 \end{cases}$$

◇ 30. Use your machine to find the a matrix in reduced row echelon form equivalent to the augmented matrix for the system in Exercise 2.3.29. and use it to give the general solution of the system. Compare your answer to that found above.

2.4   Computing Inverses of Matrices

We have seen that some square matrices have inverses and, if we are given two matrices, we can easily check to see if they are inverses of each other, but we have not yet considered a systematic method for finding inverses. Let us try to find the inverse of the matrix

$$A = \begin{pmatrix} 1 & 2 & -1 \\ -1 & -1 & 0 \\ 2 & 5 & -4 \end{pmatrix}$$

If A has an inverse, it will also be a 3×3 matrix, say

$$A^{-1} = B = \begin{pmatrix} p & t & x \\ q & u & y \\ r & v & z \end{pmatrix}$$

for some numbers p, q, r, t, u, v, x, y, and z. Writing out the condition $AB = I$ gives

$$\begin{pmatrix} 1 & 2 & -1 \\ -1 & -1 & 0 \\ 2 & 5 & -4 \end{pmatrix} \begin{pmatrix} p & t & x \\ q & u & y \\ r & v & z \end{pmatrix} = \begin{pmatrix} 1 & 0 & 0 \\ 0 & 1 & 0 \\ 0 & 0 & 1 \end{pmatrix}$$

which gives a linear system with nine equations in the nine unknowns:

$$\begin{cases}
\begin{aligned}
p + 2q - r & & & & & & = 1 \\
-p - q + 0r & & & & & & = 0 \\
2p + 5q - 4r & & & & & & = 0 \\
& t + 2u - v & & & & & = 0 \\
& -t - u + 0v & & & & & = 1 \\
& 2t + 5u - 4v & & & & & = 0 \\
& & x + 2y - z & & & & = 0 \\
& & -x - y + 0z & & & & = 0 \\
& & 2x + 5y - 4z & & & & = 1
\end{aligned}
\end{cases}$$

Actually, however, this system can be described better as three linear systems of three equations in three unknowns, each:

$$\begin{cases}
p + 2q - r = 1 \\
-p - q + 0r = 0 \\
2p + 5q - 4r = 0
\end{cases}$$

$$\begin{cases} t & + & 2u & - & v & = & 0 \\ -t & - & u & + & 0v & = & 1 \\ 2t & + & 5u & - & 4v & = & 0 \end{cases}$$

$$\begin{cases} x & + & 2y & - & z & = & 0 \\ -x & - & y & + & 0z & = & 0 \\ 2x & + & 5y & - & 4z & = & 1 \end{cases}$$

Moreover, the coefficient matrix of each of these systems is the same, namely the matrix A that we are trying to find the inverse of. Since we have noticed that the row operations necessary to solve a system do not depend on the right hand sides, only on the coefficient matrix, we will do the same row operations to solve each of these three systems. Thus we can solve them at the same time by augmenting the coefficient matrix with each of the three right hand sides.

$$\left(\begin{array}{ccc|c|c|c} 1 & 2 & -1 & 1 & 0 & 0 \\ -1 & -1 & 0 & 0 & 1 & 0 \\ 2 & 5 & -4 & 0 & 0 & 1 \end{array} \right)$$

When we have finished the reduction, the values for p, q and r will be in the first column on the right, the values for t, u, and v will be in the second column, and the values for x, y and z will be in the last column. More typically, we would remember the structure of this matrix and only use one vertical line to remind us of the augmentation:

$$\left(\begin{array}{ccc|ccc} 1 & 2 & -1 & 1 & 0 & 0 \\ -1 & -1 & 0 & 0 & 1 & 0 \\ 2 & 5 & -4 & 0 & 0 & 1 \end{array} \right)$$

We begin by replacing the second row by the second plus the first and the third row by the third minus 2 times the first.

$$\left(\begin{array}{ccc|ccc} ① & 2 & -1 & 1 & 0 & 0 \\ -1 & -1 & 0 & 0 & 1 & 0 \\ 2 & 5 & -4 & 0 & 0 & 1 \end{array} \right) \longrightarrow \left(\begin{array}{ccc|ccc} 1 & 2 & -1 & 1 & 0 & 0 \\ 0 & 1 & -1 & 1 & 1 & 0 \\ 0 & 1 & -2 & -2 & 0 & 1 \end{array} \right)$$

Now replace the third row by the third row minus the second.

$$\left(\begin{array}{ccc|ccc} 1 & 2 & -1 & 1 & 0 & 0 \\ 0 & ① & -1 & 1 & 1 & 0 \\ 0 & 1 & -2 & -2 & 0 & 1 \end{array} \right) \longrightarrow \left(\begin{array}{ccc|ccc} 1 & 2 & -1 & 1 & 0 & 0 \\ 0 & 1 & -1 & 1 & 1 & 0 \\ 0 & 0 & -1 & -3 & -1 & 1 \end{array} \right)$$

Multiplying the third row by -1,

$$\left(\begin{array}{ccc|ccc} 1 & 2 & -1 & 1 & 0 & 0 \\ 0 & 1 & -1 & 1 & 1 & 0 \\ 0 & 0 & -1 & -3 & -1 & 1 \end{array}\right) \longrightarrow \left(\begin{array}{ccc|ccc} 1 & 2 & -1 & 1 & 0 & 0 \\ 0 & 1 & -1 & 1 & 1 & 0 \\ 0 & 0 & 1 & 3 & 1 & -1 \end{array}\right)$$

Replacing the first row by the first plus the third and the second row by the second plus the third, we get

$$\left(\begin{array}{ccc|ccc} 1 & 2 & -1 & 1 & 0 & 0 \\ 0 & 1 & -1 & 1 & 1 & 0 \\ 0 & 0 & ① & 3 & 1 & -1 \end{array}\right) \longrightarrow \left(\begin{array}{ccc|ccc} 1 & 2 & 0 & 4 & 1 & -1 \\ 0 & 1 & 0 & 4 & 2 & -1 \\ 0 & 0 & 1 & 3 & 1 & -1 \end{array}\right)$$

and finally, replacing the first row by the first minus 2 times the second row, we get

$$\left(\begin{array}{ccc|ccc} 1 & 2 & 0 & 4 & 1 & -1 \\ 0 & ① & 0 & 4 & 2 & -1 \\ 0 & 0 & 1 & 3 & 1 & -1 \end{array}\right) \longrightarrow \left(\begin{array}{ccc|ccc} 1 & 0 & 0 & -4 & -3 & 1 \\ 0 & 1 & 0 & 4 & 2 & -1 \\ 0 & 0 & 1 & 3 & 1 & -1 \end{array}\right)$$

Thus, we find $p = -4$, $q = 4$, $r = 3$, $t = -3$, etc. Indeed, because of the arrangement of the unknowns, we find

$$B = \left(\begin{array}{ccc} p & t & x \\ q & u & y \\ r & v & z \end{array}\right) = \left(\begin{array}{ccc} -4 & -3 & 1 \\ 4 & 2 & -1 \\ 3 & 1 & -1 \end{array}\right)$$

It is easily checked that

$$\left(\begin{array}{ccc} 1 & 2 & -1 \\ -1 & -1 & 0 \\ 2 & 5 & -4 \end{array}\right)\left(\begin{array}{ccc} -4 & -3 & 1 \\ 4 & 2 & -1 \\ 3 & 1 & -1 \end{array}\right) = \left(\begin{array}{ccc} 1 & 0 & 0 \\ 0 & 1 & 0 \\ 0 & 0 & 1 \end{array}\right)$$

(as we set the system up) and also that it is easily checked that

$$\left(\begin{array}{ccc} -4 & -3 & 1 \\ 4 & 2 & -1 \\ 3 & 1 & -1 \end{array}\right)\left(\begin{array}{ccc} 1 & 2 & -1 \\ -1 & -1 & 0 \\ 2 & 5 & -4 \end{array}\right) = \left(\begin{array}{ccc} 1 & 0 & 0 \\ 0 & 1 & 0 \\ 0 & 0 & 1 \end{array}\right)$$

so that $B = A^{-1}$. (In fact, we will see in Section 3.8 that if A is a square matrix and we find a square matrix B so that $AB = I$, then $BA = I$ also, so this procedure always finds A^{-1} if it exists.)

Looking back at our work, we see that the original augmented matrix was of the form $(A|I)$ and we did row operations to get the matrix on the left to be I. When this occurs, the matrix on the right is A^{-1}, so the process can be described by

$$(A|I) \longrightarrow \cdots \longrightarrow (I|A^{-1})$$

Not all matrices are invertible, and if the matrix is not invertible, the elimination algorithm will discover that fact when the system of equations we are trying to solve becomes obviously inconsistent because there is a row in which the coefficient matrix has all zeros. Based on our plan for finding inverses, one way to describe the condition for invertibility is that A is invertible if and only if it is row equivalent to the identity. Indeed, if A is row equivalent to the identity, then the left block of the large augmented matrix can be converted to I by row operations and the right block will be the inverse. If A is not row equivalent to the identity, then we can never get the form we need for the inverse and the system of equations $AB = I$ is not solvable, so A is not invertible.

EXAMPLE 2.17

Find the inverse of

$$C = \begin{pmatrix} 1 & -1 & 1 & -1 \\ -1 & 2 & -4 & 2 \\ 2 & -1 & -2 & 0 \\ 1 & -2 & 3 & -1 \end{pmatrix}$$

SOLUTION We set up the four systems of four equations in four unknowns as

$$\left(\begin{array}{rrrr|rrrr} 1 & -1 & 1 & -1 & 1 & 0 & 0 & 0 \\ -1 & 2 & -4 & 2 & 0 & 1 & 0 & 0 \\ 2 & -1 & -2 & 0 & 0 & 0 & 1 & 0 \\ 1 & -2 & 3 & -1 & 0 & 0 & 0 & 1 \end{array} \right)$$

We now do row operations until the matrix C has been reduced to the identity.

$$\left(\begin{array}{rrrr|rrrr} \textcircled{1} & -1 & 1 & -1 & 1 & 0 & 0 & 0 \\ -1 & 2 & -4 & 2 & 0 & 1 & 0 & 0 \\ 2 & -1 & -2 & 0 & 0 & 0 & 1 & 0 \\ 1 & -2 & 3 & -1 & 0 & 0 & 0 & 1 \end{array} \right)$$

$$\longrightarrow \left(\begin{array}{rrrr|rrrr} 1 & -1 & 1 & -1 & 1 & 0 & 0 & 0 \\ 0 & \textcircled{1} & -3 & 1 & 1 & 1 & 0 & 0 \\ 0 & 1 & -4 & 2 & -2 & 0 & 1 & 0 \\ 0 & -1 & 2 & 0 & -1 & 0 & 0 & 1 \end{array} \right)$$

$$
\longrightarrow
\left(\begin{array}{cccc|cccc}
1 & -1 & 1 & -1 & 1 & 0 & 0 & 0 \\
0 & 1 & -3 & 1 & 1 & 1 & 0 & 0 \\
0 & 0 & \boxed{-1} & 1 & -3 & -1 & 1 & 0 \\
0 & 0 & -1 & 1 & 0 & 1 & 0 & 1
\end{array}\right)
$$

$$
\longrightarrow
\left(\begin{array}{cccc|cccc}
1 & -1 & 1 & -1 & 1 & 0 & 0 & 0 \\
0 & 1 & -3 & 1 & 1 & 1 & 0 & 0 \\
0 & 0 & -1 & 1 & -3 & -1 & 1 & 0 \\
0 & 0 & 0 & 0 & 3 & 2 & -1 & 1
\end{array}\right)
$$

This shows that C is not invertible: the last equation of the first system is $0p + 0q + 0r + 0s = 3$ which has no solution, so there is no matrix D so that $CD = I$ and the matrix C is not invertible. ∎

MATLAB finds the inverse of the matrix A with the command `inv(A)`. If the matrix is not invertible, MATLAB will probably find a matrix as the inverse but issue a warning that the matrix is `ill-conditioned` or `close to singular`. If you get a warning like this, you should determine in another way (e.g. by methods from Section 3.8) that the matrix is actually invertible and multply the matrix and its "inverse" and compare the result with the identity.

Although the following theorem is usually not the most efficient way to solve a system of linear equations, it expresses an important relationship between systems of equations and the invertibility of the coefficient matrix: it says that if the coefficient matrix of a system is invertible, the system has a unique solution and it gives a formula for the solution. You can see that it is the analogue of the approach used to solve a single linear equation in one variable.

THEOREM 2.18

If $AX = b$ is a system of n equations in n unknowns and the coefficient matrix A is invertible, then $X = A^{-1}b$ is a solution of the system and it is the only solution.

PROOF First, we must show that $X = A^{-1}b$ is a solution:

$$
AX = A\left(A^{-1}b\right) = AA^{-1}b = Ib = b
$$

We must also show that there a no other solutions! If Y is a solution of the system, then $AY = b$. Multiplying by A^{-1}, this gives $A^{-1}(AY) = A^{-1}b$. The left side is $IY = Y$, so we see $Y = A^{-1}b$ for any solution of the system. ∎

Notice that we used *both* $AA^{-1} = I$ and $A^{-1}A = I$ in the proof.

Exercises 2.4

Find the inverses of the following matrices or say it is not invertible. (Check your answers!)

1. $\begin{pmatrix} 1 & 0 \\ 3 & -1 \end{pmatrix}$

2. $\begin{pmatrix} 1 & 0 & 0 \\ -1 & 1 & 0 \\ 2 & -1 & 1 \end{pmatrix}$

3. $\begin{pmatrix} 1 & 2 & -1 & -3 \\ 0 & -2 & -2 & 0 \\ 0 & 0 & 1 & -2 \\ 0 & 0 & 0 & 1 \end{pmatrix}$

4. $\begin{pmatrix} 1 & 3 \\ 1 & 2 \end{pmatrix}$

5. $\begin{pmatrix} 2 & -1 \\ 3 & 1 \end{pmatrix}$

6. $\begin{pmatrix} 4 & -2 \\ -2 & 1 \end{pmatrix}$

7. $\begin{pmatrix} 1 & -1 & 2 \\ 2 & -3 & 3 \\ 3 & -2 & 5 \end{pmatrix}$

8. $\begin{pmatrix} 1 & 1 & -1 \\ -1 & 0 & 4 \\ 2 & 3 & 1 \end{pmatrix}$

9. $\begin{pmatrix} 1 & -2 & 1 \\ -1 & 3 & 0 \\ -2 & 6 & -1 \end{pmatrix}$

10. $\begin{pmatrix} 1 & -2 & 1 \\ -1 & 3 & 0 \\ -1 & 4 & -1 \end{pmatrix}$

11. $\begin{pmatrix} 1 & -2 & 1 \\ -1 & 3 & 0 \\ -1 & 4 & 1 \end{pmatrix}$

12. $\begin{pmatrix} 1 & -1 & 1 & 1 \\ 0 & -1 & 6 & 7 \\ -1 & 1 & 0 & 2 \\ 0 & -1 & 5 & 5 \end{pmatrix}$

13. $\begin{pmatrix} 1 & -1 & 0 & 1 \\ 1 & -2 & 2 & 0 \\ 2 & -3 & 3 & 3 \\ 1 & 0 & -3 & .5 \end{pmatrix}$

14. $\begin{pmatrix} 1 & -1 & 1 & 1 \\ -1 & 3 & 0 & -2 \\ 1 & 1 & 3 & 1 \\ 2 & -8 & 0 & 5 \end{pmatrix}$

15. $\begin{pmatrix} 1 & -1 & 1 & 2 \\ -1 & 3 & 1 & -8 \\ 2 & -1 & 3 & 2 \\ 1 & -2 & 0 & 5 \end{pmatrix}$

16. If A is an $m \times n$ matrix, we say B is a *left inverse* of A if $BA = I$ and we say C is a *right inverse* of A if $AC = I$. Observing that identity matrices are square, consideration of sizes shows that if B or C exist, they must be $n \times m$ matrices.

(a) Find left and right inverses (or say if they do not exist) for the matrix

$$A = \begin{pmatrix} 1 & -1 \\ 1 & -2 \\ 2 & -3 \end{pmatrix}$$

(b) Find left and right inverses (or say if they do not exist) for the matrix

$$E = \begin{pmatrix} 1 & 1 & 0 \\ 2 & 1 & -2 \end{pmatrix}$$

(c) Can you propose a general statement about when left and right inverses of an $m \times n$ matrix exist?

2.5 Gaussian Elimination as Matrix Factorization

A more modern point of view on the process of Gaussian Elimination is to view it as a factorization of the coefficient matrix or the augmented matrix. In this section, we will focus on relating the row operations we have done to the factorization of matrices as $A = LE$ where E is in reduced row echelon form, and L is an invertible matrix.

One interpretation of Gaussian elimination is that we have factored the augmented matrix $(A \mid b)$ as $(A \mid b) = LE$ where L is an invertible matrix and E is a reduced row echelon matrix. We usually throw L away and keep E as it represents the solution of the original system.

To see the connection between matrix factorization and Gaussian elimination, we view each row operation as multiplication by an elementary matrix. Each of the three sorts of row operations is associated with a particular kind of elementary matrix. Multiplication of the ith row by α is accomplished by multiplying by a diagonal matrix whose (i, i)–entry is α and the other diagonal entries are 1. For example, multiplication of the 2nd row of a $3 \times n$ matrix by 5 is accomplished as

$$\begin{pmatrix} 1 & 0 & 0 \\ 0 & 5 & 0 \\ 0 & 0 & 1 \end{pmatrix} \begin{pmatrix} 1 & 2 & 3 & 4 \\ 5 & 6 & 7 & 8 \\ 9 & 10 & 11 & 12 \end{pmatrix} = \begin{pmatrix} 1 & 2 & 3 & 4 \\ 25 & 30 & 35 & 40 \\ 9 & 10 & 11 & 12 \end{pmatrix}$$

Note that the matrix

$$\begin{pmatrix} 1 & 0 & 0 \\ 0 & 5 & 0 \\ 0 & 0 & 1 \end{pmatrix}$$

is obtained from the identity by multiplying the second row by 5. To interchange the ith and j^{th} rows, multiply by the matrix obtained from the identity by interchanging the ith and j^{th} rows. For example, interchanging the second and third rows of a

$3 \times n$ matrix is achieved by multiplying

$$
\begin{pmatrix} 1 & 0 & 0 \\ 0 & 0 & 1 \\ 0 & 1 & 0 \end{pmatrix}
\begin{pmatrix} 1 & 2 & 3 & 4 \\ 5 & 6 & 7 & 8 \\ 9 & 10 & 11 & 12 \end{pmatrix}
=
\begin{pmatrix} 1 & 2 & 3 & 4 \\ 9 & 10 & 11 & 12 \\ 5 & 6 & 7 & 8 \end{pmatrix}
$$

Finally, to add a multiple of one row to another, multiply by the matrix obtained from the identity matrix by doing that row operation. For example to add -9 times the first row to the last, multiply as

$$
\begin{pmatrix} 1 & 0 & 0 \\ 0 & 1 & 0 \\ -9 & 0 & 1 \end{pmatrix}
\begin{pmatrix} 1 & 2 & 3 & 4 \\ 5 & 6 & 7 & 8 \\ 9 & 10 & 11 & 12 \end{pmatrix}
=
\begin{pmatrix} 1 & 2 & 3 & 4 \\ 5 & 6 & 7 & 8 \\ 0 & -8 & -16 & -24 \end{pmatrix}
$$

All elementary matrices are invertible, indeed the inverse of an elementary matrix is the elementary matrix that undoes the row operation accomplished by the first. To get a matrix into reduced row echelon form, we must do a series of row operations, that is we multiply on the left by a succession of elementary matrices. For example, if A is a matrix and E is the result obtained by doing a series of row operations, we can see the relation by assuming that the row operations are obtained by multiplying by L_1, L_2, etc. Then we have

$$
E = L_k L_{k-1} \cdots L_3 L_2 L_1 A
$$

In fact, this shows how to factor A as an invertible matrix times a row echelon matrix:

$$
A = \left(L_k L_{k-1} \cdots L_3 L_2 L_1 \right)^{-1} E = \left(L_1^{-1} L_2^{-1} L_3^{-1} \cdots L_{k-1}^{-1} L_k^{-1} \right) E
$$

This gives the idea of the proof of the theorem below. Since row operations can be thought of as multiplication by an elementary matrix, two matrices are row equivalent if one can be obtained from the other by multiplying it on the left by a product of elementary matrices, so this result can be stated as "Every matrix is row equivalent to a reduced row echelon matrix."

THEOREM 2.19

If A is an $m \times n$ matrix, there is a unique $m \times n$ reduced row echelon matrix E and an invertible matrix L so that $A = LE$.

Just as it was convenient to combine several row operations into a single step, it is convenient to multiply several of the elementary matrices together. It is not difficult to check, for example, that the product of the matrix giving "multiply the first row by 3 and subtract from the second row", with the matrix giving "multiply the first row by -2 and subtract from the third", and the matrix giving "multiply the first row by 4 and subtract from the fourth row" looks like the combination of them:

$$\begin{pmatrix} 1 & 0 & 0 & 0 \\ 0 & 1 & 0 & 0 \\ 0 & 0 & 1 & 0 \\ -4 & 0 & 0 & 1 \end{pmatrix} \begin{pmatrix} 1 & 0 & 0 & 0 \\ 0 & 1 & 0 & 0 \\ 2 & 0 & 1 & 0 \\ 0 & 0 & 0 & 1 \end{pmatrix} \begin{pmatrix} 1 & 0 & 0 & 0 \\ -3 & 1 & 0 & 0 \\ 0 & 0 & 1 & 0 \\ 0 & 0 & 0 & 1 \end{pmatrix} = \begin{pmatrix} 1 & 0 & 0 & 0 \\ -3 & 1 & 0 & 0 \\ 2 & 0 & 1 & 0 \\ -4 & 0 & 0 & 1 \end{pmatrix}$$

In this sense, the creation of the matrices to do the row reduction is really a recording of the process of the row operations. The following example illustrates the procedure.

EXAMPLE 2.20

In Example 2.7, we row reduced the matrix

$$\begin{pmatrix} 1 & -2 & 1 & -1 & 1 \\ 2 & -4 & 3 & -1 & 4 \\ -1 & 3 & -2 & 4 & 7 \\ 1 & -4 & 4 & -4 & -7 \end{pmatrix}$$

We will follow the row reduction through the process as before, except this time, we record the matrix L as it is created.

Using the first row we get

$$\begin{pmatrix} 1 & 0 & 0 & 0 \\ -2 & 1 & 0 & 0 \\ 1 & 0 & 1 & 0 \\ -1 & 0 & 0 & 1 \end{pmatrix} \begin{pmatrix} 1 & -2 & 1 & -1 & 1 \\ 2 & -4 & 3 & -1 & 4 \\ -1 & 3 & -2 & 4 & 7 \\ 1 & -4 & 4 & -4 & -7 \end{pmatrix} = \begin{pmatrix} 1 & -2 & 1 & -1 & 1 \\ 0 & 0 & 1 & 1 & 2 \\ 0 & 1 & -1 & 3 & 8 \\ 0 & -2 & 3 & -3 & -8 \end{pmatrix}$$

Interchanging the second and third rows, we get

$$\begin{pmatrix} 1 & 0 & 0 & 0 \\ 0 & 0 & 1 & 0 \\ 0 & 1 & 0 & 0 \\ 0 & 0 & 0 & 1 \end{pmatrix} \begin{pmatrix} 1 & -2 & 1 & -1 & 1 \\ 0 & 0 & 1 & 1 & 2 \\ 0 & 1 & -1 & 3 & 8 \\ 0 & -2 & 3 & -3 & -8 \end{pmatrix} = \begin{pmatrix} 1 & -2 & 1 & -1 & 1 \\ 0 & 1 & -1 & 3 & 8 \\ 0 & 0 & 1 & 1 & 2 \\ 0 & -2 & 3 & -3 & -8 \end{pmatrix}$$

so

$$\begin{pmatrix} 1 & 0 & 0 & 0 \\ 0 & 0 & 1 & 0 \\ 0 & 1 & 0 & 0 \\ 0 & 0 & 0 & 1 \end{pmatrix} \begin{pmatrix} 1 & 0 & 0 & 0 \\ -2 & 1 & 0 & 0 \\ 1 & 0 & 1 & 0 \\ -1 & 0 & 0 & 1 \end{pmatrix} \begin{pmatrix} 1 & -2 & 1 & -1 & 1 \\ 2 & -4 & 3 & -1 & 4 \\ -1 & 3 & -2 & 4 & 7 \\ 1 & -4 & 4 & -4 & -7 \end{pmatrix}$$

$$= \begin{pmatrix} 1 & 0 & 0 & 0 \\ 1 & 0 & 1 & 0 \\ -2 & 1 & 0 & 0 \\ -1 & 0 & 0 & 1 \end{pmatrix} \begin{pmatrix} 1 & -2 & 1 & -1 & 1 \\ 2 & -4 & 3 & -1 & 4 \\ -1 & 3 & -2 & 4 & 7 \\ 1 & -4 & 4 & -4 & -7 \end{pmatrix}$$

$$= \begin{pmatrix} 1 & -2 & 1 & -1 & 1 \\ 0 & 1 & -1 & 3 & 8 \\ 0 & 0 & 1 & 1 & 2 \\ 0 & -2 & 3 & -3 & -8 \end{pmatrix}$$

Using the second row, we get

$$\begin{pmatrix} 1 & 0 & 0 & 0 \\ 0 & 1 & 0 & 0 \\ 0 & 0 & 1 & 0 \\ 0 & 2 & 0 & 1 \end{pmatrix} \begin{pmatrix} 1 & -2 & 1 & -1 & 1 \\ 0 & 1 & -1 & 3 & 8 \\ 0 & 0 & 1 & 1 & 2 \\ 0 & -2 & 3 & -3 & -8 \end{pmatrix} = \begin{pmatrix} 1 & -2 & 1 & -1 & 1 \\ 0 & 1 & -1 & 3 & 8 \\ 0 & 0 & 1 & 1 & 2 \\ 0 & 0 & 1 & 3 & 8 \end{pmatrix}$$

or

$$\begin{pmatrix} 1 & 0 & 0 & 0 \\ 0 & 1 & 0 & 0 \\ 0 & 0 & 1 & 0 \\ 0 & 2 & 0 & 1 \end{pmatrix} \begin{pmatrix} 1 & 0 & 0 & 0 \\ 1 & 0 & 1 & 0 \\ -2 & 1 & 0 & 0 \\ -1 & 0 & 0 & 1 \end{pmatrix} \begin{pmatrix} 1 & -2 & 1 & -1 & 1 \\ 2 & -4 & 3 & -1 & 4 \\ -1 & 3 & -2 & 4 & 7 \\ 1 & -4 & 4 & -4 & -7 \end{pmatrix}$$

$$= \begin{pmatrix} 1 & 0 & 0 & 0 \\ 1 & 0 & 1 & 0 \\ -2 & 1 & 0 & 0 \\ 1 & 0 & 2 & 1 \end{pmatrix} \begin{pmatrix} 1 & -2 & 1 & -1 & 1 \\ 2 & -4 & 3 & -1 & 4 \\ -1 & 3 & -2 & 4 & 7 \\ 1 & -4 & 4 & -4 & -7 \end{pmatrix}$$

$$= \begin{pmatrix} 1 & -2 & 1 & -1 & 1 \\ 0 & 1 & -1 & 3 & 8 \\ 0 & 0 & 1 & 1 & 2 \\ 0 & 0 & 1 & 3 & 8 \end{pmatrix}$$

Using the third row, we get

$$\begin{pmatrix} 1 & 0 & 0 & 0 \\ 0 & 1 & 0 & 0 \\ 0 & 0 & 1 & 0 \\ 0 & 0 & -1 & 1 \end{pmatrix} \begin{pmatrix} 1 & -2 & 1 & -1 & 1 \\ 0 & 1 & -1 & 3 & 8 \\ 0 & 0 & 1 & 1 & 2 \\ 0 & 0 & 1 & 3 & 8 \end{pmatrix} = \begin{pmatrix} 1 & -2 & 1 & -1 & 1 \\ 0 & 1 & -1 & 3 & 8 \\ 0 & 0 & 1 & 1 & 2 \\ 0 & 0 & 0 & 2 & 6 \end{pmatrix}$$

or

$$\begin{pmatrix} 1 & 0 & 0 & 0 \\ 0 & 1 & 0 & 0 \\ 0 & 0 & 1 & 0 \\ 0 & 0 & -1 & 1 \end{pmatrix} \begin{pmatrix} 1 & 0 & 0 & 0 \\ 1 & 0 & 1 & 0 \\ -2 & 1 & 0 & 0 \\ 1 & 0 & 2 & 1 \end{pmatrix} \begin{pmatrix} 1 & -2 & 1 & -1 & 1 \\ 2 & -4 & 3 & -1 & 4 \\ -1 & 3 & -2 & 4 & 7 \\ 1 & -4 & 4 & -4 & -7 \end{pmatrix}$$

$$= \begin{pmatrix} 1 & 0 & 0 & 0 \\ 1 & 0 & 1 & 0 \\ -2 & 1 & 0 & 0 \\ 3 & -1 & 2 & 1 \end{pmatrix} \begin{pmatrix} 1 & -2 & 1 & -1 & 1 \\ 2 & -4 & 3 & -1 & 4 \\ -1 & 3 & -2 & 4 & 7 \\ 1 & -4 & 4 & -4 & -7 \end{pmatrix}$$

$$= \begin{pmatrix} 1 & -2 & 1 & -1 & 1 \\ 0 & 1 & -1 & 3 & 8 \\ 0 & 0 & 1 & 1 & 2 \\ 0 & 0 & 0 & 2 & 6 \end{pmatrix}$$

Since the latter matrix is a row echelon matrix, we could use the work so far to give a factorization of A as a product of a permuted lower triangular matrix and a row echelon matrix:

$$\begin{pmatrix} 1 & -2 & 1 & -1 & 1 \\ 2 & -4 & 3 & -1 & 4 \\ -1 & 3 & -2 & 4 & 7 \\ 1 & -4 & 4 & -4 & -7 \end{pmatrix} = \begin{pmatrix} 1 & 0 & 0 & 0 \\ 1 & 0 & 1 & 0 \\ -2 & 1 & 0 & 0 \\ 3 & -1 & 2 & 1 \end{pmatrix}^{-1} \begin{pmatrix} 1 & -2 & 1 & -1 & 1 \\ 0 & 1 & -1 & 3 & 8 \\ 0 & 0 & 1 & 1 & 2 \\ 0 & 0 & 0 & 2 & 6 \end{pmatrix}$$

or

$$\begin{pmatrix} 1 & -2 & 1 & -1 & 1 \\ 2 & -4 & 3 & -1 & 4 \\ -1 & 3 & -2 & 4 & 7 \\ 1 & -4 & 4 & -4 & -7 \end{pmatrix} = \begin{pmatrix} 1 & 0 & 0 & 0 \\ 2 & 0 & 1 & 0 \\ -1 & 1 & 0 & 0 \\ 1 & -2 & 1 & 1 \end{pmatrix} \begin{pmatrix} 1 & -2 & 1 & -1 & 1 \\ 0 & 1 & -1 & 3 & 8 \\ 0 & 0 & 1 & 1 & 2 \\ 0 & 0 & 0 & 2 & 6 \end{pmatrix}$$

However, we may want to continue to get a factorization into reduced row echelon form. We had

$$\begin{pmatrix} 1 & 0 & 0 & 0 \\ 1 & 0 & 1 & 0 \\ -2 & 1 & 0 & 0 \\ 3 & -1 & 2 & 1 \end{pmatrix} \begin{pmatrix} 1 & -2 & 1 & -1 & 1 \\ 2 & -4 & 3 & -1 & 4 \\ -1 & 3 & -2 & 4 & 7 \\ 1 & -4 & 4 & -4 & -7 \end{pmatrix} = \begin{pmatrix} 1 & -2 & 1 & -1 & 1 \\ 0 & 1 & -1 & 3 & 8 \\ 0 & 0 & 1 & 1 & 2 \\ 0 & 0 & 0 & 2 & 6 \end{pmatrix}$$

Dividing the last row by 2, we get

$$\begin{pmatrix} 1 & 0 & 0 & 0 \\ 0 & 1 & 0 & 0 \\ 0 & 0 & 1 & 0 \\ 0 & 0 & 0 & .5 \end{pmatrix} \begin{pmatrix} 1 & -2 & 1 & -1 & 1 \\ 0 & 1 & -1 & 3 & 8 \\ 0 & 0 & 1 & 1 & 2 \\ 0 & 0 & 0 & 2 & 6 \end{pmatrix} = \begin{pmatrix} 1 & -2 & 1 & -1 & 1 \\ 0 & 1 & -1 & 3 & 8 \\ 0 & 0 & 1 & 1 & 2 \\ 0 & 0 & 0 & 1 & 3 \end{pmatrix}$$

so

$$\begin{pmatrix} 1 & 0 & 0 & 0 \\ 0 & 1 & 0 & 0 \\ 0 & 0 & 1 & 0 \\ 0 & 0 & 0 & .5 \end{pmatrix} \begin{pmatrix} 1 & 0 & 0 & 0 \\ 1 & 0 & 1 & 0 \\ -2 & 1 & 0 & 0 \\ 3 & -1 & 2 & 1 \end{pmatrix} \begin{pmatrix} 1 & -2 & 1 & -1 & 1 \\ 2 & -4 & 3 & -1 & 4 \\ -1 & 3 & -2 & 4 & 7 \\ 1 & -4 & 4 & -4 & -7 \end{pmatrix}$$

$$= \begin{pmatrix} 1 & 0 & 0 & 0 \\ 0 & 1 & 0 & 0 \\ 0 & 0 & 1 & 0 \\ 0 & 0 & 0 & .5 \end{pmatrix} \begin{pmatrix} 1 & -2 & 1 & -1 & 1 \\ 0 & 1 & -1 & 3 & 8 \\ 0 & 0 & 1 & 1 & 2 \\ 0 & 0 & 0 & 2 & 6 \end{pmatrix} = \begin{pmatrix} 1 & -2 & 1 & -1 & 1 \\ 0 & 1 & -1 & 3 & 8 \\ 0 & 0 & 1 & 1 & 2 \\ 0 & 0 & 0 & 1 & 3 \end{pmatrix}$$

which means

$$\begin{pmatrix} 1 & 0 & 0 & 0 \\ 1 & 0 & 1 & 0 \\ -2 & 1 & 0 & 0 \\ 1.5 & -.5 & 1 & .5 \end{pmatrix} \begin{pmatrix} 1 & -2 & 1 & -1 & 1 \\ 2 & -4 & 3 & -1 & 4 \\ -1 & 3 & -2 & 4 & 7 \\ 1 & -4 & 4 & -4 & -7 \end{pmatrix} = \begin{pmatrix} 1 & -2 & 1 & -1 & 1 \\ 0 & 1 & -1 & 3 & 8 \\ 0 & 0 & 1 & 1 & 2 \\ 0 & 0 & 0 & 1 & 3 \end{pmatrix}$$

Using the fourth row, we get

$$\begin{pmatrix} 1 & 0 & 0 & 1 \\ 0 & 1 & 0 & -3 \\ 0 & 0 & 1 & -1 \\ 0 & 0 & 0 & 1 \end{pmatrix} \begin{pmatrix} 1 & -2 & 1 & -1 & 1 \\ 0 & 1 & -1 & 3 & 8 \\ 0 & 0 & 1 & 1 & 2 \\ 0 & 0 & 0 & 1 & 3 \end{pmatrix} = \begin{pmatrix} 1 & -2 & 1 & 0 & 4 \\ 0 & 1 & -1 & 0 & -1 \\ 0 & 0 & 1 & 0 & -1 \\ 0 & 0 & 0 & 1 & 3 \end{pmatrix}$$

so

$$\begin{pmatrix} 1 & 0 & 0 & 1 \\ 0 & 1 & 0 & -3 \\ 0 & 0 & 1 & -1 \\ 0 & 0 & 0 & 1 \end{pmatrix} \begin{pmatrix} 1 & 0 & 0 & 0 \\ 1 & 0 & 1 & 0 \\ -2 & 1 & 0 & 0 \\ 1.5 & -.5 & 1 & .5 \end{pmatrix} \begin{pmatrix} 1 & -2 & 1 & -1 & 1 \\ 2 & -4 & 3 & -1 & 4 \\ -1 & 3 & -2 & 4 & 7 \\ 1 & -4 & 4 & -4 & -7 \end{pmatrix}$$

$$= \begin{pmatrix} 1 & 0 & 0 & 1 \\ 0 & 1 & 0 & -3 \\ 0 & 0 & 1 & -1 \\ 0 & 0 & 0 & 1 \end{pmatrix} \begin{pmatrix} 1 & -2 & 1 & -1 & 1 \\ 0 & 1 & -1 & 3 & 8 \\ 0 & 0 & 1 & 1 & 2 \\ 0 & 0 & 0 & 1 & 3 \end{pmatrix} = \begin{pmatrix} 1 & -2 & 1 & 0 & 4 \\ 0 & 1 & -1 & 0 & -1 \\ 0 & 0 & 1 & 0 & -1 \\ 0 & 0 & 0 & 1 & 3 \end{pmatrix}$$

or

$$\begin{pmatrix} 2.5 & -.5 & 1 & .5 \\ -3.5 & 1.5 & -2 & -1.5 \\ -3.5 & 1.5 & -1 & -.5 \\ 1.5 & -.5 & 1 & .5 \end{pmatrix} \begin{pmatrix} 1 & -2 & 1 & -1 & 1 \\ 2 & -4 & 3 & -1 & 4 \\ -1 & 3 & -2 & 4 & 7 \\ 1 & -4 & 4 & -4 & -7 \end{pmatrix} = \begin{pmatrix} 1 & -2 & 1 & 0 & 4 \\ 0 & 1 & -1 & 0 & -1 \\ 0 & 0 & 1 & 0 & -1 \\ 0 & 0 & 0 & 1 & 3 \end{pmatrix}$$

Using the third row, we get

$$\begin{pmatrix} 1 & 0 & -1 & 0 \\ 0 & 1 & 1 & 0 \\ 0 & 0 & 1 & 0 \\ 0 & 0 & 0 & 1 \end{pmatrix} \begin{pmatrix} 1 & -2 & 1 & 0 & 4 \\ 0 & 1 & -1 & 0 & -1 \\ 0 & 0 & 1 & 0 & -1 \\ 0 & 0 & 0 & 1 & 3 \end{pmatrix} = \begin{pmatrix} 1 & -2 & 0 & 0 & 5 \\ 0 & 1 & 0 & 0 & -2 \\ 0 & 0 & 1 & 0 & -1 \\ 0 & 0 & 0 & 1 & 3 \end{pmatrix}$$

so

$$\begin{pmatrix} 1 & 0 & -1 & 0 \\ 0 & 1 & 1 & 0 \\ 0 & 0 & 1 & 0 \\ 0 & 0 & 0 & 1 \end{pmatrix} \begin{pmatrix} 2.5 & -.5 & 1 & .5 \\ -3.5 & 1.5 & -2 & -1.5 \\ -3.5 & 1.5 & -1 & -.5 \\ 1.5 & -.5 & 1 & .5 \end{pmatrix} \begin{pmatrix} 1 & -2 & 1 & -1 & 1 \\ 2 & -4 & 3 & -1 & 4 \\ -1 & 3 & -2 & 4 & 7 \\ 1 & -4 & 4 & -4 & -7 \end{pmatrix}$$

$$= \begin{pmatrix} 1 & 0 & -1 & 0 \\ 0 & 1 & 1 & 0 \\ 0 & 0 & 1 & 0 \\ 0 & 0 & 0 & 1 \end{pmatrix} \begin{pmatrix} 1 & -2 & 1 & 0 & 4 \\ 0 & 1 & -1 & 0 & -1 \\ 0 & 0 & 1 & 0 & -1 \\ 0 & 0 & 0 & 1 & 3 \end{pmatrix} = \begin{pmatrix} 1 & -2 & 0 & 0 & 5 \\ 0 & 1 & 0 & 0 & -2 \\ 0 & 0 & 1 & 0 & -1 \\ 0 & 0 & 0 & 1 & 3 \end{pmatrix}$$

or

$$\begin{pmatrix} 6 & -2 & 2 & 1 \\ -7 & 3 & -3 & -2 \\ -3.5 & 1.5 & -1 & -.5 \\ 1.5 & -.5 & 1 & .5 \end{pmatrix} \begin{pmatrix} 1 & -2 & 1 & -1 & 1 \\ 2 & -4 & 3 & -1 & 4 \\ -1 & 3 & -2 & 4 & 7 \\ 1 & -4 & 4 & -4 & -7 \end{pmatrix} = \begin{pmatrix} 1 & -2 & 0 & 0 & 5 \\ 0 & 1 & 0 & 0 & -2 \\ 0 & 0 & 1 & 0 & -1 \\ 0 & 0 & 0 & 1 & 3 \end{pmatrix}$$

Finally, using the second row, we get

$$\begin{pmatrix} 1 & 2 & 0 & 0 \\ 0 & 1 & 0 & 0 \\ 0 & 0 & 1 & 0 \\ 0 & 0 & 0 & 1 \end{pmatrix} \begin{pmatrix} 1 & -2 & 0 & 0 & 5 \\ 0 & 1 & 0 & 0 & -2 \\ 0 & 0 & 1 & 0 & -1 \\ 0 & 0 & 0 & 1 & 3 \end{pmatrix} = \begin{pmatrix} 1 & 0 & 0 & 0 & 1 \\ 0 & 1 & 0 & 0 & -2 \\ 0 & 0 & 1 & 0 & -1 \\ 0 & 0 & 0 & 1 & 3 \end{pmatrix}$$

which gives

$$\begin{pmatrix} 1 & 2 & 0 & 0 \\ 0 & 1 & 0 & 0 \\ 0 & 0 & 1 & 0 \\ 0 & 0 & 0 & 1 \end{pmatrix} \begin{pmatrix} 6 & -2 & 2 & 1 \\ -7 & 3 & -3 & -2 \\ -3.5 & 1.5 & -1 & -.5 \\ 1.5 & -.5 & 1 & .5 \end{pmatrix} \begin{pmatrix} 1 & -2 & 1 & -1 & 1 \\ 2 & -4 & 3 & -1 & 4 \\ -1 & 3 & -2 & 4 & 7 \\ 1 & -4 & 4 & -4 & -7 \end{pmatrix}$$

$$
= \begin{pmatrix} 1 & 2 & 0 & 0 \\ 0 & 1 & 0 & 0 \\ 0 & 0 & 1 & 0 \\ 0 & 0 & 0 & 1 \end{pmatrix} \begin{pmatrix} 1 & -2 & 0 & 0 & 5 \\ 0 & 1 & 0 & 0 & -2 \\ 0 & 0 & 1 & 0 & -1 \\ 0 & 0 & 0 & 1 & 3 \end{pmatrix} = \begin{pmatrix} 1 & 0 & 0 & 0 & 1 \\ 0 & 1 & 0 & 0 & -2 \\ 0 & 0 & 1 & 0 & -1 \\ 0 & 0 & 0 & 1 & 3 \end{pmatrix}
$$

or

$$
\begin{pmatrix} -8 & 4 & -4 & -3 \\ -7 & 3 & -3 & -2 \\ -3.5 & 1.5 & -1 & -.5 \\ 1.5 & -.5 & 1 & .5 \end{pmatrix} \begin{pmatrix} 1 & -2 & 1 & -1 & 1 \\ 2 & -4 & 3 & -1 & 4 \\ -1 & 3 & -2 & 4 & 7 \\ 1 & -4 & 4 & -4 & -7 \end{pmatrix} = \begin{pmatrix} 1 & 0 & 0 & 0 & 1 \\ 0 & 1 & 0 & 0 & -2 \\ 0 & 0 & 1 & 0 & -1 \\ 0 & 0 & 0 & 1 & 3 \end{pmatrix}
$$

Since the latter matrix is in reduced row echelon form, this gives the factorization we were looking for. Rewriting it as

$$
\begin{pmatrix} 1 & -2 & 1 & -1 & 1 \\ 2 & -4 & 3 & -1 & 4 \\ -1 & 3 & -2 & 4 & 7 \\ 1 & -4 & 4 & -4 & -7 \end{pmatrix} = \begin{pmatrix} -8 & 4 & -4 & -3 \\ -7 & 3 & -3 & -2 \\ -3.5 & 1.5 & -1 & -.5 \\ 1.5 & -.5 & 1 & .5 \end{pmatrix}^{-1} \begin{pmatrix} 1 & 0 & 0 & 0 & 1 \\ 0 & 1 & 0 & 0 & -2 \\ 0 & 0 & 1 & 0 & -1 \\ 0 & 0 & 0 & 1 & 3 \end{pmatrix}
$$

then we get $A = LE$ where

$$
L = \begin{pmatrix} -8 & 4 & -4 & -3 \\ -7 & 3 & -3 & -2 \\ -3.5 & 1.5 & -1 & -.5 \\ 1.5 & -.5 & 1 & .5 \end{pmatrix}^{-1} = \begin{pmatrix} 1 & -2 & 1 & -1 \\ 2 & -4 & 3 & -1 \\ -1 & 3 & -2 & 4 \\ 1 & -4 & 4 & -4 \end{pmatrix}
$$

and

$$
E = \begin{pmatrix} 1 & 0 & 0 & 0 & 1 \\ 0 & 1 & 0 & 0 & -2 \\ 0 & 0 & 1 & 0 & -1 \\ 0 & 0 & 0 & 1 & 3 \end{pmatrix}
$$

$\square$

In practice, we will not usually calculate the invertible matrix L but will do row operations to calculate E and use the fact that $A = LE$ for some invertible L without knowing L. An exception to this practice would be if we expect to have several systems with the same coefficient matrix, but with several different righthand sides. In iterative applications, this is frequently the case, and in such a situation, the righthand sides are found recursively, so are not available at the beginning to augment the coefficient matrix. We will find further uses of the factorization implicit in row reduction in Section 3.7.

Exercises 2.5

1. What row operation is accomplished if you multiply a 3×5 matrix on the left by the matrix

$$L_1 = \begin{pmatrix} 1 & 0 & 0 \\ 0 & 1 & 0 \\ -2 & 0 & 1 \end{pmatrix}$$

What if you multiply by

$$L_2 = \begin{pmatrix} 1 & 0 & 0 \\ 0 & 1 & 3 \\ 0 & 0 & 1 \end{pmatrix}$$

What if you multiply by

$$L_3 = \begin{pmatrix} -4 & 0 & 0 \\ 0 & 1 & 0 \\ 0 & 0 & 1 \end{pmatrix}$$

2. If A is a 4×5 matrix, what matrix should you multiply A on the left by to interchange the second and fourth rows of A? What should you multiply A by to replace the third row by 5 times the third row? What should you multiply A by to replace the second row by the second row plus 3 times the first row?

Find a factorization of each of the following matrices as LE where L is a square invertible matrix and E is in reduced row echelon form:

3. $A = \begin{pmatrix} 1 & 3 \\ 2 & 5 \end{pmatrix}$

4. $B = \begin{pmatrix} 1 & 2 & 3 \\ 4 & 5 & 6 \end{pmatrix}$

5. $C = \begin{pmatrix} 1 & -1 & 1 \\ -1 & 3 & 3 \\ 2 & 0 & 3 \end{pmatrix}$

6. $D = \begin{pmatrix} 1 & -1 & 2 & 1 \\ -2 & 2 & -3 & -5 \\ 1 & -1 & 1 & 6 \\ 2 & -2 & 5 & -7 \end{pmatrix}$

7. $E = \begin{pmatrix} 1 & 0 & -1 & 2 \\ -1 & 1 & 3 & -1 \\ 2 & 0 & 0 & 5 \\ 1 & 1 & 1 & 3 \end{pmatrix}$

8. $F = \begin{pmatrix} -2 & 0 & 3 & -1 & -3 \\ 2 & 1 & -1 & 1 & 2 \\ 6 & 2 & -5 & 6 & 1 \\ 4 & 1 & -4 & 3 & 3 \end{pmatrix}$

9. If A is an invertible matrix and $A = LE$ where L is invertible and E is in reduced row echelon form, show that $L = A$ and $E = I$. (See Exercise 2.5.3..)

2.6   Determinants

The determinant is a scalar valued function of square matrices that is important in theoretical discussions of invertibility and solution of linear equations. Its principal numerical importance is in the change of variables formula for multiple integrals, so we will touch on it only briefly.

We will define it recursively by the *Laplace expansion formula* and state some of its important properties, but we will omit most of the proofs.

DEFINITION *For A an $n \times n$ matrix, the* determinant *of A, $\det(A)$, is a number computed as follows: for 2×2 matrices,*

$$\det \begin{pmatrix} a & b \\ c & d \end{pmatrix} = ad - bc$$

For an $n \times n$ matrix with $n > 2$, for any i, $1 \le i \le n$,

$$\det(A) = (-1)^{i+1} a_{i1} \det(\hat{A}_{i1}) + (-1)^{i+2} a_{i2} \det(\hat{A}_{i2}) + \cdots + (-1)^{i+n} a_{in} \det(\hat{A}_{in})$$

where $\hat{A}_{ij}$ is the $(n-1) \times (n-1)$ matrix obtained by deleting the ith row and the jth column of A.

The formula in the definition is called the Laplace expansion formula for expansion along the ith row or *expansion by minors* along the ith row. It is, of course, a recursive definition, in which the determinant of a 4×4 matrix is to be computed in terms of four 3×3 determinants and each of these is to be computed by three 2×2 determinants for which we have the formula. MATLAB computes the determinant of the matrix A with the command det(A).

For this to be an acceptable definition, we should prove that the Laplace expansion gives the same answer no matter which row is chosen for the expansion; rather than give a proof, we'll just accept this as true. A similar column expansion also can be used to compute the determinant.

EXAMPLE 2.21

Find the determinant of the matrix

$$\begin{pmatrix} 3 & -2 & 1 \\ 2 & 5 & -1 \\ -1 & -2 & 4 \end{pmatrix}$$

SOLUTION We will expand along the second row, that is, $i = 2$ in the definition.

$$\det \begin{pmatrix} 3 & -2 & 1 \\ 2 & 5 & -1 \\ -1 & -2 & 4 \end{pmatrix} = 2(-1)^{2+1} \det \begin{pmatrix} -2 & 1 \\ -2 & 4 \end{pmatrix}$$

$$+ 5(-1)^{2+2} \det \begin{pmatrix} 3 & 1 \\ -1 & 4 \end{pmatrix} + (-1)(-1)^{2+3} \det \begin{pmatrix} 3 & -2 \\ -1 & -2 \end{pmatrix}$$

Thus, the single 3×3 determinant we needed to compute has been replaced by three 2×2 determinants. These determinants are computed directly from the definition.

$$= (-2)\left((-2)(4) - (1)(-2)\right) + (5)\left((3)(4) - (1)(-1)\right) + (1)\left((3)(-2) - (-2)(-1)\right)$$

$$= (-2)(-6) + (5)(13) + (1)(-8) = 69$$

$\square$

There are a few easy general principles that can be proved.

COROLLARY 2.22

If A is an upper or lower triangular $n \times n$ matrix,

$$\det(A) = a_{11}a_{22} \cdots a_{nn}$$

the product of the diagonal entries.

PROOF   Consider the lower triangular matrix

$$A = \begin{pmatrix} a_{11} & 0 & 0 & \cdots & 0 \\ a_{21} & a_{22} & 0 & \cdots & 0 \\ a_{31} & a_{32} & a_{33} & & 0 \\ & & & \ddots & \\ a_{n1} & a_{n2} & a_{n3} & \cdots & a_{nn} \end{pmatrix}$$

Expanding along the first row, we see all the terms but the first are zero, so we get

$$\det(A) = (-1)^{1+1} a_{11} \det \begin{pmatrix} a_{22} & 0 & \cdots & 0 \\ a_{32} & a_{33} & & 0 \\ & & \ddots & \\ a_{n2} & a_{n3} & \cdots & a_{nn} \end{pmatrix}$$

Expanding this matrix along its first row, we get

$$\det(A) = (-1)^{1+1}a_{11}\left((-1)^{1+1}a_{22}\det\begin{pmatrix} a_{33} & & 0 \\ & \ddots & \\ a_{n3} & \cdots & a_{nn} \end{pmatrix}\right)$$

Continuing, after $n-1$ such expansions, we see that

$$\det(A) = a_{11}a_{22}\cdots a_{nn}$$

as we wished to show.

The proof for upper triangular matrices is similar except we expand along the last row of the matrix at each step. ∎

EXAMPLE 2.23

As an illustration of using the Laplace expansion formula, we will develop a formula for 3×3 matrices that is analogous to the formula in the definition for 2×2 matrices. Expanding along the first row, we get

$$\det\begin{pmatrix} a & b & c \\ d & e & f \\ g & h & j \end{pmatrix}$$

$$= a(-1)^{1+1}\det\begin{pmatrix} e & f \\ h & j \end{pmatrix} + b(-1)^{1+2}\det\begin{pmatrix} d & f \\ g & j \end{pmatrix} + c(-1)^{1+3}\det\begin{pmatrix} d & e \\ g & h \end{pmatrix}$$

$$= a(ej - fh) - b(dj - fg) + c(dh - eg)$$

$$= aej + bfg + cdh - afh - bdj - ceg$$

This formula works for finding the determinant of any 3×3 and can be memorized if you wish. A mnemonic for remembering the formula is to notice the three positive products are the products in the matrix moving down and to the right and the three negative products are the products in the matrix moving down and to the left. Figure 2.3 illustrates the mnemonic: it shows the basketweave mnemonic for the 2×2 formula and the basketweave mnemonic for the 3×3 formula. For the 3×3 formula, it is sometimes helpful to copy the first two columns to the right of the matrix to line up the arrows as has been done in the figure.

WARNING! This basket weave suggests a pattern for finding the determinant of an arbitrary square matrix, but this generalization to 4×4 and larger matrices is

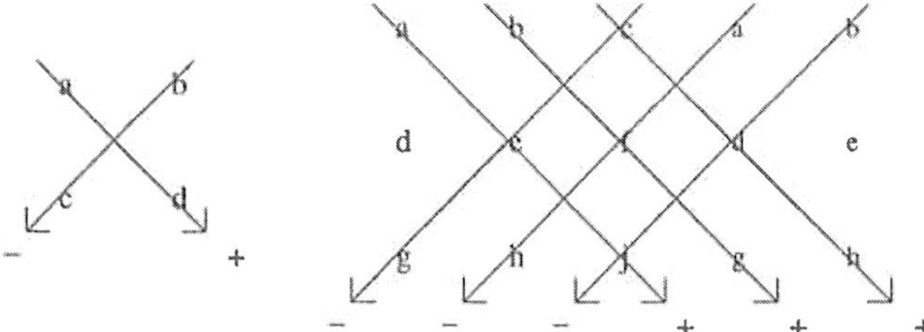

FIGURE 2.3
   Mnemonic for the determinant of 2×2 and 3×3 matrices

incorrect! In fact, the determinant of an $n \times n$ matrix is a sum of $n!$ products of n terms each, with one term in each product from each row and one term from each column. Half of the resulting terms have a $+$ in the sum for the determinant and half have a $-$. For example, expanding the determinant of a 4×4 matrix by the Laplace expansion will show that it is the sum of $4! = 24$ terms each the product of 4 entries of the matrix. A basket weave would only include 8 of the 24 terms and thus cannot be correct.   ☐

You should try the basketweave mnemonic on the problem of Example 2.21.

Because the number of terms in computing a determinant from the Laplace expansion is so large for even moderate sized matrices, using row operations as in Gaussian elimination to find determinants is a more practical approach to computing them than using the Laplace expansion. The following theorem tells how.

Before we can state the theorem, however, we must discuss the notation of block matrices we will use. Instead of thinking about a matrix as consisting of its entries, we can think of the matrix as being built from its rows. Viewed this way, we write the matrix as a column of row vectors. For example, the matrix

$$A = \begin{pmatrix} 1 & 2 & 3 \\ 4 & 5 & 6 \\ 7 & 8 & 9 \end{pmatrix}$$

can be written as

$$A = \begin{pmatrix} R_1 \\ R_2 \\ R_3 \end{pmatrix}$$

where $R_1 = \begin{pmatrix} 1 & 2 & 3 \end{pmatrix}$, $R_2 = \begin{pmatrix} 4 & 5 & 6 \end{pmatrix}$, and $R_3 = \begin{pmatrix} 7 & 8 & 9 \end{pmatrix}$. By writing it in this way, we say we have *blocked* the matrix A in rows. The advantage of this point of view is that we can then more readily focus on the rows as entities in themselves. This topic is explored in greater generality in Appendix B.

THEOREM 2.24

Suppose A is an $n \times n$ matrix. Write A as a block matrix where each row is a block

$$A = \begin{pmatrix} R_1 \\ R_2 \\ \vdots \\ R_n \end{pmatrix}$$

The following properties hold:

1. *If two rows of A are interchanged, the determinant changes signs, that is, for $i \neq j$*

$$\det \begin{pmatrix} R_1 \\ R_2 \\ \vdots \\ R_i \\ \vdots \\ R_j \\ \vdots \\ R_n \end{pmatrix} = -\det \begin{pmatrix} R_1 \\ R_2 \\ \vdots \\ R_j \\ \vdots \\ R_i \\ \vdots \\ R_n \end{pmatrix}$$

2. *If two rows of A are the same, then $\det(A) = 0$.*

3. *The determinant is linear in each row of A separately, that is,*

$$\det \begin{pmatrix} R_1 \\ R_2 \\ \vdots \\ \alpha R_j \\ \vdots \\ R_n \end{pmatrix} = \alpha \det \begin{pmatrix} R_1 \\ R_2 \\ \vdots \\ R_j \\ \vdots \\ R_n \end{pmatrix}$$

and

$$\det \begin{pmatrix} R_1 \\ R_2 \\ \vdots \\ S_j + T_j \\ \vdots \\ R_n \end{pmatrix} = \det \begin{pmatrix} R_1 \\ R_2 \\ \vdots \\ S_j \\ \vdots \\ R_n \end{pmatrix} + \det \begin{pmatrix} R_1 \\ R_2 \\ \vdots \\ T_j \\ \vdots \\ R_n \end{pmatrix}$$

4. *If a multiple of one row is added to another row, the determinant does not change, that is, for $i \neq j$*

$$\det \begin{pmatrix} R_1 \\ R_2 \\ \vdots \\ R_i + \alpha R_j \\ \vdots \\ R_j \\ \vdots \\ R_n \end{pmatrix} = \det \begin{pmatrix} R_1 \\ R_2 \\ \vdots \\ R_i \\ \vdots \\ R_j \\ \vdots \\ R_n \end{pmatrix}$$

5. $\det I = 1.$

While each of the row operations mentioned above, when applied to a system of equations, gives a system equivalent system to the original one, the theorem says some of these row operations alter the determinant. In particular, interchanging two rows of a matrix multiplies the determinant by -1 and multiplying a row by a number multiplies the determinant by that number. Replacing a row by it plus a multiple of another row does not change the value of the determinant. To use row operations to find determinants, we need to keep track of the changes in the determinant caused by the row operations we do.

EXAMPLE 2.25

Find the determinant of the matrix

$$\begin{pmatrix} 3 & -2 & 1 \\ 2 & 5 & -1 \\ -1 & -2 & 4 \end{pmatrix}$$

SOLUTION This is the problem of Example 2.21, but this time we will use the properties described in Theorem 2.24. The row operations will be easier if we use the -1 in the last row as a pivot. Using property 1. of the Theorem, we interchange the first and third row which changes the sign of the determinant.

$$\det \begin{pmatrix} 3 & -2 & 1 \\ 2 & 5 & -1 \\ -1 & -2 & 4 \end{pmatrix} = -\det \begin{pmatrix} -1 & -2 & 4 \\ 2 & 5 & -1 \\ 3 & -2 & 1 \end{pmatrix}$$

Now, use property 4. of the Theorem to replace the second row by the second row plus 2 times the first row (which does not change the determinant).

$$-\det \begin{pmatrix} -1 & -2 & 4 \\ 2 & 5 & -1 \\ 3 & -2 & 1 \end{pmatrix} = -\det \begin{pmatrix} -1 & -2 & 4 \\ 0 & 1 & 7 \\ 3 & -2 & 1 \end{pmatrix}$$

Similarly, we use property 4. of the Theorem to replace the third row by the third plus 3 times the first row (which does not change the determinant).

$$-\det \begin{pmatrix} -1 & -2 & 4 \\ 0 & 1 & 7 \\ 3 & -2 & 1 \end{pmatrix} = -\det \begin{pmatrix} -1 & -2 & 4 \\ 0 & 1 & 7 \\ 0 & -8 & 13 \end{pmatrix}$$

Now replace the third row by the third plus 8 times the second.

$$-\det \begin{pmatrix} -1 & -2 & 4 \\ 0 & 1 & 7 \\ 0 & -8 & 13 \end{pmatrix} = -\det \begin{pmatrix} -1 & -2 & 4 \\ 0 & 1 & 7 \\ 0 & 0 & 69 \end{pmatrix}$$

Now by Corollary 2.22, the last determinant can be evaluated easily and we find

$$\det \begin{pmatrix} 3 & -2 & 1 \\ 2 & 5 & -1 \\ -1 & -2 & 4 \end{pmatrix} = (-1)(-1)(1)(69) = 69$$

as before. ☐

EXAMPLE 2.26

Find the determinant of the matrix

$$\begin{pmatrix} 1 & -2 & 1 & -1 \\ 2 & -4 & 3 & -1 \\ -1 & 3 & -2 & 4 \\ 1 & -4 & 4 & -4 \end{pmatrix}$$

SOLUTION We will use row operations as described in Theorem 2.24, this time at a somewhat more typical pace.

$$\det \begin{pmatrix} 1 & -2 & 1 & -1 \\ 2 & -4 & 3 & -1 \\ -1 & 3 & -2 & 4 \\ 1 & -4 & 4 & -4 \end{pmatrix} = \det \begin{pmatrix} 1 & -2 & 1 & -1 \\ 0 & 0 & 1 & 1 \\ 0 & 1 & -1 & 3 \\ 0 & -2 & 3 & -3 \end{pmatrix}$$

because we have replaced each of rows 2, 3, and 4 by themselves plus a multiple of the first row. Now, interchanging the second and third rows and then replacing rows by themselves plus multiples of another row gives

$$= -\det \begin{pmatrix} 1 & -2 & 1 & -1 \\ 0 & 1 & -1 & 3 \\ 0 & 0 & 1 & 1 \\ 0 & -2 & 3 & -3 \end{pmatrix} = -\det \begin{pmatrix} 1 & -2 & 1 & -1 \\ 0 & 1 & -1 & 3 \\ 0 & 0 & 1 & 1 \\ 0 & 0 & 1 & 3 \end{pmatrix}$$

$$= -\det \begin{pmatrix} 1 & -2 & 1 & -1 \\ 0 & 1 & -1 & 3 \\ 0 & 0 & 1 & 1 \\ 0 & 0 & 0 & 2 \end{pmatrix} = -(1)(1)(1)(2) = -2$$

where the last step follows from Corollary 2.22.   ☐

In general, you should expect that doing row operations to find determinants will be faster than the Laplace expansion for matrices 4×4 or larger.

The following two theorems are important facts about determinants but their proofs are beyond the scope of this book. Since the determinant of a matrix and its transpose are the same, determinants can be computed with column operations as well as row operations.

THEOREM 2.27

If A is an $n \times n$ matrix, $\det(A^t) = \det(A)$.

COROLLARY 2.28

If A is an $n \times n$ complex matrix, $\det(A') = \overline{\det(A)}$.

PROOF   The complex conjugate of a sum or product is the sum or product of the conjugates; since the determinant of a matrix is a sum of products of the entries, putting the conjugates on the entries is the same as conjugating the whole sum. Since A' is the conjugate of the transpose and $\det(A^t) = \det(A)$, the determinant of A' is the conjugate of $\det(A)$.   ∎

THEOREM 2.29

If A and B are $n \times n$ matrices, then $\det(AB) = \det(A)\det(B)$.

COROLLARY 2.30

If A is invertible, $\det(A^{-1}) = 1/\det(A)$

PROOF   We have $AA^{-1} = I$, so

$$1 = \det(I) = \det(AA^{-1}) = \det(A)\det(A^{-1})$$

∎

In particular, if A is invertible, $\det(A)$ is non-zero. The converse is also true and the following theorem gives a "formula" for the inverse of a matrix using determinants. The matrix in the formula is sometimes called the *classical adjoint* or the *adjugate* of A. Although it is appealing to have a formula for the inverse of a matrix, the computation of the inverse is more easily carried out by row operations as in Section 2.4. This formula is of mostly theoretical interest.

THEOREM 2.31

If A is an $n \times n$ matrix and $\det(A)$ is not zero, then A is invertible and A^{-1} is given by

$$\frac{1}{\det(A)}\begin{pmatrix} (-1)^{1+1}\det(\hat{A}_{11}) & (-1)^{2+1}\det(\hat{A}_{21}) & \cdots & (-1)^{n+1}\det(\hat{A}_{n1}) \\ (-1)^{1+2}\det(\hat{A}_{12}) & (-1)^{2+2}\det(\hat{A}_{22}) & \cdots & (-1)^{n+2}\det(\hat{A}_{n2}) \\ \vdots & & \ddots & \vdots \\ (-1)^{1+n}\det(\hat{A}_{1n}) & (-1)^{2+n}\det(\hat{A}_{2n}) & \cdots & (-1)^{n+n}\det(\hat{A}_{nn}) \end{pmatrix}$$

Recall that $\hat{A}_{ij}$ is the $(n-1) \times (n-1)$ matrix obtained by deleting the ith row and the jth column of A. If we call the expression $(-1)^{i+j} \det(\hat{A}_{ij})$ the *cofactor* of a_{ij}, then the matrix in the theorem is the *transpose* of the matrix of cofactors. The proof consists of multiplying A times the matrix in the theorem and interpreting each of the sums for the entry of the product as a row expansion for a determinant. It is not difficult, but we will omit it.

COROLLARY 2.32

If A is an $n \times n$ matrix, A is invertible if and only if $\det(A) \neq 0$.

PROOF   In the formula above, the only thing that can prevent the formula from giving the inverse is if the final step of dividing by $\det(A)$ is impossible because this number is zero. Thus if the $\det(A)$ is non-zero, the matrix A is invertible.

If $\det(A)$ is zero, then the matrix is not invertible; indeed, if it were,

$$1 = \det(AA^{-1}) = \det(A)\det(A^{-1}) = 0$$

which is a contradiction.   ∎

EXAMPLE 2.33

Let's use the adjugate formula above to find the inverse of

$$A = \begin{pmatrix} 2 & -1 & 3 \\ 1 & 0 & -2 \\ 3 & 2 & -1 \end{pmatrix}$$

We need to find the determinant of A and the nine 2×2 determinants of the submatrices of A to find the entries of the adjugate matrix.

$$\det(\hat{A}_{11}) = \det \begin{pmatrix} 0 & -2 \\ 2 & -1 \end{pmatrix} = 4$$

$$\det(\hat{A}_{12}) = \det \begin{pmatrix} 1 & -2 \\ 3 & -1 \end{pmatrix} = 5$$

$$\det(\hat{A}_{13}) = \det \begin{pmatrix} 1 & 0 \\ 3 & 2 \end{pmatrix} = 2$$

$$\det(\hat{A}_{21}) = \det \begin{pmatrix} -1 & 3 \\ 2 & -1 \end{pmatrix} = -5$$

$$\det(\hat{A}_{22}) = \det \begin{pmatrix} 2 & 3 \\ 3 & -1 \end{pmatrix} = -11$$

$$\det(\hat{A}_{23}) = \det \begin{pmatrix} 2 & -1 \\ 3 & 2 \end{pmatrix} = 7$$

$$\det(\hat{A}_{31}) = \det \begin{pmatrix} -1 & 3 \\ 0 & -2 \end{pmatrix} = 2$$

$$\det(\hat{A}_{32}) = \det \begin{pmatrix} 2 & 3 \\ 1 & -2 \end{pmatrix} = -7$$

$$\det(\hat{A}_{33}) = \det \begin{pmatrix} 2 & -1 \\ 1 & 0 \end{pmatrix} = 1$$

Using the formula of Example 2.23, the determinant of A is

$$(2)(0)(-1)+(-1)(-2)(3)+(3)(1)(2)-(3)(0)(3)-(-1)(1)(-1)-(2)(-2)(2) = 19$$

Therefore, according to the formula of Theorem 2.31, the inverse of A is

$$A^{-1} = \frac{1}{19} \begin{pmatrix} 4 & 5 & 2 \\ -5 & -11 & 7 \\ 2 & -7 & 1 \end{pmatrix}$$

As usual, to check this answer, we multiply the matrix A by the matrix obtained above and, indeed, it is A^{-1}. $\square$

Combining the formula for the inverse given in Theorem 2.31 with the observation that a system $AX = b$ of n equations in n unknowns has a unique solution if A is invertible gives the famous "Cramer's Rule" which is a formula for the solution of the system. This formula, however, is again mostly of theoretical interest because the amount of effort needed to solve a system of n equations in n unknowns is much more using Cramer's Rule than that using Gaussian elimination when $n \geq 4$. Indeed, if one were to use a mainframe computer that does 10^9 multiplications each second, to solve a system of 20 equations in 20 unknowns using Cramer's Rule (computing the determinants by minors), it would require more than *1000 years* of computer time, whereas MATLAB running on personal computer can do such a problem using Gaussian elimination in less than a second.

COROLLARY 2.34   *Cramer's Rule*

If A is an $n \times n$ matrix and $\det(A)$ is not zero, then the unique solution of $AX = b$ is $x_j = \det(B_j)/\det(A)$ where B_j is the matrix obtained by replacing the jth column of A by b, that is

$$B_j = \begin{pmatrix} C_1 & \cdots & C_{j-1} & b & C_{j+1} & \cdots & C_n \end{pmatrix}$$

where the columns of A are $C_1, C_2, \cdots, C_n$.

PROOF   If A is an $n \times n$ matrix and $\det(A) \neq 0$, then A is invertible and the unique solution of $AX = b$ is $X = A^{-1}b$. Using the formula for the adjugate matrix in Theorem 2.31 and noting that from $X = A^{-1}b$, then x_j is the jth row of A^{-1} times b, so

$$x_j = \frac{1}{\det(A)} \left((-1)^{1+j} \det(\hat{A}_{1j})b_1 + (-1)^{2+j} \det(\hat{A}_{2j})b_2 + \cdots + (-1)^{n+j} \det(\hat{A}_{nj})b_n \right)$$

Now the expression in parentheses is the expansion along the jth column of the matrix whose j^{th} column is b and whose other columns agree with A, that is, it is the determinant of B_j. In other words,

$$x_j = \frac{\det(B_j)}{\det(A)}$$

∎

EXAMPLE 2.35

Use Cramer's Rule to find the solution of the system

$$\begin{cases} 2x & + & 3y & - & z & = & 4 \\ x & - & 2y & + & 3z & = & 1 \\ x & + & 4y & - & 2z & = & -3 \end{cases}$$

SOLUTION Cramer's Rule will apply if the determinant of the coefficient matrix is non-zero. We will check this first, using the formula of Example 2.23.

$$\det \begin{pmatrix} 2 & 3 & -1 \\ 1 & -2 & 3 \\ 1 & 4 & -2 \end{pmatrix} = 8 + 9 - 4 - 24 + 6 - 2 = -7$$

Cramer's rule says, replacing the first column of A by b,

$$x = \frac{1}{-7} \det \begin{pmatrix} 4 & 3 & -1 \\ 1 & -2 & 3 \\ -3 & 4 & -2 \end{pmatrix} = \frac{-51}{-7} = \frac{51}{7}$$

replacing the second column of A by b,

$$y = \frac{1}{-7} \det \begin{pmatrix} 2 & 4 & -1 \\ 1 & 1 & 3 \\ 1 & -3 & -2 \end{pmatrix} = \frac{38}{-7} = -\frac{38}{7}$$

and, replacing the third column of A by b,

$$z = \frac{1}{-7} \det \begin{pmatrix} 2 & 3 & 4 \\ 1 & -2 & 1 \\ 1 & 4 & -3 \end{pmatrix} = \frac{40}{-7} = -\frac{40}{7}$$

As usual, we can check these answers by putting these values back into the equations and they are in fact correct. $\square$

Exercises 2.6

Find the determinants of the following matrices.

1. $A = \begin{pmatrix} -2 & 0 & 1 \\ 3 & 4 & -2 \\ 5 & 3 & -1 \end{pmatrix}$

2. $B = \begin{pmatrix} 1 & 2 & 0 & 1 \\ -2 & -2 & -1 & 1 \\ 2 & 2 & 4 & 0 \\ 3 & 6 & 3 & 3 \end{pmatrix}$

3. $C = \begin{pmatrix} 1 & 1 & -2 & 1 \\ -1 & 4 & 3 & -2 \\ 0 & 5 & 3 & 0 \\ -2 & 3 & 7 & 1 \end{pmatrix}$

4. $D = \begin{pmatrix} 0 & 0 & 3 & -1 \\ 1 & 2 & -1 & 1 \\ -2 & -2 & 3 & -5 \\ 1 & 6 & -8 & -4 \end{pmatrix}$

5. $E = \begin{pmatrix} 3 & -1 & 0 & -1 \\ 0 & 3 & -1 & -7 \\ -2 & -4 & 1 & 11 \\ -1 & 2 & 3 & 0 \end{pmatrix}$

6. $F = \begin{pmatrix} 0 & 0 & -4 & 0 & 0 \\ 0 & 1 & 3 & -1 & 1 \\ 0 & -2 & -1 & 4 & -3 \\ 2 & -1 & 4 & 3 & 0 \\ -2 & 3 & -9 & -11 & 10 \end{pmatrix}$

7. $G = \begin{pmatrix} -1 & 3 & -4 & 2 \\ 0 & -1 & 1 & 3 \\ 3 & 0 & -2 & -1 \\ 2 & -1 & 4 & 3 \end{pmatrix}$

9. $K = \begin{pmatrix} 1 & -1 & 2 & 1 & 0 \\ 0 & 2 & 1 & 3 & 1 \\ -1 & 1 & 4 & -1 & 1 \\ 0 & 1 & -1 & 0 & 3 \\ 2 & 0 & -1 & 1 & 1 \end{pmatrix}$

8. $H = \begin{pmatrix} -1 & 3 & -4 & 2 & 1 \\ 2 & 0 & 1 & -1 & 3 \\ 1 & 1 & 0 & -2 & -1 \\ 1 & 0 & -1 & 4 & 3 \\ 1 & 2 & 0 & -5 & 0 \end{pmatrix}$

10. $L = \begin{pmatrix} 0 & 1 & -1 & 2 & 1 \\ 1 & 0 & 1 & 1 & -1 \\ 2 & 1 & 1 & -1 & 1 \\ 0 & -1 & 1 & 1 & 1 \\ 2 & 1 & -2 & -3 & 1 \end{pmatrix}$

Use the formula in Theorem 2.31 to find the inverses of the following matrices. Check your answers by multiplying your answer by the given matrix.

11. $\begin{pmatrix} 3 & -1 \\ 4 & 7 \end{pmatrix}$

13. $\begin{pmatrix} 3 & 1 \\ 5 & -2 \end{pmatrix}$

12. $\begin{pmatrix} 2 & 1 & -1 \\ -2 & 3 & 0 \\ 1 & 2 & -1 \end{pmatrix}$

14. $\begin{pmatrix} 1 & -1 & 3 \\ 4 & 0 & -1 \\ -3 & 1 & 2 \end{pmatrix}$

Use Cramer's Rule to solve the systems. Check your answers by substituting into the original system.

15. $\begin{cases} 2x - 3y = 5 \\ x + 4y = -2 \end{cases}$

17. $\begin{cases} 3x + 4y = -7 \\ 2x - 3y = 2 \end{cases}$

16. $\begin{cases} 2x + y + 3z = 6 \\ x - 2y + z = -1 \\ 2x - 3y + z = 2 \end{cases}$

18. $\begin{cases} -y + z = -4 \\ -x + 3y + 3z = 4 \\ 2x + 3z = 1 \end{cases}$

2.7 Projects on Linear Systems

Linear systems are important because they come up in so many different applications. Typical applications involve solution of thousands of equations in thousands of unknowns, and the current bounds on the size of problems that can be solved in practice are on the order of a few million equations in a few million unknowns. Even problems where the sizes are measured in hundreds of equations are beyond the scope of this book, but the theory that applies to the solution of systems of all sizes is the same.

In this section, we introduce a few elementary applications of interest for engineering and science that involve the solution of linear systems. These are meant to indicate some typical applications rather than to be an exhaustive list. We could just as easily have included examples from traffic flow, analysis of chemical processes, economic models, or control theory, but we cannot include all examples.

2.7.1 Nodal Analysis of of Simple Circuits

We will illustrate the application of linear algebra to circuit analysis with the nodal analysis of simple circuits. Ohm's law states that the voltage drop across a resistor is proportional to the resistance and to the current flowing through the resistor: $v_2 - v_1 = Ri$, where the voltages v_k are measured in volts (V), the resistance R in ohms (Ω), and the current i in amps (A). For example, if the current through a 28 ohm resistor is 4 amps, the voltage drop across the resistor is $4 \cdot 28 = 112$ volts. We will consider only circuits consisting of current sources and resistors connected at "nodes". Since current flow has a direction, the current in any circuit element has a magnitude and a direction that we indicate by a sign; a current of $+3$ amps and -3 amps have the same magnitude but opposite directions in the circuit. Besides Ohm's law, we will use Kirchhoff's current law for the analysis of the circuits we consider: Kirchhoff's current law says that

the algebraic sum of the currents leaving any node of the circuit is zero

For example, in Figure 2.4, currents i_1 and i_2 are flowing into the node and i_3 and i_4 are flowing out of the node. Kirchhoff's current law says $-i_1 - i_2 + i_3 + i_4 = 0$.

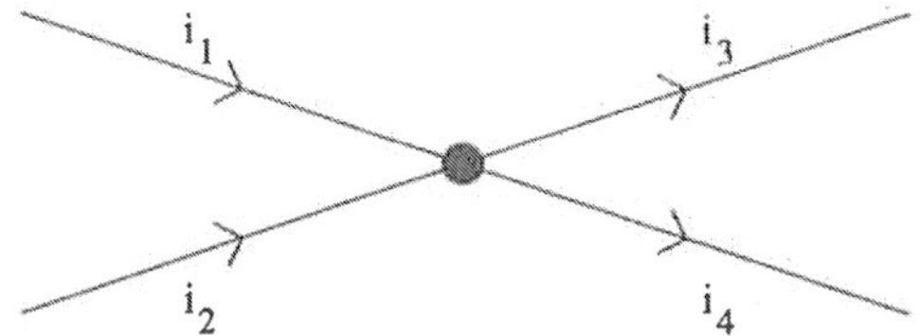

FIGURE 2.4

Current flow through a node

If the current i_1 is 4A, the current in i_2 is 3A, and the current in i_3 is 12A, then the current in i_4 is

$$i_4 = i_1 + i_2 - i_3 = 4 + 3 - 12 = -5A$$

which means that, actually, the current in the circuit element carrying i_4 is flowing into the node. (In particular, you don't need to know in advance which way the current is flowing to find the currents; the algebra will find out for you based on your assumptions, whatever they are.)

We can use Kirchhoff's current law to find the voltages at the nodes of a circuit, relative to the "reference node". That is, Ohm's law refers only to voltage drops, so the difference in voltages is what is important; the analysis yields the differences between the voltages at each of the nodes and the reference node. Frequently, if a "ground" is present, the "ground" is chosen as the reference node and one assumes the voltage at that node is 0; the differences between the ground node and the other nodes are referred to as the node voltages.

Figure 2.5 shows the symbols used in circuit diagrams for some circuit elements. In drawing a circuit, it would sometimes be confusing to draw, as a single node, all the physical elements that make up that node. For example, in a circuit with a "ground", each of the "ground"'s represents a connection to the single node G even

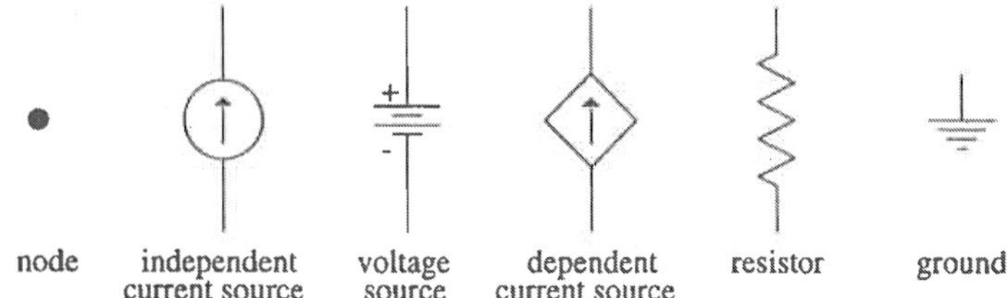

FIGURE 2.5

Some symbols for circuit diagrams

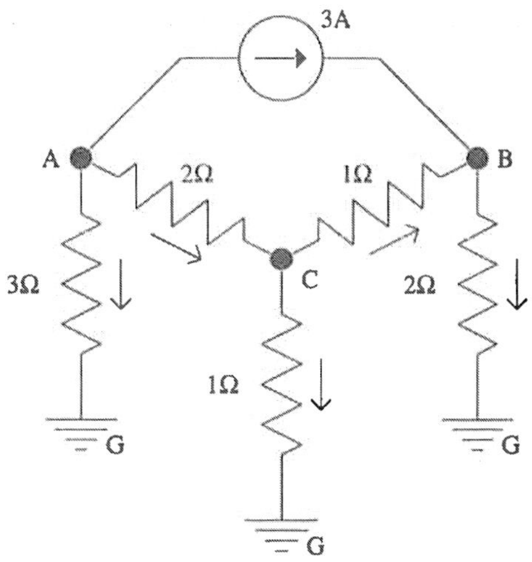

FIGURE 2.6

A simple circuit

though, for convenience, they may be drawn separately.

In the circuit in Figure 2.6, taking the reference node (G) to be the ground at the bottom of the diagram, we get the following equations from applying Kirchhoff's current law to each of the remaining nodes:

$$\begin{cases} i_{AG} + i_{AC} + 3 &= 0 \qquad (Node A) \\ i_{BC} + i_{BG} - 3 &= 0 \qquad (Node B) \\ -i_{AC} - i_{BC} + i_{CG} &= 0 \qquad (Node C) \end{cases} \qquad (2.7.3)$$

We do not have enough equations to solve for all of the currents directly, but applying Ohm's law to each of the resistors, we get a system involving the node voltages that we can solve:

$$V_A = V_A - V_G = 3i_{AG}$$

$$V_A - V_C = 2i_{AC}$$

$$V_B - V_C = 1i_{BC}$$

$$V_B = V_B - V_G = 2i_{BG}$$

$$V_C = V_C - V_G = 1i_{CG}$$

which gives

$$\begin{cases} \frac{1}{3}V_A + \frac{1}{2}(V_A - V_C) + 3 = 0 & (Node\,A) \\ \frac{1}{1}(V_B - V_C) + \frac{1}{2}V_B - 3 = 0 & (Node\,B) \\ -\frac{1}{2}(V_A - V_C) - \frac{1}{1}(V_B - V_C) + \frac{1}{1}V_C = 0 & (Node\,C) \end{cases}$$

Rewriting this in the more standard format, we get

$$\begin{cases} \frac{5}{6}V_A \quad\quad\quad - \frac{1}{2}V_C = -3 \\ \quad\quad \frac{3}{2}V_B - V_C = 3 \\ -\frac{1}{2}V_A - V_B + \frac{5}{2}V_C = 0 \end{cases}$$

and solving we get $V_A = -3.5217$, $V_B = 2.0870$, and $V_C = 0.1304$. Using Ohm's law we can find the currents in each of the resistors: $i_{AC} = -1.8261$, $i_{AG} = -1.1739$, $i_{BC} = 1.9565$, $i_{BG} = 1.0435$, and $i_{CG} = 0.1304$.

In addition to passive elements such as the resistors in the circuit above, simple circuits can be built with transistors which act as voltage dependent current sources, that is, the current produced depends on the voltage drop elsewhere in the circuit. For example, if the voltage across the resistor R_1 is 5V and the transfer ratio of the transistor is .04, then the current produced by the transistor is $5 \cdot .04 = .2A$.

2.7.2   Allocation of Internal Costs

A profitable business tries to set prices on its products so that all its costs are covered (and a little more added in for profit). Clearly, in order to do that, the company must know what its costs are. If the company makes only one product, the situation is simple, the price on that one product must cover all the costs of the company. Already if the company has two products a problem arises; the prices on the products must be set so that the total costs are met, but that does not help in determining what the prices are on each product. Of course, if a decision is made that makes the price of one of the products high relative to the prices of its competitors, then it will lose business on that product, and thereby might actually not meet its costs.

We should realize that the setting of prices and the *policy* for allocating costs are management decisions, not mathematics problems. Our role in the process is to convert the *policy* for allocating costs into a *mathematical model*, analyze the model mathematically, then report back to management on whether this model reflects their policy decision accurately. To emphasize the role of management, consider the very common grocery marketing practice of selling eggs at (say) 49 cents a dozen when they actually pay (say) 63 cents a dozen for them. They do this consciously to achieve a certain goal: bring customers into the store who will buy steak at (say) \$5.95 a pound that costs them (say) \$2.67 a pound. While we are not going to try to make this decision, we can advise management on what the costs actually are, facts that should be a part of the marketing decisions.

In addition, we should realize that many of the services that departments perform can be purchased: rather than having a cafeteria in which the food is cooked by company employees, a catering company could be hired to provide meals to employees at a reasonable cost or rather than having the accounting department set up and run retirement and health plans for the employees, the company could hire a benefits administration firm to set up and run these programs. Before a decision can be made on "outsourcing", an accurate assessment of the costs involved in doing the job internally must be made so that they can be compared with the cost to have the job done externally.

> **The Problem**: If a company has several products, how do we allocate
> the total costs of the company to each of these products so that we know
> how much each of the products "actually costs" to produce?

Again, we must realize that the actual attribution of the costs is also a management decision: Suppose a company produces forgings and castings and has one janitor to clean up the factory. If the president wishes, the wages of the janitor can all be added to the "cost of producing forgings". Such a decision might make the price of the company's forgings high and the price of its castings low, but that is a *policy* decision. Our role is to help the president carry out this policy, or more likely, help compare the effects of two proposed policies.

One reasonable model of cost allocation is to find out how much effort the janitor spends cleaning up the part of the plant where forgings are produced and how much where castings are produced, then divide the cost of the janitor proportionately.

Unfortunately, this simple statement of policy does not solve the problem, for not all parts of the company make products for sale and all parts of the company do use the janitor. For the purposes of the model we will develop, we divide departments of the company into two types, the *service departments* and the *production departments*. The service departments, like the janitor, the company cafeteria, the accounting department, supply a service to the departments of the company internally, but do not supply a service or product to customers outside the company. The production departments supply products outside the company, bring in revenue, and it is to these departments that we want to attribute the total costs of the company, including the cost of the service departments. Thus, while the accounting department has costs, like salary, it has no direct customer generated revenue to pay these costs. It writes checks for the production departments, so should get its costs covered by the production departments. A complicating factor is that the accounting department writes checks for itself and for the janitor as well, so the service departments provide service not only to the production departments, but also to the service departments. Similarly, we divide the costs into two categories, the *direct costs* that are paid to someone outside the company, and the *indirect costs* that a department incurs by using the service of one of the service departments. In some sense, the direct costs are "real" and the indirect costs are "funny money". What we really intend is to use the concept of indirect costs to distribute the total direct costs of the company, both those paid by the production departments and those paid by the service departments, among the production departments to arrive at an "actual cost" of producing each product. We will assume that the service departments do not try to make a profit on the work they do.

EXAMPLE 2.36

The Star Chain Company makes two types of chain: roller chain and timing chain. Besides the roller chain and timing chain departments, there is the accounting department, the food services department, and the security and maintenance department. The direct costs of the various departments are 40 (thousand dollars per month) for the accounting department, 60 for the food services department, 100 for the security and maintenance department, 500 for the roller chain department, and 300 for the timing chain department. Thus, the total direct costs for the company are 1000.

department	direct costs	acct.	food	sec.
accounting	40	10%	10%	10%
food	60	20%	10%	10%
sec. & main.	100	10%	10%	5%
roller	500	40%	40%	45%
timing	300	20%	30%	30%

We want to allocate part of the 1000 to the roller chain department and the rest to the timing chain department in order to determine their "actual costs". The efforts of the various service departments are used by the departments in proportions given in the following table. For example, the 45% in the right column of the table means that 45% of the output of the security and maintenance department is used by the roller chain department. In our model of this problem, we will try to allocate 45% of the total costs of the security and maintenance department to the roller chain department.   □

Our model will try to allocate the total direct costs of the company among the production departments (in this case, the roller and timing chain departments) by setting up linear equations that express the relationships between the costs of running the various departments. To begin, we set up our notation and give our assumptions:

Suppose an organization has m service departments (numbered 1 to m) and n production departments (numbered $m + 1$ to $m + n$). Let d_i denote the direct costs of department i, for $1 \le i \le m + n$, and let c_i denote the total costs of department i, for $1 \le i \le m + n$. That is, c_i is the sum of d_i and the indirect costs allocated to department i due to its use of other departments' services. Let s_{ij} denote the share of the total cost of department j that will be charged to department i, for $1 \le i \le m + n$ and $1 \le j \le m$. Implicit in the notation for the cost sharing, that is, $1 \le j \le m$ rather than $1 \le j \le m + n$, is the assumption that costs of the service departments are shared, the costs for the production departments are not. We further require that $0 \le s_{ij} \le 1$ and that, for each j with $1 \le j \le m$,

$$s_{1j} + s_{2j} + \ldots + s_{m+n,j} = 1$$

This means that the portion of the costs of department j charged to department i are between none ($s_{ij} = 0$) and all ($s_{ij} = 1$) and that all of the costs of department j are charged to some department. Our problem is to determine the total costs of all the departments, that is, we want to determine $c_1, c_2, \ldots, c_{m+n}$.

The intuition behind the choice of variables suggests that the following system of equations should be satisfied:

$$(\dagger) \quad \begin{cases} c_1 &= d_1 + s_{11}c_1 + s_{12}c_2 + \cdots + s_{1m}c_m \\ c_2 &= d_2 + s_{21}c_1 + s_{22}c_2 + \cdots + s_{2m}c_m \\ &\vdots \\ c_m &= d_m + s_{m1}c_1 + s_{m2}c_2 + \cdots + s_{mm}c_m \\ c_{m+1} &= d_{m+1} + s_{m+1,1}c_1 + s_{m+1,2}c_2 + \cdots + s_{m+1,m}c_m \\ &\vdots \\ c_{m+n} &= d_{m+n} + s_{m+n,1}c_1 + s_{m+n,2}c_2 + \cdots + s_{m+n,m}c_m \end{cases}$$

Each equation corresponds to a particular department and asserts that the total costs for that department are its direct costs plus its indirect costs, that is, its share of the total costs of the service departments who provide services to it. Since the direct costs (d_i) and the proportions (s_{ij}) of costs to be allocated are assumed to be known, this is a system of $m + n$ equations in $m + n$ unknowns (c_i). Examination of the system, however, shows that actually, the first m equations involve only the first m unknowns and the last n equations give the last n unknowns (the production department costs) in terms of the first m unknowns (the service department costs).

EXAMPLE 2.36, continued. For the Star Chain Company, numbering the departments in the order accounting, food, security, roller chain, timing chain; these equations become

$$\begin{cases} c_1 = 40 + .1c_1 + .1c_2 + .1c_3 \\ c_2 = 60 + .2c_1 + .1c_2 + .1c_3 \\ c_3 = 100 + .1c_1 + .1c_2 + .05c_3 \\ c_4 = 500 + .4c_1 + .4c_2 + .45c_3 \\ c_5 = 300 + .2c_1 + .3c_2 + .3c_3 \end{cases}$$

Rewriting the first three equations, we get

$$\begin{cases} .9c_1 - .1c_2 - .1c_3 = 40 \\ -.2c_1 + .9c_2 - .1c_3 = 60 \\ -.1c_1 - .1c_2 + .95c_3 = 100 \end{cases}$$

which has the solution $c_1 = 68.68$, $c_2 = 95.54$, and $c_3 = 122.55$. Putting these values into the last two equations, we find that the total costs of the roller chain department are $c_4 = 620.84$ and total costs of the timing chain department are $c_5 = 379.16$. For this example, the combined total costs of the production departments is 1000 which is also the combined direct costs for the entire company, so we have achieved our goal of allocating all the direct costs among the production departments. $\square$

Was this balance a lucky accident or will it always happen this way? Are there similar problems in which some of the costs are negative? These are some of the questions we must answer if we are to evaluate our model. Clearly, if the balance between the total costs allocated to production and the total direct costs is an accident, then the model does not achieve its stated goal. Even if the balance always occurs, if sometimes the costs come out negative, we will lose faith in the model because a department should not get useful output for (less than) nothing. In fact these things are not accidental, we will analyze the model and then prove a theorem that says, under some reasonable conditions, all costs are positive and the company's total direct costs are allocated among the production departments.

The system of equations (†) will be easier to understand if we express it in terms of block matrices. Let $C_s = (c_1, c_2, \ldots, c_m)$ be the column vector of total costs of the service departments and let $C_p = (c_{m+1}, c_{m+2}, \ldots, c_{m+n})$ be the column vector of total costs of the production departments. Similarly, let D_s and D_p be the column vectors of direct costs of the service and production departments and let S_s and S_p be the $m \times m$ and $n \times m$ matrices whose entries are the proportions of the costs of the service departments to be allocated to the service and production departments. With this notation, the system (†) becomes

$$(\ddagger) \qquad \begin{pmatrix} C_s \\ C_p \end{pmatrix} = \begin{pmatrix} D_s \\ D_p \end{pmatrix} + \begin{pmatrix} S_s \\ S_p \end{pmatrix} C_s$$

Isolating the part of the system involving only the service departments, we get the equation

$$C_s = D_s + S_s C_s$$

In other words,

$$\begin{aligned} C_s - S_s C_s &= D_s \\ (I - S_s) C_s &= D_s \end{aligned}$$

and if $I - S_s$ is invertible,

$$C_s = (I - S_s)^{-1} D_s$$

This implies that

$$C_p = D_p + S_p (I - S_s)^{-1} D_s$$

so the system is solved.

It is clear from this calculation that the applicability of the model will depend on the existence and nature of $(I - S_s)^{-1}$. We are now ready to state the theorem justifying the model. This is really a theorem about $(I - S_s)^{-1}$ and its proof is beyond the scope of this book.

THEOREM 2.37

Let the notation be as above. Suppose

(a)   for each i, we have $d_i \geq 0$,

(b)   for each j, there is at least one i with $m + 1 \leq i \leq m + n$ for which $s_{ij} > 0$.

Then

(1) $(I - S_s)^{-1}$ exists and has all non-negative entries,

(2) $C_s = (I - S_s)^{-1} D_s$ and $C_p = D_p + S_p(I - S_s)^{-1}D_s$,

(3)   for each i, we have $c_i \geq 0$,

(4)

$$\sum_{i=m+1}^{m+n} c_i = \sum_{i=1}^{m+n} d_i$$

Let's examine the meaning of each of the statements. The assumption (a) is that all the direct costs are non-negative; this is clearly a sensible assumption since, as noted above, we cannot expect to get something for (less than) nothing. Assumption (b) says that every service department should provide service to some production department. This assumption makes sense in that the focus of the company is on the production departments, so a service department should provide service to one of them. On the other hand, it is easy to imagine that it would be convenient to have departments in a company that served production by serving the service departments, for example, it is conceivable that a company might have a data processing department that served only the accounting and inventory departments and no production departments. In this case, the theorem above would not (directly) apply. We have two choices: we could create another model (and another theorem) in which there were three classes of departments, production, departments serving production departments and service departments not serving production departments, or, more easily, we could incorporate, for the purposes of the model only, departments not directly serving production departments into those that were. For example, we could, for the purposes of analysis only, incorporate the data processing and the accounting departments, and the new (fictitious) department would then fit into the current model.

Parts (1) and (2) of the conclusion reiterate the formulas given above, assert that they make mathematical sense, and give further information about them. Part (3) of the conclusion says that the computed costs satisfy the minimum demand of reasonableness, that they be non-negative. Part (4) says that the model presented solves the problem: it asserts that the sum of the total costs of the production departments as computed by the model equal the total direct costs of the company.

2.7.3 Linear Optimization Problems

Companies have limited resources and must choose how to allocate them in their business. For example, a company may have a grinder that can be used in the production of timing chain or roller chain. Someone must decide how much time the grinder should be used to produce roller chain and how much to produce timing chain. Since the profit in making roller chain and timing chain may be different, this should affect our decision. However, it may not be reasonable to only use the grinder to produce timing chain, because it may be that the heat treating furnace can only produce enough timing chain parts to keep the grinder busy half the time. Thus, we would like to allocate all of our resources in such a way as to make the largest profit.

In this section, we will consider optimization problems that have linear constraints and the outcome depends on the inputs linearly; these problems are usually called *linear programming problems*.

In many situations, the assumption of linearity is a good approximation: if one bolt making machine can produce 1000 bolts per hour, three machines together should be able to produce 3000 bolts per hour; if the company makes a profit of $100 selling 1000 bolts, we might expect it could make about $500 profit selling 5000 bolts. In the latter case, we are assuming that economies of scale and the strain on our resources in moving production from 1000 bolts to 5000 bolts are insignificant and we are assuming that the extra supply of bolts would not distort the market enough to significantly change the price we can sell them for. (If these assumptions are not appropriate, then we do not have a linear problem.)

To illustrate the type of problem we are talking about and to see how to set up the problem, we will consider the following example.

EXAMPLE 2.38

The Widget Manufacturing Co. makes regular and deluxe widgets. Regular widgets require 1 hour in the casting department, whereas deluxe widgets need 2 hours. Regular widgets spend 3 hours in the shaping department, and deluxe widgets only 2 hours. In addition, deluxe widgets spend 2 hours in the finishing department. Departmental capacities (in process hours per day) are 16 hours for the casting department, 24 hours for the shaping department, and 14 hours for the finishing department. Each regular widget yields $300 profit and each deluxe widget $400 profit. How many of each variety should WMC make each day to maximize their profits?

Set-Up: We need to decide how many of each kind of widgets to manufacture, so we will let r be the number of regular widgets and d the number of deluxe widgets to be made each day. Clearly, since making $d = -3$ widgets does not make sense, we must have

$$\begin{cases} r & \geq & 0 \\ d & \geq & 0 \end{cases}$$

and we want to maximize our profit

$$300r + 400d$$

Obviously, to make a large profit, we should make lots of both kinds of widgets. However, we are constrained from making an unlimited number of widgets by our plant capacity. Indeed, the capacity of the casting department means

$$r + 2d \leq 16$$

the capacity of the shaping department means

$$3r + 2d \leq 24$$

and the capacity of the finishing department means

$$2d \leq 14$$

To put it all together, we wish to maximize the *objective function*

$$\text{Maximize } 300r + 400d$$

subject to the constraints

$$\begin{cases} r & + & 2d & \leq & 16 \\ 3r & + & 2d & \leq & 24 \\ & & 2d & \leq & 14 \\ r & & & \geq & 0 \\ & & d & \geq & 0 \end{cases}$$

It is probably helpful to try to visualize the problem geometrically. We can plot everything in the problem in the (r, d)–plane. In this plane, the set of points satisfying each of the constraint inequalities is a half plane bounded by the line whose equation is the constraint with equality in place of inequality. In Figure 2.7, lines for each of the constraints are plotted and the set of points satisfying all the

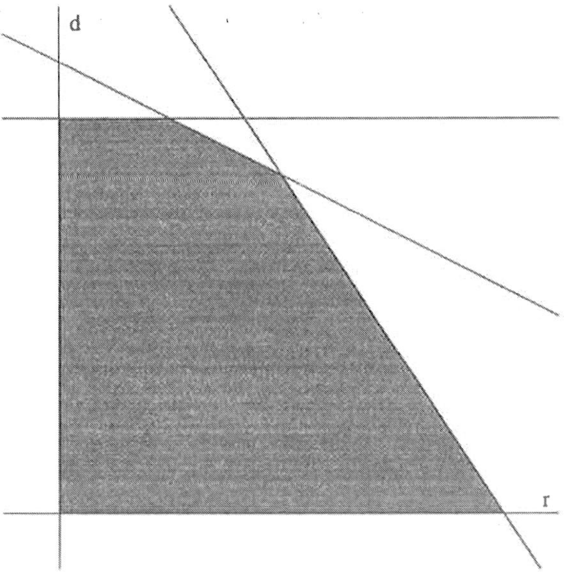

FIGURE 2.7

Feasible region for Widget Man. Co.

inequalities, called the *feasible region*, is shaded. (Each point satisfying all the constraint inequalities is called a *feasible point*.)

Thus, we are looking for the point in the shaded region at which the function $300r + 400d$ is maximized; this point will correspond to the number of regular (r) widgets and the number of deluxe widgets (d) that should be made to maximize the profit. Recall from calculus that the level curves of a function of two variables are the curves in the plane at which the function values are equal to a particular value. Figure 2.8, shows the feasible region for the problem together with the level curves $300r + 400d = -400$, $300r + 400d = 0$, $300r + 400d = 400$, $\cdots$, $300r + 400d = 5200$, $300r + 400d = 5600$, and $300r + 400d = 6000$. Since the level curve $300r + 400d = 2800$ intersects the feasible region in a large number of points, there are many choices of targets for manufacturing that will lead to a profit of \$2800. For example, WMC has the plant capacity to make $r = 0$ regular widgets and $d = 7$ deluxe widgets, $r = 4$ regular widgets and $d = 4$ deluxe widgets, or $r = 6\frac{2}{3}$ regular widgets and $d = 2$ deluxe widgets and make a \$2800 profit. The figure also makes it clear that the maximum profit that can be made is \$3600, and this can be achieved within the plant capacity only by making $r = 4$ regular widgets and $d = 6$ deluxe widgets. ☐

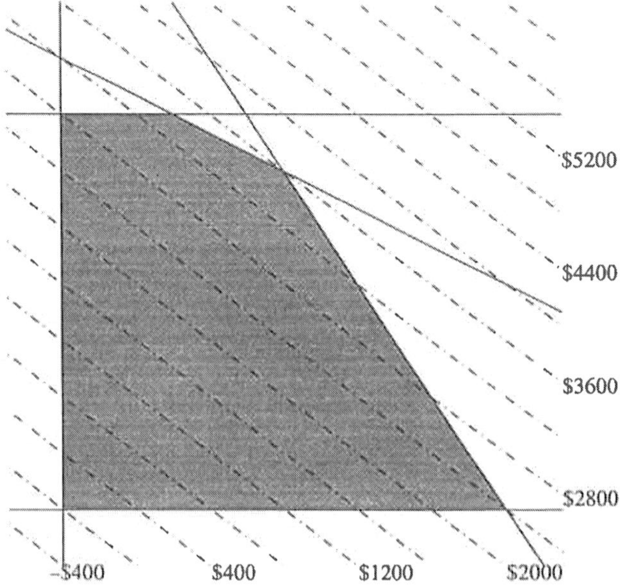

FIGURE 2.8

Profit level curves for Widget Manufacturing Co.

Notice that the maximum profit is made at one of the corners of the feasible region. In fact, under fairly general hypotheses on the nature of the problem, this is *always* the case! Thus, one strategy for solving this type of problem is to find *all* the corners of the feasible region and the values of the objective function at those corners and choosing the point at which the objective function is maximized.

In order to implement this strategy, we need to identify the corners and be able to find them. The first difficulty is that we are unaccustomed to dealing with inequalities; we are much more comfortable dealing with equalities. In the example above, the inequality

$$r + 2d \le 16$$

expresses the constraint on production due to the resources of the casting department. If c represents the number of unused hours in the casting department, this inequality is equivalent to

$$r + 2d + c = 16 \quad \text{where} \quad c \ge 0$$

The variable c is called a *slack variable* because it represents the slack in the constraint corresponding to the casting department. Introducing the variable c allows us to replace the inequality constraint by an equality together with the simple non-negativity condition $c \geq 0$. Introducing slack variables s and f representing the unused capacity of the shaping department and the finishing department, the linear programming problem becomes

$$\text{Maximize } 300r + 400d$$

subject to the constraints

$$\begin{cases} r + 2d + c & = 16 \\ 3r + 2d \quad\quad + s & = 24 \\ 2d \quad\quad\quad + f & = 14 \end{cases}$$

where

$$r \geq 0,\ d \geq 0,\ c \geq 0,\ s \geq 0,\ f \geq 0$$

Thus, the feasible region for this problem is the set of solutions of the above system of three equations in five unknowns for which each of the variables is non-negative. We expect that there are infinitely many solutions of the given system and we hope that at least some of them have each of the variables non-negative (otherwise, the set of constraint inequalities cannot be solved and there are no points of the feasible region).

Now each corner of the feasible region is the intersection of two boundary lines and each boundary line corresponds to the set of points at which one of the variables is 0. For example, the line $r + 2d = 16$ corresponds to no excess capacity in the casting department, that is, the slack variable c is 0. To find each corner of the feasible region, we can find the solutions of the system

$$\begin{cases} r + 2d + c & = 16 \\ 3r + 2d \quad\quad + s & = 24 \\ 2d \quad\quad\quad + f & = 14 \end{cases}$$

in which some pair of variables is zero. Care must be used however, since there may be a number of solutions that do not correspond to points in the feasible region; these must be ignored. In Example 2.38, the lines $3r + 2d = 24$ and $2d = 14$ intersect at $r = 10/3$ and $d = 7$. This point does not satisfy the constraint $r + 2d \leq 16$, indeed, when the solution is found using the slack variables, the slack variable corresponding to this point is negative ($c = -4/3$). If the lines bounding the feasible region do not intersect at all, as $2d = 14$ and $d = 0$ in that example, then there will be no solution for the corresponding system involving the slack variables.

EXAMPLE 2.38, continued. We can solve the problem of finding the number of regular and deluxe widgets that the Widget Manufacturing Company should try to make in order to maximize their profits by choosing all possible combinations of pairs of variables to be zero and solving the resulting systems.

If $s = 0$ and $f = 0$, we get the system

$$\begin{cases} r + 2d + c & = 16 \\ 3r + 2d + s & = 24 \\ 2d + f & = 14 \\ s & = 0 \\ f & = 0 \end{cases}$$

which has solution $r = 3\frac{1}{3}$, $d = 7$, and $c = -1\frac{1}{3}$. Since $c = -1\frac{1}{3} < 0$, this is not a feasible point.

If $c = 0$ and $f = 0$, we get the system

$$\begin{cases} r + 2d + c & = 16 \\ 3r + 2d + s & = 24 \\ 2d + f & = 14 \\ c & = 0 \\ f & = 0 \end{cases}$$

which has solution $r = 2$, $d = 7$, and $s = 4$. This is a feasible point, so is a corner of the feasible region.

If $d = 0$ and $f = 0$, we get the system

$$\begin{cases} r + 2d + c & = 16 \\ 3r + 2d + s & = 24 \\ 2d + f & = 14 \\ d & = 0 \\ f & = 0 \end{cases}$$

which is inconsistent: there are no points corresponding to $d = 0$ and $f = 0$.

Continuing in this way, we get solutions

$$\begin{array}{lllll} r = 0 & d = 7 & c = 2 & s = 10 & f = 0; \\ r = 4 & d = 6 & c = 0 & s = 0 & f = 2; \\ r = 8 & d = 0 & c = 8 & s = 0 & f = 14; \\ r = 0 & d = 12 & c = -8 & s = 0 & f = -10; \\ r = 16 & d = 0 & c = 0 & s = -24 & f = 14; \\ r = 0 & d = 8 & c = 0 & s = 8 & f = -2; \\ r = 0 & d = 0 & c = 16 & s = 24 & f = 14 \end{array}$$

Of these, $r = 0$, $d = 12$; $r = 16$, $d = 0$; and $r = 0$, $d = 8$ are also not feasible points because some of the slack variables are negative. Thus, we have altogether five corner points of the feasible region: $r = 2$, $d = 7$; $r = 0$, $d = 7$; $r = 4$, $d = 6$; $r = 8$, $d = 0$; and $r = 0$, $d = 0$. To see which is optimal, we only need evaluate the objective function at each of these points:

$$\begin{array}{lll}
r = 2 & d = 7 & 300r + 400d = \$3400; \\
r = 0 & d = 7 & 300r + 400d = \$2800; \\
r = 4 & d = 6 & 300r + 400d = \$3600; \\
r = 8 & d = 0 & 300r + 400d = \$2400; \\
r = 0 & d = 0 & 300r + 400d = \$0
\end{array}$$

As in the graphical solution, we see that the Widget Manufacturing Company should try to make 4 regular and 6 deluxe widgets per day, and if they do so, they will make \$3600 profit per day. Since $c = s = 0$ and $f = 2$ for this solution, we see that the casting and shaping departments will be working to capacity, but the finishing department will have excess capacity. $\Box$

In general, it is customary to set up linear programming problems to seek non-negative solutions. If we have such a problem with m variables and n inequality constraints, we will need to add n slack variables and we will get a system of n equations in $m + n$ variables for which the feasible region will be the set of solutions with all the variables non-negative. The maximum value of the objective function will occur at a corner point of the feasible region. Since the corner points are all solutions of the system with m of the variables equal to zero, the naive strategy is to find all solutions of the system with m of the variables equal to zero and choose the feasible solution for which the objective function is largest.

In Example 2.38, we had a problem with two variables (r and d) and three constraints, so we added three slack variables (c, s, and f) and ended up with a system of three equations in five unknowns. We found all solutions of this system with two of the variables equal to zero (there were ten such solutions, counting the case with no solutions) and five of these corresponded to feasible solutions. We then chose the one of these for which the objective function was largest.

For this problem, the naive strategy was possible to carry out. In larger problems, however, since the number of ways to set m variables to zero among $m + n$ variables is

$$\frac{(m + n)!}{m!n!}$$

this is no longer a practical strategy. (Recall that $k! = 1 \cdot 2 \cdot 3 \cdots k$.) Linear programming, as a part of operations research, has been developed over the past half

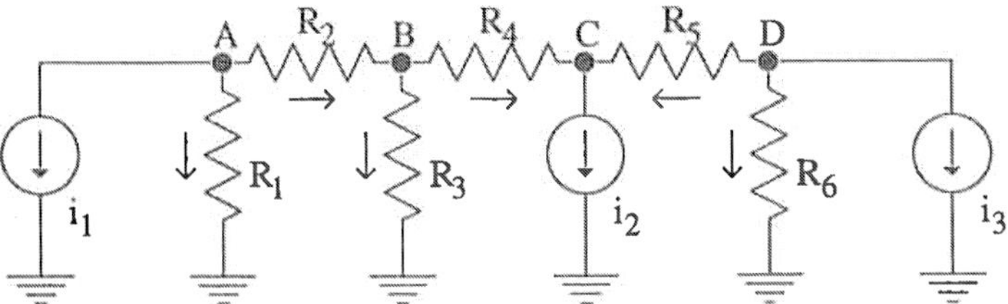

FIGURE 2.9

Circuit A

century and other strategies for their solution are practical. The most common of these, developed by Dantzig in the late 1940's, is the *simplex algorithm*. Problems involving thousands of variables are routinely solved using the simplex algorithm. The simplex algorithm works by more carefully selecting which corners are important and only solving the systems corresponding to important corners. More recently, *interior point methods* such as *Karmarkar's algorithm* or the *ellipsoidal algorithm* have been developed. Although these are faster, in principle, for very large problems, they are not as widely used as the simplex algorithm. Discussions of the simplex algorithm and interior point methods for the solution of linear programming problems are beyond the scope of this book, so we will be content to attack only very small problems for which the naive strategy is possible.

Exercises 2.7

1. In the discussion above, we found the currents in the circuit using the system (2.7.3) that did not include the equation obtained from Kirchhoff's current law at the ground node. (Recall that all "ground"s constitute a single node which we used as the reference node.) Show that the system obtained from including the ground node is equivalent to the system (2.7.3), that is, that adding this equation does not help. Can you explain why this is so for all circuits?

2. For the circuit of Figure 2.8, find the currents in each of the resistors and the voltages at each of the nodes. (Use the "ground" as the reference node and assume $V_G = 0$.) $i_1 = 20A$, $i_2 = 5A$, $i_3 = 10A$, $R_1 = 10\Omega$, $R_2 = 20\Omega$, $R_3 = 30\Omega$, $R_4 = 40\Omega$, $R_5 = 5\Omega$, $R_6 = 100\Omega$.

3. In the circuit of problem 2.7.2., find the currents in the circuit if the currents in each of the independent current sources is doubled to $i_1 = 40A$, $i_2 = 10A$, $i_3 = 20A$.

4. In the circuit of problem 2.7.2., what should the current source i_2 be so that no current flows through resistor R_3?

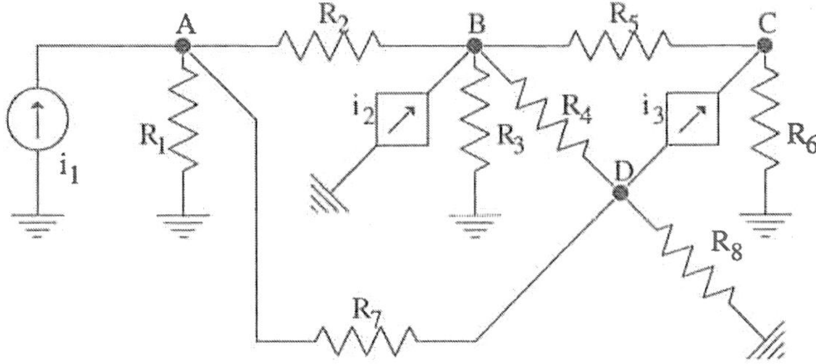

FIGURE 2.10

Circuit B

5. For the circuit of Figure 2.10, find the currents in each of the resistors and the voltages at each of the nodes. (Use the "ground" as the reference node and assume $V_G = 0$.) $i_1 = .1$, $i_2 = .4V_A$, $i_3 = .1(V_B - V_C)$, $R_1 = R_4 = 250\Omega$, $R_2 = R_5 = 1,000\Omega$, $R_3 = R_8 = 100\Omega$, $R_6 = 10\Omega$, $R_7 = 28\Omega$.

6. In the circuit of problem 2.7.5., find the currents in the circuit if the current in the independent current source is doubled to $i_1 = .2$.

---

7. The Acme Manufacturing Company makes molded plastic products. They have divided their organization into 6 departments: the toy department, the kitchen gadget department, and the novelty department, which are production departments, and the accounting department, the maintenance department, and the security department, which are service departments. Their activities and costs are shown in the table below. Direct costs are given in thousands of dollars and the "20%" in the upper right corner of the table means that 20% of the security departments efforts are providing service to the accounting department.

department	direct costs	acct.	main.	sec.
accounting	20	10%	10%	20%
maintenance	100	10%	20%	10%
security	40	10%	10%	0%
kitchen	50	20%	20%	30%
novelty	20	20%	20%	20%
toy	100	30%	20%	20%

(a) Find the total costs for each department and verify that the combined total costs of the production departments are equal to the total direct costs for all departments.

(b) Al's Protection Company (A. Capone, president) has offered their services to Acme for $60,000 per year. Should the company eliminate their security department and buy protection from Al's company instead? Why?

8. Yolanda, a chemical engineer, and Zeke, an industrial engineer, share an office, a secretary, and a computer programmer in their consulting businesses. The secretary is paid $8 an hour and works 25 percent of the time for the computer programmer, 30 percent of the time for Yolanda, and 45 percent of the time for Zeke. The computer programmer is paid $13 an hour and works 20 percent of the time for the secretary, 20 percent of the time doing maintenance on her own programs, 35 percent of the time for Yolanda, and 25 percent of the time for Zeke. They split the other office expenses equally, each paying $3 an hour. If Yolanda wants to earn a salary of $30 an hour and Zeke wants to earn $25 an hour, how much should they each charge their customers?

9. The Top Brass Manufacturing Company makes brass clothing accessories. They have divided their organization into 4 departments: the button department and the buckle department, which are production departments, and the personnel department and the accounting department, which are service departments. Their activities and costs are shown in the table below. Direct costs are given in thousands of dollars and the "40%" in the upper right corner of the table means that 40% of the accounting department's efforts are providing service to the personnel department.

department	direct costs	pers.	acct.
personnel	30	40%	40%
accounting	20	20%	20%
button	50	30%	20%
buckle	40	10%	20%

   Find the total costs for each department.

10. Pinkham Pharmaceuticals makes and sells penicillin, potassium chloride (KCl) in aqueous solution, and Lydia Pinkham's Pink Pills for Pale People. In addition, the company consists of the production services department, the accounting department, the advertising department, and the product development department. Ms. Pinkham, CEO, has found that the direct costs of the penicillin department are 600 (thousand dollars per month); the potassium chloride department, 240; the pink pill department, 180; the production services department, 300; the accounting department, 120; the advertising department, 500; and the development department, 280. The following table represents use of the various departments services by the other departments; she assigned the costs of the advertising and development departments not on the basis of reports from her accounting staff, rather on the basis of her expectation of the benefits from that department's operation. Find the total costs for each department.

department	prod.	acct.	adv.	dev.
penicillin	40%	10%	20%	30%
KCl	20%	15%	30%	20%
pills	10%	30%	45%	10%
prod. ser.	10%	10%	0%	30%
accounting	5%	5%	0%	0%
advert.	5%	20%	0%	0%
develop.	10%	10%	5%	10%

11. The Boilermaker Iron Company makes assorted castings and forgings. They have divided their organization into 4 departments: the casting department, the forging department, the accounting department, and the maintenance department. Some of their activities and costs are shown in the table below.

department	direct costs	acct.	main.
accounting	50	10%	10%
maintenance	100	20%	10%
casting	750	60%	30%
forging	300	10%	50%

Unfortunately, this company does not directly fit our model because the casting department does 2/3 of its work for outside customers but does 1/3 of its work making castings for the forging department, in other words, it is neither a purely production department nor a purely service department. Modify the model or the way you set the problem up, explaining what you are doing, and find the total costs that should be passed on to the customers of the forging department and the outside customers of the casting department.

---

12. Find the number of systems that would need to be solved to use the naive strategy described in the section to solve a small linear programming problem in which there are 18 unknowns and 27 constraints.

13. Cruisin' Craig Andrews has his own rig and is a contract hauler out of West Lafayette. His truck can carry up to 39000 pounds and up to 2100 cubic feet of cargo. He has two potential customers wanting containers shipped to Denver; the Acme Bird Trap company and the Better Biscuit Baking company. Acme has 1200 containers each of which weighs 40 pounds and is 2 cubic feet in volume. Better Biscuit has 600 containers each of which weighs 50 pounds and is 3 cubic feet in volume. If Acme is willing to pay $2.00 per container and Better Biscuits is willing to pay $2.60 per container for shipping, how many containers from each company should Craig take to maximize his income?

14. Steve's Cycle Shop makes unicycles, bicycles, and tricycles. Each unicycle uses 2 work–hours from the wheel department, spends 1 work–hour in the assembly department, and 3 work–hours in the finishing department. Each bicycle needs 4 work–hours from the wheel department, 3 work–hours in the assembly department, and 5 work–hours in the finishing department. Each tricycle needs 5 work–hours from the wheel department, 4 work–hours in the assembly department, and 7 work–hours in the finishing department. The capacity of the wheel department is 270 work–hours per week, the capacity of the assembly department is 200 work–hours per week, and the capacity of the finishing department is 360 work–hours per week. The company makes $19 profit on each unicycle, $34 profit on each bicycle, and $43 profit on each tricycle. How many of each kind of cycle should they make if they wish to maximize their profit? Steve, the president of the company, may decide that to preserve their prestige and their competitive position,

they should make all three kinds of cycle. Is there a feasible corner point that includes all three cycles? If so, how much less profit would the company make if they chose this product mix?

15. The Grocery Tricycle Company, one of Steve's major competitors in the tricycle market (see Exercise 2.7.14.), has gone out of business and the price on tricycles has increased. Steve now estimates that he can make a $49 profit on each tricycle. How does this change the product mix they should make in their shop?

3

Subspaces and Bases

3.1   A Statics Example

Blocks B_1, B_2, B_3, and B_4 rest on a frictionless platform connected by springs whose spring constants are k_1, k_2, and k_3. In Figure 3.1, the blocks are shown in equilibrium with all springs at their natural length and no external forces acting. In Figure 3.2, the blocks are shown in equilibrium in a different position with external forces f_1, f_2, f_3, and f_4 applied. The tick marks in the figures indicate the location of the blocks in their natural and different positions. The displacements of the four blocks from their natural position are denoted by d_1, d_2, d_3, and d_4, where the displacements and forces are measured positive to the right. We want to find the displacements of the blocks when the system with the applied forces is in static equilibrium.

To find these displacements, we set up balance equations for each of the blocks: in equilibrium, the total of the external forces and the forces exerted by the springs on each block must be zero. The balance equations, one for each block, are

$$\begin{cases} f_1 + k_1(d_2 - d_1) &= 0 \\ f_2 - k_1(d_2 - d_1) + k_2(d_3 - d_2) &= 0 \\ f_3 - k_2(d_3 - d_2) + k_3(d_4 - d_3) &= 0 \\ f_4 - k_3(d_4 - d_3) &= 0 \end{cases}$$

The first equation says that the sum of forces on B_1 must be zero. The equation is that the external force plus the force exerted on B_1 by the spring (which will be to the right if $d_1 < d_2$, the displacement of the first block is less than the displacement of the second block, and to the left if $d_1 > d_2$, the displacement of the first block is greater than the displacement of the second) equals zero. Similarly, the second equation says that the sum of the forces on B_2 must be zero. The equation is that

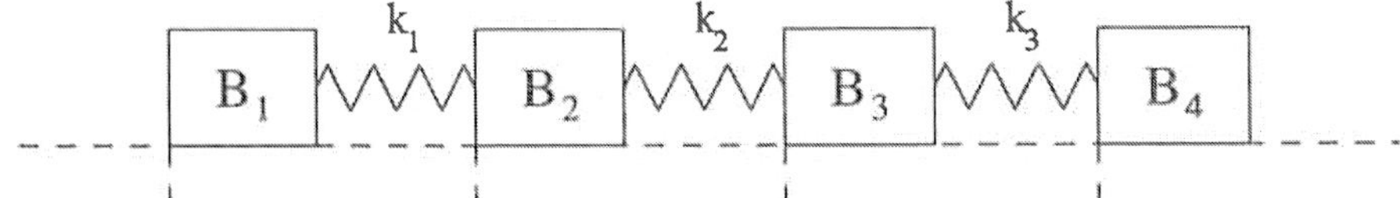

FIGURE 3.1

Blocks in natural position

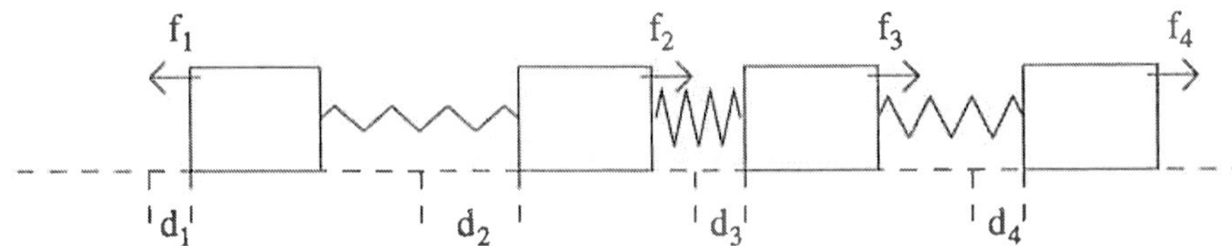

FIGURE 3.2

Blocks, with forces, in equilibrium

the external force on B_2 plus the force exerted on B_2 by the first spring (which will
be to the left if $d_1 < d_2$ and to the right if $d_1 > d_2$) plus the force exerted on B_2 by
the middle spring (which will be to the right if $d_2 < d_3$ and to the left if $d_2 > d_3$) is
zero. In the same way, the remaining two equations say that the sums of the forces
on blocks B_3 and B_4 are zero.

Rewriting this system, we see that it is a linear system of four equations in the
four unknown displacements.

$$
\begin{cases}
-k_1 d_1 + k_2 d_2 & = -f_1 \\
k_1 d_1 - (k_1 + k_2)d_2 + k_2 d_3 & = -f_2 \\
k_2 d_2 - (k_2 + k_3)d_3 + k_3 d_4 & = -f_3 \\
k_3 d_3 - k_3 d_4 & = -f_4
\end{cases}
$$

Specifically, if $k_1 = 3$, $k_2 = 5$, and $k_3 = .4$ and the external forces are $f_1 = -2$,
$f_2 = -1$, $f_3 = 0$, and $f_4 = 5$, we get the system

$$
\begin{cases}
-3d_1 + 3d_2 & = 2 \\
3d_1 - 8d_2 + 5d_3 & = 1 \\
5d_2 - 5.4d_3 + .4d_4 & = 0 \\
.4d_3 - .4d_4 & = -5
\end{cases}
$$

Alternatively, if $f_1 = -2$, $f_2 = -1$, $f_3 = 0$, and $f_4 = 3$, we get the system

$$\begin{cases} -3d_1 &+& 3d_2 & & & & & = & 2 \\ 3d_1 &-& 8d_2 &+& 5d_3 & & & = & 1 \\ & & 5d_2 &-& 5.4d_3 &+& .4d_4 & = & 0 \\ & & & & .4d_3 &-& .4d_4 & = & -3 \end{cases}$$

In the last chapter, we found a way to solve systems of equations like these. In this chapter, we will investigate more fully the nature of the solutions of these and other systems like them. In particular, we will see whether there is an equilibrium solution for every set of applied forces, if not, we will try to determine when there is a solution, and in this case, we will investigate how many solutions there are and how they are related to each other. To understand the answers, we must develop terminology to describe them and find the relationships that underly the various concepts we introduce. In particular, we will see that there are sets of vectors, called *subspaces*, that are at the heart of the answers to our questions. In this chapter, we will learn how to use *bases* to describe subspaces efficiently, how to determine how large they are (*dimension*), and we will discover how some of the subspaces associated with a matrix are related to each other.

3.2   Vector Spaces and Subspaces

We think of vectors geometrically and algebraically in terms of our experience with $\mathbf{R}^n$, especially $\mathbf{R}^3$. Sometimes it is helpful to think about mathematical objects more in terms of their important abstract qualities, unencumbered by other features that may be present in a particular setting but irrelevant for the work of the moment. On the other hand, sometimes thinking about mathematical issues purely in the abstract forces us to leave our intuition behind and we have no way to think about them. One of the goals of this book is to help you move from the concrete to the abstract and be comfortable in both settings. In this section, we want to pick out several features of vectors in $\mathbf{R}^n$ that are important for our work and identify them. Most of what follows will depend only on these abstract features, not on particular facts about $\mathbf{R}^n$, but we will not emphasize this excessively. Most of what we need can be accomplished by thinking about subspaces of $\mathbf{R}^n$, and it is in this context that we will think about the results of linear algebra.

Vectors in $\mathbf{R}^n$ have several algebraic features: they can be added and multiplied by numbers and these operations satisfy the properties below:

1. $u + v = v + u$ for all vectors u and v.

2. $u + (v + w) = (u + v) + w$ for all vectors u, v, and w.

3. The vector 0 satisfies $v + 0 = v$ for all vectors v.

4. For each vector v, the vector $-v$ satisfies $v + (-v) = 0$.

5. $1v = v$ for all vectors v.

6. $\alpha(\beta v) = (\alpha\beta)v$ for all vectors v and all numbers α and β.

7. $(\alpha + \beta)v = \alpha v + \beta v$ for all vectors v and all numbers α and β.

8. $\alpha(u + v) = \alpha u + \alpha v$ for all vectors u and v and all numbers α.

It is these algebraic properties of $\mathbf{R}^n$ that are fundamental for our work. But there are other systems that satisfy these rules of combination, including many of importance in engineering. For example, if u, v, and w represent matrices these properties remain true. Similarly, if u, v, and w represent polynomials of degree less or equal to N, a set important for approximation problems, or if u, v, and w represent continuous real valued functions defined on the set $0 \leq t \leq 2\pi$, a set important for Fourier analysis and signal processing, these properties remain true because the sum $u + v$ of the continuous functions u and v is again a continuous function, as is αu, and these notions of vector addition and scalar multiplication have the eight properties listed above. To abstract these features, we call a set with addition and scalar multiplication with these properties a *vector space*; that is, vector spaces are assumed to have these algebraic properties but are not assumed to have specific other properties. Whatever can be discovered about vector spaces using only these features must be true for all vector spaces, not just our favorite examples $\mathbf{R}^n$.

Notice that scalar multiplication is an essential feature of the structure of a vector space. We will understand that the relevant set of scalars, that is, the relevant numbers, for use in the vector space $\mathbf{R}^n$ is the set of real numbers. Similarly, we will understand that the relevant set of scalars for use in the vector space $\mathbf{C}^n$ is the set of complex numbers. More generally, in discussing any vector space, the set of scalars being used must be understood as a part of the definition of the space.

DEFINITION *If $\mathcal{V}$ is $\mathbf{R}^n$, $\mathbf{C}^n$, or other vector space and $\mathcal{W}$ is a non–empty subset of $\mathcal{V}$, we say $\mathcal{W}$ is a* subspace *of $\mathcal{V}$ if*

> *$v + w$ is in $\mathcal{W}$ for all v and w in $\mathcal{W}$*
> *and αw is in $\mathcal{W}$ for all w in $\mathcal{W}$ and all numbers α.*

The first of these properties says the sum of vectors in the set is again in the set, so this property says that $\mathcal{W}$ is *closed under addition*. The second property says that the product of a number and a vector in the set again lies in the set, so this property says that $\mathcal{W}$ is *closed under scalar multiplication*. Notice that the second property implies that subspaces containing a non–zero vector contain infinitely many vectors: if $w \neq 0$ belongs to $\mathcal{W}$, then so does $2w$, $3w$, $-17.6w$, $\sqrt{2}w$, etc., and these are all different. These two properties of a subset of $\mathbf{R}^n$ guarantee that the subset will, itself, be a vector space. In this book, if it helps your intuition, you may think of *vector space* as being synonymous with *subspace of* $\mathbf{R}^n$ although, as noted above, there are many vector spaces of importance in engineering that are not subspaces of $\mathbf{R}^n$.

EXAMPLE 3.1

If $\mathcal{V}$ is $\mathbf{R}^n$, or other vector space, then $\mathcal{V}$ itself and $\{0\}$ are subspaces of $\mathcal{V}$. The subspace $\{0\}$ is called the *zero subspace* and both subspaces are called *trivial subspaces*. To check that $\{0\}$ is a subspace, we need only check that $0 + 0 = 0$, so $\{0\}$ is closed under addition, and that $\alpha 0 = 0$ for all numbers α so $\{0\}$ is closed under scalar multiplication. Of course, $\mathcal{V}$ itself is, by definition, closed under addition and scalar multiplication. $\quad\square$

EXAMPLE 3.2

In $\mathbf{R}^2$, the set of solutions of $3x + 2y = 0$ is a subspace, but the set of solutions of $3x + 2y = 5$ is not a subspace.

SOLUTION To formalize slightly, let $\mathcal{W} = \{(x, y) : 3x + 2y = 0\}$ and let $\mathcal{Z} = \{(x, y) : 3x + 2y = 5\}$. The notation $\{\,\cdot\, : \,\cdot\,\}$ means that $\mathcal{W}$ is the set of objects of the form noted before the colon, vectors of the form (x, y) in this case, that satisfy the property noted after the colon, satisfy the equation $3x + 2y = 0$ in this case. To see whether a vector such as $(4, -6)$ belongs to the set $\mathcal{W}$, we must see if it satisfies the given condition: in this case, since $3(4) + 2(-6) = 0$, the condition is satisfied and the vector $(4, -6)$ is in $\mathcal{W}$. On the other hand, the vector $(-2, 4)$ is not in $\mathcal{W}$ because $3(-2) + 2(4) = 2 \neq 0$ and the condition is not satisfied.

To see that $\mathcal{W}$ is a subspace, we need to show that it is closed under addition and scalar multiplication. Suppose (x_1, y_1) and (x_2, y_2) are vectors in $\mathcal{W}$, that is, $3x_1 + 2y_1 = 0$ and $3x_2 + 2y_2 = 0$. Their sum is $(x_1 + x_2, y_1 + y_2)$ and to see if this vector is in $\mathcal{W}$, we need to see if it satisfies the condition that defines $\mathcal{W}$:

$$3(x_1 + x_2) + 2(y_1 + y_2) = 3x_1 + 3x_2 + 2y_1 + 2y_2$$

$$= (3x_1 + 2y_1) + (3x_2 + 2y_2) = 0 + 0 = 0$$

so this vector does satisfy the condition. This means $\mathcal{W}$ is closed under addition. To see that $\mathcal{W}$ is closed under scalar multiplication, we need to check that for every number α, $\alpha(x_1, y_1)$ is in $\mathcal{W}$ whenever (x_1, y_1) is. Since $\alpha(x_1, y_1) = (\alpha x_1, \alpha y_1)$ we need to see if this vector satisfies the equation

$$3(\alpha x_1) + 2(\alpha y_1) = 3\alpha x_1 + 2\alpha y_1 = \alpha(3x_1 + 2y_1) = \alpha \cdot 0 = 0$$

so $\mathcal{W}$ is closed under scalar multiplication as well, and is a subspace.

To see that $\mathcal{Z}$ is a not subspace, we need only provide one example that $\mathcal{Z}$ is not closed under addition *or* not closed under scalar multiplication. It is easily checked that $(1, 1)$ is in $\mathcal{Z}$ and that $(-3, 7)$ is in $\mathcal{Z}$. However, $(1, 1) + (-3, 7) = (-2, 8)$ and $3 \cdot (-2) + 2 \cdot 8 = 10 \neq 5$ so $(-2, 8)$ does not satisfy the defining condition for $\mathcal{Z}$ and it is not in $\mathcal{Z}$. Therefore, $\mathcal{Z}$ is not closed under addition and $\mathcal{Z}$ is not a subspace of $\mathbf{R}^2$.   ⬚

This example can be generalized to a statement about all subspaces.

PROPOSITION 3.3

If $\mathcal{W}$ is a subspace, then 0 is a vector in $\mathcal{W}$.

PROOF   Suppose v is a vector in $\mathcal{W}$. Then, since $\mathcal{W}$ is closed under scalar multiplication, $0v = 0$ is in $\mathcal{W}$.   ∎

The subset $\mathcal{Z}$ of the example above does not contain the zero vector $(0, 0)$ so it cannot be a subspace. In fact, later, we will see that every subspace of $\mathbf{R}^2$ is either one of the trivial subspaces $\{0\}$ and $\mathbf{R}^2$, or is a line through the origin. Of course, the subspace $\mathcal{W}$ in the example above is a line through the origin while $\mathcal{Z}$ is a line not passing through the origin.

EXAMPLE 3.4

If A is an $m \times n$ matrix, show that the set $\mathcal{N}(A) = \{X \in \mathbf{R}^n : AX = 0\}$ is a subspace.

SOLUTION This is the content of *(a)* of Theorem 2.2. If X_1 and X_2 are in $\mathcal{N}(A)$, that is $AX_1 = 0$ and $AX_2 = 0$, then $A(X_1 + X_2) = AX_1 + AX_2 = 0 + 0 = 0$ so $X_1 + X_2$ is in $\mathcal{N}(A)$ and $\mathcal{N}(A)$ is closed under addition. If α is a number, then $A(\alpha X_1) = \alpha AX_1 = \alpha 0 = 0$ so αX_1 is in $\mathcal{N}(A)$ and $\mathcal{N}(A)$ is closed under scalar multiplication. Since $\mathcal{N}(A)$ is closed under addition and scalar multiplication, it is a subspace of $\mathbf{R}^n$.

Note that this is a generalization of the previous example: there $A = \begin{pmatrix} 3 & 2 \end{pmatrix}$.

It is even a generalization of the first example: $\mathbf{R}^n = \{X \in \mathbf{R}^n : 0X = 0\}$ and $\{0\} = \{X \in \mathbf{R}^n : IX = 0\}$ □

The subspace of this example will be important in our work later; we formalize it in the following definition.

DEFINITION *If A is an $m \times n$ matrix, the subspace*

$$\mathcal{N}(A) = \{X \in \mathbf{R}^n : AX = 0\}$$

is called the nullspace *of A, the* solution space *of $AX = 0$, or the* kernel *of A.*

EXAMPLE 3.5

Consider the system

$$\begin{cases} 2x & - & 4y & = & a \\ -x & + & 2y & = & b \end{cases}$$

Let $\mathcal{W}$ be the set of righthand sides for which the system is solvable, that is, let

$$\mathcal{W} = \{\begin{pmatrix} a \\ b \end{pmatrix} : \begin{cases} 2x & - & 4y & = & a \\ -x & + & 2y & = & b \end{cases} \text{ has a solution}\}$$

Show that $\mathcal{W}$ is a subspace of $\mathbf{R}^2$.

SOLUTION Before beginning the "solution" of the problem, let's explore a little. First of all, note that if $a = 0$ and $b = 0$, so that the system is

$$\begin{cases} 2x & - & 4y & = & 0 \\ -x & + & 2y & = & 0 \end{cases}$$

then $x = 0$, $y = 0$ is a solution, so $(0, 0)$ satisfies the condition to be in the set $\mathcal{W}$. Also, for $a = 4$, $b = -2$, so that the system is

$$\begin{cases} 2x & - & 4y & = & 4 \\ -x & + & 2y & = & -2 \end{cases}$$

then $x = 6$, $y = 2$ is a solution, so $(4, -2)$ also satisfies the defining condition and is in $\mathcal{W}$. On the other hand, if $a = 1$ and $b = 1$, then the system becomes

$$\begin{cases} 2x & - & 4y & = & 1 \\ -x & + & 2y & = & 1 \end{cases}$$

and the system has no solution: indeed if $-x + 2y = 1$ as in the second equation, then, multiplying by -2, we see $2x - 4y = -2 \neq 1$. This means $(1, 1)$ does not satisfy the defining condition for belonging to $\mathcal{W}$, so it is not in $\mathcal{W}$.

One way to solve this problem is to note that, as above, if the system is

$$\begin{cases} 2x & - & 4y & = & a \\ -x & + & 2y & = & b \end{cases}$$

and (x, y) is a solution to the second equation, then $-x + 2y = b$ and multiplying by -2, this forces $2x - 4y = -2b$. Thus if $a \neq -2b$ the system has no solution. On the other hand, if $a = -2b$ so the system is

$$\begin{cases} 2x & - & 4y & = & -2b \\ -x & + & 2y & = & b \end{cases}$$

then $x = -b$, $y = 0$ is a solution of the system, so $(-2b, b)$ is in $\mathcal{W}$ for all numbers b.

This means that the set $\mathcal{W}$ can be more easily described as

$$\mathcal{W} = \left\{ \begin{pmatrix} a \\ b \end{pmatrix} : a = -2b \right\}$$

that is, $\mathcal{W}$ is the set of multiples of $(-2, 1)$ which is the line through the origin and $(-2, 1)$ and this set is a subspace. $\square$

EXAMPLE 3.6

If A is an $m \times n$ matrix, prove that the set $\mathcal{R}(A) = \{b \in \mathbf{R}^m : AX = b \text{ for some } X \text{ in } \mathbf{R}^n\}$ is a subspace.

SOLUTION Notice that this is a generalization of the previous example, for

$$A = \begin{pmatrix} 2 & -4 \\ -1 & 2 \end{pmatrix}$$

Suppose b_1 and b_2 are in $\mathcal{R}(A)$, that is, there are X_1 and X_2 in $\mathbf{R}^n$ so that $AX_1 = b_1$ and $AX_2 = b_2$. We need to find a vector X in $\mathbf{R}^n$ so that $AX = b_1 + b_2$. We claim $X = X_1 + X_2$ is such a vector: $A(X_1 + X_2) = AX_1 + AX_2 = b_1 + b_2$, so $b_1 + b_2$ is in $\mathcal{R}(A)$ and $\mathcal{R}(A)$ is closed under addition. Similarly, if α is a number and b_1 is in $\mathcal{R}(A)$ because $AX_1 = b_1$, then $A(\alpha X_1) = \alpha A X_1 = \alpha b_1$, so αb_1 is also in $\mathcal{R}(A)$ and $\mathcal{R}(A)$ is closed under scalar multiplication. Thus, $\mathcal{R}(A)$ is a subspace of $\mathbf{R}^m$. $\Box$

The set in this example will also be important in our future work and we formalize it in the following definition.

DEFINITION *If A is an $m \times n$ matrix, the subspace*

$$\mathcal{R}(A) = \{b \in \mathbf{R}^m : AX = b \text{ for some } X \text{ in } \mathbf{R}^n\}$$

is called the range *of A.*

In the next two sections, we will learn about ways to describe subspaces more efficiently. After finding "standard" descriptions of subspaces, we will apply our understanding to the problem of solving systems of equations and be able to answer the questions asked in the introduction to our study of linear systems. The subspaces $\mathcal{N}(A)$ and $\mathcal{R}(A)$ introduced in the examples will play a major role in this understanding.

Exercises 3.2

1. Let $\mathcal{W} = \{(s, 0, t, 2s + 3t) : s, t \text{ real}\}$. Show that $\mathcal{W}$ is a subspace of $\mathbf{R}^4$.
2. Let $\mathcal{U} = \{(s, 3, t, 2s + 3t) : s, t \text{ real}\}$. Show that $\mathcal{U}$ is not a subspace of $\mathbf{R}^4$.
3. Show that the system

$$\begin{cases} w + 2x - y + z &= 2 \\ 3x - 2y + 2z &= 4 \\ 2w - x + y - z &= -1 \end{cases}$$

   has infinitely many solutions but that the set of solutions of this system does *not* form a subspace of $\mathbf{R}^4$.
4. Show that the set $\mathcal{W}$ of Exercise 3.2.1. is the range of the matrix

$$A = \begin{pmatrix} 1 & 0 \\ 0 & 0 \\ 0 & 1 \\ 2 & 3 \end{pmatrix}$$

   which means that $\mathcal{W}$ is a subspace by Example 3.6.

5. Let $\mathcal{Z} = \{(2s + t, s, -s + 3t, t) : s, t \text{ real}\}$. Find a matrix A as in Exercise 3.2.4. to show that $\mathcal{Z}$ is a subspace of $\mathbf{R}^4$ by Example 3.6.

6. Let $\mathcal{V} = \{(s - t, -s + 3t, s, t) : s, t \text{ real}\}$. Show that $\mathcal{V}$ is a subspace of $\mathbf{R}^4$ by showing it is the nullspace of the matrix

$$A = \begin{pmatrix} 1 & 0 & -1 & 1 \\ 0 & 1 & 1 & -3 \end{pmatrix}$$

and using the result of Example 3.4.

7. Let $\mathcal{U} = \{(r - 2s + t, -2r + s, r, s, t) : r, s, t \text{ real}\}$. Show that $\mathcal{U}$ is a subspace of $\mathbf{R}^5$ by showing it is the nullspace of some matrix (as in Exercise 3.2.6.) and and using the result of Example 3.4.

8. Show that the set $\mathcal{P}_3$ of polynomials of degree less than or equal to 3 is a subspace of the vector space of all polynomials. (Recall that a function p is a polynomial of degree 3 or less if there are real numbers a_0, a_1, a_2, and a_3 so that $p(x) = a_0 + a_1 x + a_2 x^2 + a_3 x^3$ for all x.)

9. Show that the set $\{p \in \mathcal{P}_3 : p(1) = 0\}$ of polynomials of degree 3 or less that have a root at $x = 1$ is a subspace of $\mathcal{P}_3$.

10. Show that the set $\{p \in \mathcal{P}_3 : p(1) = 4\}$ of polynomials of degree 3 or less that have the value 4 at $x = 1$ is not a subspace of $\mathcal{P}_3$.

11. If A is an $m \times n$ matrix and $b \neq 0$ is a vector in $\mathbf{R}^m$, show that $\{X : AX = b\}$ is not a subspace of $\mathbf{R}^n$.

3.3   Linear Combinations and Spanning Sets

In this section, we want to emphasize the way in which subspaces can be built from a few vectors. This building process must be based on the ways that vectors can be combined which, in a vector space, are limited to addition and multiplication by numbers. Thus, the following definition is natural, it just describes the way a vector can be built from others in a vector space.

DEFINITION *If v_1, v_2, $\cdots$, v_k, and w are vectors in the vector space $\mathcal{V}$, we say w* is a linear combination of v_1, v_2, $\cdots$, v_k *if there are numbers c_1, c_2, $\cdots$, c_k so that*

$$w = c_1 v_1 + c_2 v_2 + \cdots + c_k v_k$$

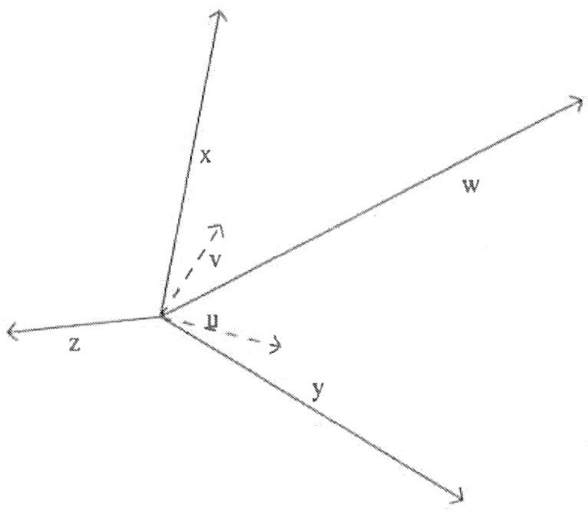

FIGURE 3.3

Linear combinations of u and v

For example, the vector $(-5, 4, 8)$ is a linear combination of $(1, 2, -1), (1, -1, 3)$, and $(2, 0, -1)$ because

$$\begin{pmatrix} -5 \\ 4 \\ 8 \end{pmatrix} = 3 \begin{pmatrix} 1 \\ 2 \\ -1 \end{pmatrix} + 2 \begin{pmatrix} 1 \\ -1 \\ 3 \end{pmatrix} - 5 \begin{pmatrix} 2 \\ 0 \\ -1 \end{pmatrix}$$

We will describe answers to most problems in linear algebra in terms of linear combinations.

If $u = (4, -1)$ and $v = (2, 3)$ were especially interesting vectors in the plane for some problem, we might choose to write other vectors as a linear combination of them. Figure 3.3 illustrates the linear combinations $w = (14, 7) = 2u + 3v$, $x = (2, 10) = -u + 3v$, $y = (10, -6) = 3u - v$, and $z = (-5, -.5) = -u - .5v$. It is not difficult to imagine that every vector in the plane can be written as a linear combination of these two vectors.

THEOREM 3.7

If $v_1, v_2, \cdots, v_k$ are vectors in the vector space V, the set of linear combinations of $v_1, v_2, \cdots, v_k$ is a subspace of V.

PROOF Let W be the set of linear combinations of $v_1, v_2, \cdots, v_k$, that is

$$W = \{w : w = c_1 v_1 + c_2 v_2 + \cdots + c_k v_k \text{ for some scalars } c_1, c_2, \cdots, c_k\}$$

We need to show $\mathcal{W}$ is closed under addition and scalar multiplication. Suppose w and x are in $\mathcal{W}$. Then there are scalars so that $w = c_1 v_1 + c_2 v_2 + \cdots + c_k v_k$ and $x = d_1 v_1 + d_2 v_2 + \cdots + d_k v_k$. then

$$w + x = (c_1 v_1 + c_2 v_2 + \cdots + c_k v_k) + (d_1 v_1 + d_2 v_2 + \cdots + d_k v_k)$$
$$= (c_1 + d_1)v_1 + (c_2 + d_2)v_2 + \cdots + (c_k + d_k)v_k$$

so $w + x$ is also a linear combination of the v's and $w + x$ is in $\mathcal{W}$. Similarly, if a is a scalar and w is in $\mathcal{W}$, then

$$aw = a(c_1 v_1 + c_2 v_2 + \cdots + c_k v_k)$$
$$= (ac_1)v_1 + (ac_2)v_2 + \cdots + (ac_k)v_k$$

which is a linear combination of the v's and aw is in $\mathcal{W}$. Since $\mathcal{W}$ is closed under addition and scalar multiplication, it is a subspace of $\mathcal{V}$.   ∎

DEFINITION *If v_1, v_2, $\cdots$, v_k are vectors in the vector space $\mathcal{V}$, the span of v_1, v_2, $\cdots$, v_k is the set of linear combinations of v_1, v_2, $\cdots$, v_k. If $\mathcal{W}$ is a subspace of $\mathcal{V}$ and v_1, v_2, $\cdots$, v_k are vectors in $\mathcal{W}$ such that every vector in $\mathcal{W}$ is a linear combination of these vectors, we say $\{v_1, v_2, \cdots, v_k\}$ is a spanning set for $\mathcal{W}$ or we say $\mathcal{W}$ is spanned by v_1, v_2, $\cdots$, v_k.*

EXAMPLE 3.8

The vectors $e_1 = (1, 0, 0)$, $e_2 = (0, 1, 0)$, and $e_3 = (0, 0, 1)$ span $\mathbf{R}^3$. Indeed, if v is a vector in $\mathbf{R}^3$, $v = (a_1, a_2, a_3)$ for some numbers a_1, a_2, and a_3, so $v = a_1 e_1 + a_2 e_2 + a_3 e_3$. Indeed, it is difficult to think of $\mathbf{R}^3$ without thinking of vectors as linear combinations of e_1, e_2, and e_3 because the coordinates of a vector are just the coefficients in the expression of the vector as a linear combination of these special vectors.   ☐

EXAMPLE 3.9

Two vectors u and v in $\mathbf{R}^3$ span a plane unless one is multiple of the other (see Figure 3.4). Of course the plane passes through the origin because $0u + 0v = 0$ and 0 is a linear combination of u and v.   ☐

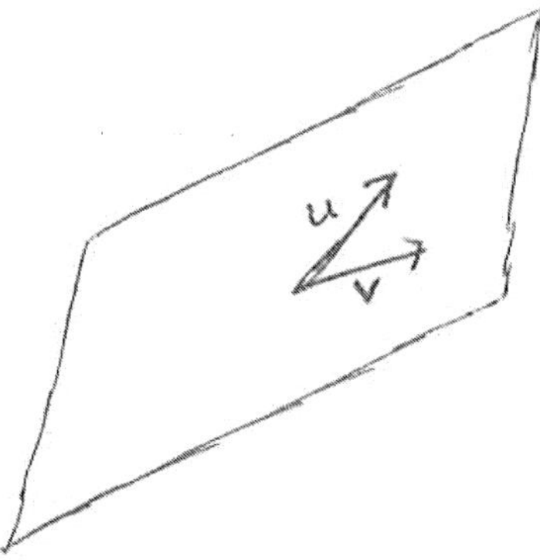

FIGURE 3.4

Span of u and v in $\mathbf{R}^3$

EXAMPLE 3.10

Find vectors that span $\mathcal{N}(A)$ the nullspace of the matrix $A = \begin{pmatrix} 1 & -1 & 2 & 1 \\ 2 & -1 & 1 & 0 \end{pmatrix}$.

SOLUTION The nullspace of A is the set of vectors in $\mathbf{R}^4$ that satisfy $AX = 0$; so first we solve this system.

$$\begin{pmatrix} 1 & -1 & 2 & 1 & | & 0 \\ 2 & -1 & 1 & 0 & | & 0 \end{pmatrix} \longrightarrow \begin{pmatrix} 1 & -1 & 2 & 1 & | & 0 \\ 0 & 1 & -3 & -2 & | & 0 \end{pmatrix}$$

$$\longrightarrow \begin{pmatrix} 1 & 0 & -1 & -1 & | & 0 \\ 0 & 1 & -3 & -2 & | & 0 \end{pmatrix}$$

so the system $AX = 0$ is equivalent to the system $w-y-z = 0$ and $x-3y-2z = 0$, or $w = y + z$ and $x = 3y + 2z$. In other words, X is a solution of the system if and only if $X = (y + z, 3y + 2z, y, z)$ for some numbers y and z. In other words, X is in $\mathcal{N}(A)$ if and only if

$$X = \begin{pmatrix} y + z \\ 3y + 2z \\ y \\ z \end{pmatrix} = y \begin{pmatrix} 1 \\ 3 \\ 1 \\ 0 \end{pmatrix} + z \begin{pmatrix} 1 \\ 2 \\ 0 \\ 1 \end{pmatrix}$$

which says $\mathcal{N}(A)$ is spanned by the vectors $(1, 3, 1, 0)$ and $(1, 2, 0, 1)$. $\quad\square$

A subspace of $\mathbf{R}^n$ spanned by one vector, that is, the set of multiples of a single vector, is just a line through the origin. If two vectors do not lie on the same line, the subspace they span is the plane through the origin that contains them (see Figure 3.3).

In general, there are many ways to express a subspace as a span. For example, if $\mathcal{W}$ is the span of $v_1 = (1, -1, 2)$, $v_2 = (2, 0, 1)$, and $v_3 = (1, -3, 5)$, then $\mathcal{W}$ is also spanned by v_1 and v_2 because v_3 is a linear combination of v_1 and v_2, namely, $v_3 = 3v_1 - v_2$, so any linear combination of v_1, v_2, and v_3 can be replaced by a linear combination of v_1 and v_2. For example, $w = (1, -5, 8)$ which can be written as $w = 2v_1 - v_2 + v_3$ can also be written as $w = 2v_1 - v_2 + (3v_1 - v_2) = 5v_1 - 2v_2$. That is, $\mathcal{W}$ is spanned by v_1 and v_2.

EXAMPLE 3.11

Suppose v_1, v_2, and v_3 span the subspace $\mathcal{W}$. If v_4 is another vector in $\mathcal{W}$, do v_1, v_2, v_3, and v_4 also span $\mathcal{W}$?

SOLUTION Yes! To say that v_1, v_2, and v_3 span the subspace $\mathcal{W}$ means that if w is any vector in $\mathcal{W}$, then there are constants so that

$$w = c_1 v_1 + c_2 v_2 + c_3 v_3$$

On the other hand, if this is so, then

$$w = c_1 v_1 + c_2 v_2 + c_3 v_3 + 0 v_4$$

which means w is a linear combination of v_1, v_2, v_3, and v_4 and they also span $\mathcal{W}$. ∎

There are two other subspaces associated with A that would appear to be important.

DEFINITION *If A is a $m \times n$ matrix, the* column space *is the subspace of $\mathbf{R}^m$ spanned by the columns of A and the* row space *is the subspace of $\mathbf{R}^n$ spanned by the rows of A.*

To think more about the column space of a matrix, we extend the notation of block matrices introduced in Section 2.6. In that section, instead of thinking about a matrix as consisting of its entries, we thought of the matrix as being built from its rows. We can just as easily think of a matrix as being built from its column vectors: it is a row of column vectors. For example, the matrix

$$A = \begin{pmatrix} 1 & 2 & 3 \\ 4 & 5 & 6 \\ 7 & 8 & 9 \end{pmatrix}$$

can be written as

$$A = \begin{pmatrix} A_1 & A_2 & A_3 \end{pmatrix}$$

where

$$A_1 = \begin{pmatrix} 1 \\ 4 \\ 7 \end{pmatrix} \qquad A_2 = \begin{pmatrix} 2 \\ 5 \\ 8 \end{pmatrix} \qquad \text{and} \quad A_3 = \begin{pmatrix} 3 \\ 6 \\ 9 \end{pmatrix}$$

We say we have *blocked* the matrix A by columns. This point of view helps us focus more easily on the columns as entities in themselves.

EXAMPLE 3.12

Let A be the matrix above, blocked in columns as

$$A = \begin{pmatrix} A_1 & A_2 & A_3 \end{pmatrix}$$

If $X = (c_1, c_2, c_3)$ is a vector in $\mathbf{R}^3$, then $AX = c_1 A_1 + c_2 A_2 + c_3 A_3$. This becomes easier to see if we write it as a matrix multiplication:

$$AX = \begin{pmatrix} A_1 & A_2 & A_3 \end{pmatrix} \begin{pmatrix} c_1 \\ c_2 \\ c_3 \end{pmatrix} = A_1 c_1 + A_2 c_2 + A_3 c_3$$

which can be rewritten in the form $c_1 A_1 + c_2 A_2 + c_3 A_3$ because the c's are numbers and therefore can be multiplied by a vector in either order. To write this out completely, we write

$$AX = \begin{pmatrix} 1 & 2 & 3 \\ 4 & 5 & 6 \\ 7 & 8 & 9 \end{pmatrix} \begin{pmatrix} c_1 \\ c_2 \\ c_3 \end{pmatrix}$$

$$= \begin{pmatrix} 1c_1 + 2c_2 + 3c_3 \\ 4c_1 + 5c_2 + 6c_3 \\ 7c_1 + 8c_2 + 9c_3 \end{pmatrix}$$

$$= c_1 \begin{pmatrix} 1 \\ 4 \\ 7 \end{pmatrix} + c_2 \begin{pmatrix} 2 \\ 5 \\ 8 \end{pmatrix} + c_3 \begin{pmatrix} 3 \\ 6 \\ 9 \end{pmatrix}$$

$$= c_1 A_1 + c_2 A_2 + c_3 A_3$$

Notice that this generalizes the work in Exercise 1.3.4. in which a similar calculation was noted for the special vectors e_1, e_2, and e_3. $\quad\Box$

In Section 3.2, we defined the *range*[1] *of A* to be the subspace of $\mathbf{R}^m$

$$\mathcal{R}(A) = \{b \in \mathbf{R}^m : AX = b \text{ for some } X \text{ in } \mathbf{R}^n\}$$

The following result says that the range of A is the same subspace as the column space of A! If A is a real matrix, then the row space of A is just the range of A'.

THEOREM 3.13

The range of A is the column space of A.

PROOF To prove the equality of these two sets, we need to prove that the range of A is a subset of the column space and the column space is a subset of the range of A. That is, we need to prove that every vector in the range of A is a linear combination of the columns of A and, conversely, that every linear combination of the columns of A is in the range.

Let $A = \begin{pmatrix} A_1 & A_2 & \cdots & A_n \end{pmatrix}$ be the block representation of A in which A_j is the jth column of A. If w is a linear combination of the columns, say $w = c_1 A_1 + c_2 A_2 + \cdots + c_n A_n$, then, as in Example 3.12

$$w = \begin{pmatrix} A_1 & A_2 & \cdots & A_n \end{pmatrix} \begin{pmatrix} c_1 \\ c_2 \\ \vdots \\ c_n \end{pmatrix}$$

that is, $AX = w$ where $X = (c_1, c_2, \cdots, c_n)$. Thus, every vector in the column space is in the range.

Conversely, this calculation also shows that if $AX = b$ for $X = (c_1, c_2, \cdots, c_n)$, then $b = c_1 A_1 + c_2 A_2 + \cdots + c_n A_n$. Thus, if b is a vector in the range of A, it is a linear combination of the columns of A. ∎

EXAMPLE 3.14

Find vectors that span $\mathcal{R}(A)$, the range of the matrix $A = \begin{pmatrix} 1 & -1 & 2 & 1 \\ 2 & -1 & 1 & 0 \end{pmatrix}$.

---

[1]The word "range" is applied to the subspace associated with A because this subspace is the range the function mapping $\mathbf{R}^n$ into $\mathbf{R}^m$ by the rule $X \to AX$.

SOLUTION Theorem 3.13 says that the range of A is the same as the column space of A, which is the subspace spanned by the columns of A. Thus, the set of column vectors

$$\left\{ \begin{pmatrix} 1 \\ 2 \end{pmatrix}, \begin{pmatrix} -1 \\ -1 \end{pmatrix}, \begin{pmatrix} 2 \\ 1 \end{pmatrix} \begin{pmatrix} 1 \\ 0 \end{pmatrix} \right\}$$

is a spanning set for the range of A.

However, another answer is that the vectors $(1, 2)$ and $(-1, -1)$ span the range of A. Clearly every vector that can be written as a linear combination of two of the column vectors can be written as a linear combination of all four of them; if $w = \alpha(1, 2) + \beta(1, -1)$, then

$$w = \alpha \begin{pmatrix} 1 \\ 2 \end{pmatrix} + \beta \begin{pmatrix} -1 \\ -1 \end{pmatrix} = \alpha \begin{pmatrix} 1 \\ 2 \end{pmatrix} + \beta \begin{pmatrix} -1 \\ -1 \end{pmatrix} + 0 \begin{pmatrix} 2 \\ 1 \end{pmatrix} + 0 \begin{pmatrix} 1 \\ 0 \end{pmatrix}$$

Conversely, since $(2, 1) = -(1, 2) - 3(-1, -1)$ and $(1, 0) = -(1, 2) - 2(-1, -1)$ every vector that can be written as a linear combination of all four columns can be written as a linear combination of the first two. For example,

$$\begin{pmatrix} 0 \\ 1 \end{pmatrix} = \begin{pmatrix} 1 \\ 2 \end{pmatrix} + 2 \begin{pmatrix} -1 \\ -1 \end{pmatrix} + \begin{pmatrix} 2 \\ 1 \end{pmatrix} - \begin{pmatrix} 1 \\ 0 \end{pmatrix}$$

but also

$$\begin{pmatrix} 0 \\ 1 \end{pmatrix} = \begin{pmatrix} 1 \\ 2 \end{pmatrix} + 2 \begin{pmatrix} -1 \\ -1 \end{pmatrix} + \left(-\begin{pmatrix} 1 \\ 2 \end{pmatrix} - 3 \begin{pmatrix} -1 \\ -1 \end{pmatrix} \right) - \left(-\begin{pmatrix} 1 \\ 2 \end{pmatrix} - 2 \begin{pmatrix} -1 \\ -1 \end{pmatrix} \right)$$

$$= \begin{pmatrix} 1 \\ 2 \end{pmatrix} + \begin{pmatrix} -1 \\ -1 \end{pmatrix}$$

□

This example shows that there are many ways to describe a subspace in terms of spanning vectors: we saw above that $\mathcal{R}(A) = \text{span}\{(1, 2), (-1, -1), (2, 1), (1, 0)\}$ and also that $\mathcal{R}(A) = \text{span}\{(1, 2), (-1, -1)\}$. Indeed, since $(1, 0) = -(1, 2) - 2(-1, -1)$ and $(0, 1) = (1, 2) + (-1, -1)$, every linear combination of $(1, 0)$ and $(0, 1)$ is a linear combination of $(1, 2)$ and $(-1, -1)$. Conversely, $(1, 2) = (1, 0) + 2(0, 1)$ and $(-1, -1) = -(1, 0) - (0, 1)$, so every linear combination of $(1, 2)$ and $(-1, -1)$ is also a linear combination of $(1, 0)$ and $(0, 1)$. This means that $\mathcal{R}(A) = \text{span}\{(1, 0), (0, 1)\}$ which is $\mathbf{R}^2$. It should be clear, and we will discuss this more later, that some spanning sets for a subspace are more convenient than others: in this case, it seems clear that $\{(1, 0), (0, 1)\}$ is a more convenient spanning set for $\mathcal{R}(A) = \mathbf{R}^2$ than $\{(1, 2), (-1, -1), (2, 1), (1, 0)\}$.

EXAMPLE 3.15

Let $B = \begin{pmatrix} 2 & 0 & -1 \\ 1 & 3 & 1 \\ 1 & 1 & 0 \end{pmatrix}$. If $X = (1, -2, 2)$, then

$$BX = \begin{pmatrix} 2 & 0 & -1 \\ 1 & 3 & 1 \\ 1 & 1 & 0 \end{pmatrix} \begin{pmatrix} 1 \\ -2 \\ 2 \end{pmatrix} = \begin{pmatrix} 0 \\ -3 \\ -1 \end{pmatrix}$$

is in the range of B. According to the Theorem above, this vector is a linear combination of the columns of B. Indeed, this is easy to see:

$$\begin{pmatrix} 0 \\ -3 \\ -1 \end{pmatrix} = 1 \begin{pmatrix} 2 \\ 1 \\ 1 \end{pmatrix} - 2 \begin{pmatrix} 0 \\ 3 \\ 1 \end{pmatrix} + 2 \begin{pmatrix} -1 \\ 1 \\ 0 \end{pmatrix}$$

Conversely, the Theorem says that if a vector is a linear combination of the columns, of B, for example, if

$$v = 2 \begin{pmatrix} 2 \\ 1 \\ 1 \end{pmatrix} + 1 \begin{pmatrix} 0 \\ 3 \\ 1 \end{pmatrix} - 3 \begin{pmatrix} -1 \\ 1 \\ 0 \end{pmatrix} = \begin{pmatrix} 7 \\ 2 \\ 3 \end{pmatrix}$$

then v is BX for some X. Again, the proof of the Theorem shows us that this is easy:

$$\begin{pmatrix} 7 \\ 2 \\ 3 \end{pmatrix} = \begin{pmatrix} 2 & 0 & -1 \\ 1 & 3 & 1 \\ 1 & 1 & 0 \end{pmatrix} \begin{pmatrix} 2 \\ 1 \\ -3 \end{pmatrix}$$

that is, that $v = BX$ for $X = (2, 1, -3)$.

Is the vector $(1, 1, -2)$ a linear combination of the columns of B? To answer this question, we must decide if there are numbers a, b, and c so that

$$a \begin{pmatrix} 2 \\ 1 \\ 1 \end{pmatrix} + b \begin{pmatrix} 0 \\ 3 \\ 1 \end{pmatrix} + c \begin{pmatrix} -1 \\ 1 \\ 0 \end{pmatrix} = \begin{pmatrix} 1 \\ 1 \\ -2 \end{pmatrix}$$

Rewriting this linear system in its matrix form, we get

$$\begin{pmatrix} 2 & 0 & -1 \\ 1 & 3 & 1 \\ 1 & 1 & 0 \end{pmatrix} \begin{pmatrix} a \\ b \\ c \end{pmatrix} = \begin{pmatrix} 1 \\ 1 \\ -2 \end{pmatrix}$$

In other words, we must decide if $(1, 1, -2)$ is in the range of B. This is precisely what the Theorem says!

Solving the system by row operations:

$$\left(\begin{array}{ccc|c} 2 & 0 & -1 & 1 \\ 1 & 3 & 1 & 1 \\ 1 & 1 & 0 & 2 \end{array} \right) \rightarrow \left(\begin{array}{ccc|c} 1 & 1 & 0 & 2 \\ 1 & 3 & 1 & 1 \\ 2 & 0 & -1 & 1 \end{array} \right)$$

$$\rightarrow \left(\begin{array}{ccc|c} 1 & 1 & 0 & 2 \\ 0 & 2 & 1 & -1 \\ 0 & -2 & -1 & -3 \end{array} \right) \rightarrow \left(\begin{array}{ccc|c} 1 & 1 & 0 & 2 \\ 0 & 2 & 1 & -1 \\ 0 & 0 & 0 & -4 \end{array} \right)$$

so the system is inconsistent and the vector $(1, 1, -2)$ is *not* a linear combination of the columns of B, that is, it is *not* in the range of B. In fact, the range of B is a plane (through the origin) in $\mathbf{R}^3$ and the point $(1, 1, -2)$ is not on this plane. $\square$

To summarize, there are four important subspaces associated with an $m \times n$ matrix A: the nullspace of A, the range (column space) of A, the nullspace of A', and the range of A'. The nullspace of A and the range of A' are subspaces of $\mathbf{R}^n$ and the nullspace of A' and the range of A are subspaces of $\mathbf{R}^m$. In the next few sections, we will discover important relationships between them.

Exercises 3.3

1. Let $u = (4, -1)$ and $v = (2, 3)$ as in Figure 3.3.
   (a) Write $p = (10, 8)$, $q = (-4, 8)$, $r = (7, 7)$ as linear combinations of u and v.
   (b) On a piece of graph paper, draw arrows to represent u, v, p, q, and r as in Figure 3.3.
2. In doing his linear algebra homework, Eduardo found that if $u = (2, 1, -3)$, $v = (1, -1, 2)$, and $w = (1, 5, -12)$ then $w = 2u - 3v$. Is w a linear combination of u and v?
3. In doing their linear algebra homework, Jennifer found that if $p = (-2, 4, -3)$, $q = (1, -1, 2)$, and $r = (-1, 7, 1)$, $s = (1, 1, 3)$ then $p = q + r - 2s$, and Bud found that $p = -3q + s$. Is p a linear combination of q, r, and s?
4. For $w = (4, -1, 5)$, $x = (2, -1, 1)$, $y = (1, 0, 2)$, and $z = (1, 1, -3)$, Drew found that w is a linear combination of x and y, namely, $w = x + 2y$. Write w as a linear combination of x, y, and z.
5. For $v = (4, -1, 2)$, $w = (2, 1, 4)$, $x = (1, -1, -1)$, and $y = (9, -3, 3)$, Melissa found that $y = 3v - w - x$. On the other hand, Melissa also found that $v = w + 2x$, Write y as a linear combination of w and x.

6. Write $(1, -1, 5, -5)$ as a linear combination of the vectors $(1, 1, -1, 1)$, and $(2, 1, 1, -1)$ or explain why it is not possible to do so.

7. Write $(1, -1, 1, 2)$ as a linear combination of the vectors $(1, 1, -1, 1)$, and $(2, 1, 1, -1)$ or explain why it is not possible to do so.

8. Write $(1, -3, 0, 2)$ as a linear combination of the vectors $(1, 0, 1, 1)$, $(0, 1, 1, -1)$, and $(2, 1, 1, 3)$ or explain why it is not possible to do so.

9. Write $\begin{pmatrix} -1 & 3 \\ 5 & -2 \end{pmatrix}$ as a linear combination of the matrices $\begin{pmatrix} 1 & 0 \\ 1 & -1 \end{pmatrix}$, $\begin{pmatrix} 2 & 1 \\ 0 & 1 \end{pmatrix}$, and $\begin{pmatrix} -1 & 2 \\ 1 & 1 \end{pmatrix}$.

10. Find vectors that span $\mathcal{N}(B)$ the nullspace of the matrix $B = \begin{pmatrix} 1 & 1 & 1 & -2 \\ 2 & 3 & 0 & -1 \end{pmatrix}$.

11. Find vectors that span $\mathcal{N}(C)$ the nullspace of the matrix $C = \begin{pmatrix} 1 & -1 & 0 & 2 \\ 2 & 1 & 1 & 0 \\ 3 & 0 & 1 & 2 \end{pmatrix}$.

12. Find vectors that span $\mathcal{R}(C)$, the range of the matrix $C = \begin{pmatrix} 1 & -1 & 0 & 2 \\ 2 & 1 & 1 & 0 \\ 3 & 0 & 1 & 2 \end{pmatrix}$.

13. In answering a question on her linear algebra homework, April claimed that the subspace $\mathcal{W}$ is spanned by the set $u_1 = (1, 0, 1)$ and $u_2 = (0, 1, -1)$. Michelle claimed that the subspace $\mathcal{W}$ is spanned by $v_1 = (1, 1, 0)$, $v_2 = (2, 1, 1)$, and $v_3 = (1, -1, 2)$. Do their answers agree with each other, that is, is the subspace spanned by the set $\{u_1, u_2\}$ the same as the subspace spanned by $\{v_1, v_2, v_3\}$?

14. In answering a question on his linear algebra homework, Max claimed that the subspace $\mathcal{U}$ is spanned by the set $u_1 = (1, 0, 1, 1)$, $u_2 = (0, 1, -1, 0)$ and $u_3 = (0, 0, 1, 2)$. Spike claimed that the subspace $\mathcal{U}$ is spanned by $v_1 = (1, 1, 0, 1)$, $v_2 = (2, 1, 1, 2)$, and $v_3 = (1, -1, 2, 1)$. Do they agree with each other, that is, is the subspace spanned by the set $\{u_1, u_2, u_3\}$ the same as the subspace spanned by $\{v_1, v_2, v_3\}$?

15. Let $B = \begin{pmatrix} 1 & 1 & -1 \\ 2 & -1 & 2 \\ -1 & 0 & 1 \end{pmatrix}$.

    (a) Taking $X = (1, 1, -2)$ shows that $(4, -3, -3) = BX$ is in the range of B. Show that $(4, -3, -3)$ is a linear combination of $(1, 2, -1)$, $(1, -1, 0)$, and $(-1, 2, 1)$.

    (b) The vector $v = (1, 2, -1) + 2(1, -1, 0) - (-1, 2, 1) = (4, -2, -2)$ is a linear combination of the columns of B. Find X so that $BX = v$.

    (c) Is the vector $(-2, 11, 1)$ in the subspace spanned by $(1, 2, -1)$, $(1, -1, 0)$, and $(-1, 2, 1)$?

16. Let $A = \begin{pmatrix} 1 & 0 & 1 \\ 3 & 1 & 2 \\ -1 & 1 & -2 \end{pmatrix}$

    (a) Taking $X = (1, 1, -2)$ shows that $(-1, 0, 4) = AX$ is in the range of A. Show that $(-1, 0, 4)$ is a linear combination of $(1, 3, -1)$, $(0, 1, 1)$, and $(1, 2, -2)$.
    (b) The vector $v = 2(1, 3, -1) + (0, 1, 1) - 3(1, 2, -2) = (-1, 1, 5)$ is a linear combination of the columns of A. Find X so that $AX = v$.
    (c) Is the vector $(3, 7, -5)$ in the subspace spanned by $(1, 3, -1)$, $(0, 1, 1)$, and $(1, 2, -2)$?
    (d) Show that $(1, 2, -2)$ is a linear combination of $(1, 3, -1)$ and $(0, 1, 1)$.
    (e) Show that $\mathcal{R}(A)$ is the subspace spanned by $(1, 3, -1)$ and $(0, 1, 1)$.

17.

$$\text{Let } E = \begin{pmatrix} 2 & 1 & 0 & 1 & -2 \\ 0 & 1 & 1 & 2 & 3 \\ -1 & 1 & 1 & -2 & 1 \\ 1 & 3 & 2 & 1 & 2 \end{pmatrix}$$

    (a) Find vectors that span $\mathcal{N}(E)$ the nullspace of E.
    (b) Find vectors that span $\mathcal{R}(E)$ the range of E.
    (c) Find vectors that span $\mathcal{N}(E')$ the nullspace of E'.
    (d) Find vectors that span $\mathcal{R}(E')$ the range of E'.

3.4 Linear Independence

Notice that 0 is a linear combination of any set of vectors, $0v_1 + 0v_2 + \cdots + 0v_k = 0$, but this is a trivial way of getting 0 as a linear combination of the vectors.

Sometimes there are other ways 0 can be written as a linear combination of some vectors, for example,

$$2\begin{pmatrix} 1 \\ -2 \\ 1 \end{pmatrix} + 3\begin{pmatrix} 0 \\ 1 \\ -1 \end{pmatrix} - \begin{pmatrix} 2 \\ -1 \\ -1 \end{pmatrix} = \begin{pmatrix} 0 \\ 0 \\ 0 \end{pmatrix}$$

as well as

$$0\begin{pmatrix} 1 \\ -2 \\ 1 \end{pmatrix} + 0\begin{pmatrix} 0 \\ 1 \\ -1 \end{pmatrix} + 0\begin{pmatrix} 2 \\ -1 \\ -1 \end{pmatrix} = \begin{pmatrix} 0 \\ 0 \\ 0 \end{pmatrix}$$

Other times, the only way 0 can be written as a linear combination of some vectors is in the trivial way, for example, if

$$\alpha \begin{pmatrix} 1 \\ -1 \end{pmatrix} + \beta \begin{pmatrix} 0 \\ 1 \end{pmatrix} = \begin{pmatrix} 0 \\ 0 \end{pmatrix}$$

then $\alpha = 0$ from the first component, so $\beta = 0$ from the second component. Thus, the only way 0 is a linear combination of $(1, -1)$ and $(0, 1)$ is in the trivial way. It is this distinction that the following definition addresses.

DEFINITION *The set of vectors* $\{v_1, v_2, \cdots, v_k\}$ *is said to be* linearly independent *if the only way in which*

$$c_1 v_1 + c_2 v_2 + \cdots + c_k v_k = 0$$

is for $c_1 = 0,\ c_2 = 0,\ \cdots,$ *and* $c_k = 0$. *A set of vectors that is not linearly independent is said to be* linearly dependent, *that is, a set of vectors in linearly dependent if there is an equation of the form*

$$c_1 v_1 + c_2 v_2 + \cdots + c_k v_k = 0$$

with some of the c_j*'s not zero.*

Less formally, we occasionally say $v_1,\ v_2,\ \cdots,\ v_k$ are linearly independent instead of saying the set $\{v_1, v_2, \cdots, v_k\}$ is linearly independent.

EXAMPLE 3.16

Show that if $u = (1, 1, 2)$, $v = (1, -1, 0)$, and $w = (1, 1, -1)$ the set $\{u, v, w\}$ is linearly independent in $\mathbf{R}^3$.

SOLUTION We must show that the only linear combination of u, v, and w that gives 0 is the trivial linear combination. Suppose a, b, and c satisfy $au + bv + cw = 0$, that is,

$$a \begin{pmatrix} 1 \\ 1 \\ 2 \end{pmatrix} + b \begin{pmatrix} 1 \\ -1 \\ 0 \end{pmatrix} + c \begin{pmatrix} 1 \\ 1 \\ -1 \end{pmatrix} = \begin{pmatrix} 0 \\ 0 \\ 0 \end{pmatrix}$$

Solving the system, we get

$$\left(\begin{array}{ccc|c} 1 & 1 & 1 & 0 \\ 1 & -1 & 1 & 0 \\ 2 & 0 & -1 & 0 \end{array} \right) \rightarrow \left(\begin{array}{ccc|c} 1 & 1 & 1 & 0 \\ 0 & 2 & 0 & 0 \\ 0 & -2 & -3 & 0 \end{array} \right)$$

$$\rightarrow \begin{pmatrix} 1 & 1 & 1 & | & 0 \\ 0 & 1 & 0 & | & 0 \\ 0 & -2 & -3 & | & 0 \end{pmatrix} \rightarrow \begin{pmatrix} 1 & 0 & 1 & | & 0 \\ 0 & 1 & 0 & | & 0 \\ 0 & 0 & -3 & | & 0 \end{pmatrix}$$

$$\rightarrow \begin{pmatrix} 1 & 0 & 1 & | & 0 \\ 0 & 1 & 0 & | & 0 \\ 0 & 0 & 1 & | & 0 \end{pmatrix} \rightarrow \begin{pmatrix} 1 & 0 & 0 & | & 0 \\ 0 & 1 & 0 & | & 0 \\ 0 & 0 & 1 & | & 0 \end{pmatrix}$$

which means $a = 0$, $b = 0$, and $c = 0$ is the only solution, that is, the set of vectors $\{u, v, w\}$ is linearly independent. ☐

EXAMPLE 3.17

Are the vectors $v_1 = (1, 1, 1, -1)$, $v_2 = (1, 1, -1, 0)$, and $v_3 = (1, 1, 3, -2)$ linearly independent in $\mathbf{R}^4$?

SOLUTION We must decide whether the only linear combination of v_1, v_2, and v_3 that gives 0 is the trivial linear combination. Suppose a, b, and c satisfy $av_1 + bv_2 + cv_3 = 0$, that is,

$$a \begin{pmatrix} 1 \\ 1 \\ 1 \\ -1 \end{pmatrix} + b \begin{pmatrix} 1 \\ 1 \\ -1 \\ 0 \end{pmatrix} + c \begin{pmatrix} 1 \\ 1 \\ 3 \\ -2 \end{pmatrix} = \begin{pmatrix} 0 \\ 0 \\ 0 \\ 0 \end{pmatrix}$$

Solving the system, we get

$$\begin{pmatrix} 1 & 1 & 1 & | & 0 \\ 1 & 1 & 1 & | & 0 \\ 1 & -1 & 3 & | & 0 \\ -1 & 0 & -2 & | & 0 \end{pmatrix} \rightarrow \begin{pmatrix} 1 & 1 & 1 & | & 0 \\ 0 & 0 & 0 & | & 0 \\ 0 & -2 & 2 & | & 0 \\ 0 & 1 & -1 & | & 0 \end{pmatrix} \rightarrow \begin{pmatrix} 1 & 0 & 2 & | & 0 \\ 0 & 1 & -1 & | & 0 \\ 0 & 0 & 0 & | & 0 \\ 0 & 0 & 0 & | & 0 \end{pmatrix}$$

which means $a = -2$, $b = 1$, and $c = 1$ is the one of infinitely many solutions, that is, the vectors v_1, v_2, and v_3 are linearly dependent. ☐

EXAMPLE 3.18

Suppose $\{v_1, v_2, v_3\}$ is a linearly dependent set in the subspace $\mathcal{W}$. If v_4 is another vector in $\mathcal{W}$, is $\{v_1, v_2, v_3, v_4\}$ also a linearly dependent set?

SOLUTION Yes! To say that v_1, v_2, and v_3 are linearly dependent means that there are constants, not all zero, so that

$$c_1 v_1 + c_2 v_2 + c_3 v_3 = 0 \qquad (3.4.1)$$

On the other hand, if this is so, then

$$c_1 v_1 + c_2 v_2 + c_3 v_3 + 0 v_4 = 0$$

Now these constants are not all zero (for example, if $c_2 \neq 0$ in Equation (3.4.1), then $c_2 \neq 0$ in this equation, too), so $\{v_1, v_2, v_3, v_4\}$ also a linearly dependent set.
$\square$

EXAMPLE 3.19

Suppose A is a 6×5 matrix and that $v = (1, -2, 1, 0, 3)$ is in the nullspace of A. Show that the columns of A are linearly dependent in $\mathbf{R}^6$.

SOLUTION Let A_1, A_2, A_3, A_4, and A_5 be the columns of A. In other words, blocking the matrix A by columns, we have $A = \begin{pmatrix} A_1 & A_2 & A_3 & A_4 & A_5 \end{pmatrix}$. The fact that $v = (1, -2, 1, 0, 3)$ is in the nullspace of A means $Av = 0$, or

$$\begin{pmatrix} A_1 & A_2 & A_3 & A_4 & A_5 \end{pmatrix} \begin{pmatrix} 1 \\ -2 \\ 1 \\ 0 \\ 3 \end{pmatrix} = 0$$

Multiplying this out, we get

$$(A_1)(1) + (A_2)(-2) + (A_3)(1) + (A_4)(0) + (A_5)(3) = 0$$

or, since multiplying by scalars can be done in any order, that

$$A_1 - 2A_2 + A_3 + 0A_4 + 3A_5 = 0$$

Since not all of the coefficients are zero, this says that the columns of A are linearly dependent. Note that this gives an interesting connection between the range of A and the nullspace of A, two subspaces of different spaces that do not have obvious connections. $\square$

Linear dependence leads to redundancy of linear combinations as the following example illustrates.

EXAMPLE 3.20

The vectors $v_1 = (1, -1, 2)$, $v_2 = (1, 1, -1)$, and $v_3 = (2, -4, 7)$ are linearly dependent because $3v_1 - v_2 - v_3 = 0$. The vector $w = (2, -6, 10)$ is a linear combination of v_1, v_2, and v_3 because, for example, $w = v_1 - v_2 + v_3$. In fact, because the v's are linearly dependent, w can be written as a linear combination of the v's in infinitely many different ways. We see that $w = -2v_1 + 2v_3$ and $w = 4v_1 - 2v_2$. Indeed,

$$w = (v_1 - v_2 + v_3) + \alpha(3v_1 - v_2 - v_3)$$

for any number α, and in this case, it can be shown that this is the only way. $\square$

The following theorem gives another aspect of the fact that linear dependence is associated with redundancy in forming linear combinations. It shows that if a set of vectors is linearly dependent, then every linear combination of vectors in that set is also a linear combination of vectors from a smaller set.

THEOREM 3.21

The set $\{v_1, v_2, \cdots, v_k\}$ is linearly dependent if and only if one of the vectors in the set is a linear combination of the rest.

PROOF Suppose v_j is a linear combination of the other vectors, say

$$v_j = a_1 v_1 + \cdots + a_{j-1} v_{j-1} + a_{j+1} v_{j+1} + \cdots + a_k v_k$$

Then

$$a_1 v_1 + \cdots + a_{j-1} v_{j-1} + (-1) v_j + a_{j+1} v_{j+1} + \cdots + a_k v_k = 0$$

Since not all the coefficients in this linear combination are zero (the coefficient of v_j is -1), the vectors are linearly dependent.

Conversely, if the vectors are linearly dependent, there is a non-trivial linear combination of the vectors that gives zero, say

$$a_1 v_1 + \cdots + a_{j-1} v_{j-1} + a_j v_j + a_{j+1} v_{j+1} + \cdots + a_k v_k = 0$$

where $a_j \neq 0$. Then, we can solve for v_j to get

$$v_j = -\frac{a_1}{a_j} v_1 - \cdots - \frac{a_{j-1}}{a_j} v_{j-1} - \frac{a_{j+1}}{a_j} v_{j+1} - \cdots - \frac{a_k}{a_j} v_k$$

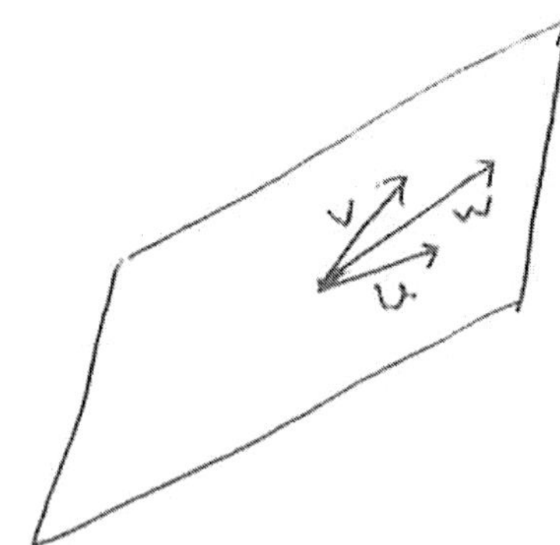

FIGURE 3.5

Span of dependent vectors in $\mathbf{R}^3$

which gives v_j as a linear combination of the others.   ∎

Notice that the theorem does not say that each of the vectors is a linear combination of the others, it says *some* vector is. The proof shows that the vectors that can be expressed as linear combinations of the others are the ones with non-zero coefficient in a linear combination that gives zero.

EXAMPLE 3.22

If three vectors u, v, and w in $\mathbf{R}^3$ are dependent, then Theorem 3.21 says one of the vectors is a linear combination of the others. While we might expect the span of three vectors in $\mathbf{R}^3$ is all of $\mathbf{R}^3$, the fact one is a linear combination of the rest means that, really, we only need two vectors to span the subspace: so it is a plane through the origin as in Figure 3.5.   ☐

EXAMPLE 3.23

Decide if the set

$$\left\{ \begin{pmatrix} 1 \\ -1 \\ 2 \end{pmatrix}, \begin{pmatrix} 1 \\ -2 \\ 3 \end{pmatrix}, \begin{pmatrix} 2 \\ -1 \\ 2 \end{pmatrix} \right\}$$

is linearly independent or dependent. If it is a dependent set, write one of the vectors as a linear combination of the rest.

SOLUTION We need to decide how many solutions the system

$$c_1 \begin{pmatrix} 1 \\ -1 \\ 2 \end{pmatrix} + c_2 \begin{pmatrix} 1 \\ -2 \\ 3 \end{pmatrix} + c_3 \begin{pmatrix} 2 \\ -1 \\ 2 \end{pmatrix} = \begin{pmatrix} 0 \\ 0 \\ 0 \end{pmatrix}$$

has. This is really a homogeneous system of three equations in three unknowns; setting it up as an augmented matrix, we get

$$\left(\begin{array}{ccc|c} 1 & 1 & 2 & 0 \\ -1 & -2 & -1 & 0 \\ 2 & 3 & 2 & 0 \end{array} \right) \longrightarrow \left(\begin{array}{ccc|c} 1 & 1 & 2 & 0 \\ 0 & -1 & 1 & 0 \\ 0 & 1 & -2 & 0 \end{array} \right)$$

$$\longrightarrow \left(\begin{array}{ccc|c} 1 & 1 & 2 & 0 \\ 0 & -1 & 1 & 0 \\ 0 & 0 & -1 & 0 \end{array} \right) \longrightarrow \cdots \longrightarrow \left(\begin{array}{ccc|c} 1 & 0 & 0 & 0 \\ 0 & -1 & 0 & 0 \\ 0 & 0 & -1 & 0 \end{array} \right)$$

so the only solution of the system is $c_1 = 0$, $c_2 = 0$, and $c_3 = 0$. Thus, the vectors are linearly independent in $\mathbf{R}^3$ and none of them can be written as a linear combination of the others. $\quad\Box$

EXAMPLE 3.24

Decide if the set

$$\left\{ \begin{pmatrix} 1 \\ 1 \\ 2 \\ -1 \end{pmatrix}, \begin{pmatrix} -1 \\ -2 \\ -1 \\ 3 \end{pmatrix}, \begin{pmatrix} 2 \\ 1 \\ 5 \\ 0 \end{pmatrix}, \begin{pmatrix} -1 \\ -3 \\ 2 \\ 12 \end{pmatrix} \right\}$$

is linearly independent or dependent. If it is a dependent set, write one of the vectors as a linear combination of the rest.

SOLUTION We need to decide how many solutions there are to the system

$$c_1 \begin{pmatrix} 1 \\ 1 \\ 2 \\ -1 \end{pmatrix} + c_2 \begin{pmatrix} -1 \\ -2 \\ -1 \\ 3 \end{pmatrix} + c_3 \begin{pmatrix} 2 \\ 1 \\ 5 \\ 0 \end{pmatrix} + c_4 \begin{pmatrix} -1 \\ -3 \\ 2 \\ 12 \end{pmatrix} = \begin{pmatrix} 0 \\ 0 \\ 0 \\ 0 \end{pmatrix}$$

Setting up the augmented matrix to solve this system of four equations in four unknowns we get

$$
\left(\begin{array}{cccc|c}
1 & -1 & 2 & -1 & 0 \\
1 & -2 & 1 & -3 & 0 \\
2 & -1 & 5 & 2 & 0 \\
-1 & 3 & 0 & 12 & 0
\end{array}\right)
\longrightarrow
\left(\begin{array}{cccc|c}
1 & -1 & 2 & -1 & 0 \\
0 & -1 & -1 & -2 & 0 \\
0 & 1 & 1 & 4 & 0 \\
0 & 2 & 2 & 11 & 0
\end{array}\right)
$$

$$
\longrightarrow
\left(\begin{array}{cccc|c}
1 & -1 & 2 & -1 & 0 \\
0 & -1 & -1 & -2 & 0 \\
0 & 0 & 0 & 2 & 0 \\
0 & 0 & 0 & 7 & 0
\end{array}\right)
\longrightarrow
\left(\begin{array}{cccc|c}
1 & -1 & 2 & 0 & 0 \\
0 & -1 & -1 & 0 & 0 \\
0 & 0 & 0 & 1 & 0 \\
0 & 0 & 0 & 0 & 0
\end{array}\right)
$$

$$
\longrightarrow
\left(\begin{array}{cccc|c}
1 & 0 & 3 & 0 & 0 \\
0 & 1 & 1 & 0 & 0 \\
0 & 0 & 0 & 1 & 0 \\
0 & 0 & 0 & 0 & 0
\end{array}\right)
$$

So there are infinitely many solutions to the system which means that the set of vectors is linearly dependent. In particular, the general solution of the system is $c_1 = -3c_3$, $c_2 = -c_3$, $c_4 = 0$ and c_3 is arbitrary. For example, choosing $c_3 = 1$, we get

$$
-3\begin{pmatrix}1 \\ 1 \\ 2 \\ -1\end{pmatrix} - \begin{pmatrix}-1 \\ -2 \\ -1 \\ 3\end{pmatrix} + \begin{pmatrix}2 \\ 1 \\ 5 \\ 0\end{pmatrix} + 0\begin{pmatrix}-1 \\ -3 \\ 2 \\ 12\end{pmatrix} = \begin{pmatrix}0 \\ 0 \\ 0 \\ 0\end{pmatrix}
$$

which shows their dependence because not all the coefficients in this linear combination are zero.

According to Theorem 3.21, one of the vectors can be written as a linear combination of the rest, for example, solving for the third vector, we get

$$
\begin{pmatrix}2 \\ 1 \\ 5 \\ 0\end{pmatrix} = 3\begin{pmatrix}1 \\ 1 \\ 2 \\ -1\end{pmatrix} + \begin{pmatrix}-1 \\ -2 \\ -1 \\ 3\end{pmatrix} + 0\begin{pmatrix}-1 \\ -3 \\ 2 \\ 12\end{pmatrix}
$$

or we could have written the first vector as

$$
\begin{pmatrix}1 \\ 1 \\ 2 \\ -1\end{pmatrix} = -\frac{1}{3}\begin{pmatrix}-1 \\ -2 \\ -1 \\ 3\end{pmatrix} + \frac{1}{3}\begin{pmatrix}2 \\ 1 \\ 5 \\ 0\end{pmatrix} + 0\begin{pmatrix}-1 \\ -3 \\ 2 \\ 12\end{pmatrix}
$$

We could also have written the second as a linear combination of the rest, but the fourth vector, $(-1, -3, 2, 12)$ is not a linear combination of the other three. The proof of shows why this is the case: to solve for the last vector, we would need to divide by the coefficient it has in linear combination for the zero vector, but this cannot be done because that coefficient is 0 for the fourth vector. ⬚

According to Theorem 3.21, if we have just two vectors, they are linearly dependent exactly when one of the vectors is a linear combination of the other, that is, one of the vectors is a multiple of the other. This means it is usually very easy to tell when two vectors are independent: $(1, -2)$ and $(-3, 6)$ are dependent because $(-3, 6) = -3(1, -2)$ whereas $(1, -2)$ and $(2, 3)$ are independent because $(2, 3) \neq \alpha(1, -2)$ and $(1, -2) \neq \beta(2, 3)$ for any numbers α or β. Geometrically, two vectors are linearly independent if they are pointing in different directions, that is, if they are not in the same line. Three vectors are considerably more difficult. We saw above that $(1, -2, 1)$, $(0, 1, -1)$, and $(2, -1, -1)$ are linearly dependent: $(2, -1, -1) = 2(1, -2, 1) + 3(0, 1, -1)$, a relationship that takes some effort to discover. Geometrically, three vectors are linearly dependent if one of the vectors is in the plane spanned by the other two, and they are independent if each vector points out of the plane determined by the other two.

Exercises 3.4

1. In doing his linear algebra homework, Eduardo found that if $u = (2, 1, -3)$, $v = (1, -1, 2)$, and $w = (1, 5, -12)$ then $2u - 3v - w = 0$. Does this calculation show that u, v, and w form a linearly dependent set, that they form a linearly independent set, or does it not show either of these?

2. In her linear algebra homework, Pauline was given that $p = (1, 1, -3)$, $q = (2, -1, 2)$, and $r = (1, -2, 5)$. She noticed that $0p + 0q + 0r = 0$. Does this calculation show that p, q, and r form a linearly dependent set, that they form a linearly independent set, or does it not show either of these?

3. In doing their linear algebra homework, Abbie and John found that if $x = (2, -1, 1)$, $y = (1, 0, 2)$, and $z = (1, 1, -3)$ then the system $c_1 x + c_2 y + c_3 z = 0$ has only one solution, namely, $c_1 = 0$, $c_2 = 0$, and $c_3 = 0$. Does this calculation show that x, y, and z form a linearly dependent set, that they form a linearly independent set, or does it not show either of these?

Decide if the following sets of vectors are linearly dependent or independent. If independent, prove that they are, if dependent, find a non-trivial linear combination of the vectors that gives zero.

4. $\{(2,3,1),\ (-1,1,2),\ (0,1,1)\}$

5. $\{(-1,2,1),\ (1,1,3),\ (1,0,1)\}$

6. $\{(1,-1,1),\ (2,1,-2),\ (3,-1,1),\ (1,4,-2)\}$

7. $\{(1,1,-1,1),\ (2,1,-1,2),\ (1,-1,1,1),\ (1,-2,2,1)\}$

8. $\{(1,1,-2,-1),\ (-1,0,4,2),\ (0,2,5,5),\ (1,2,1,2)\}$

9. $\{(0,1,1,-1),\ (1,3,1,-2),\ (2,1,0,-3),\ (3,1,-1,2),\ (2,-1,2,0)\}$

10. $\{(1,1,-1,2),\ (3,-1,1,1),\ (2,0,-1,1),\ (0,2,-3,2)\}$

11. $\left\{ \begin{pmatrix} 1 & 0 \\ 1 & -1 \end{pmatrix},\ \begin{pmatrix} 2 & -1 \\ -1 & 1 \end{pmatrix},\ \begin{pmatrix} 6 & 1 \\ -1 & 5 \end{pmatrix},\ \begin{pmatrix} 0 & 2 \\ 1 & 1 \end{pmatrix} \right\}$

Decide if the following sets of functions are linearly dependent or independent.

12. $\{x, x^2, x^3\}$
13. $\{\sin(x), \cos(x), \sin(2x), \cos(2x)\}$
14. $\{\sin^2(x), \cos^2(x), 1\}$
15. In doing his linear algebra homework, Eduardo found that if $u = (2,1,-3)$, $v = (1,-1,2)$, and $w = (1,5,-12)$, then u, v, and w form a linearly dependent set in $\mathbf{R}^3$ because $2u - 3v - w = 0$. Write one of these vectors as a linear combination of the rest.
16. The vectors $v_1 = (1,-1,2)$, $v_2 = (-1,2,-3)$, $v_3 = (1,1,-1)$, and $v_4 = (-2,3,-4)$ are linearly dependent in $\mathbf{R}^3$. Write one of the vectors as a linear combination of the rest.
17. The vector $u = (8,-7,11)$ is a linear combination of the vectors $v_1 = (1,-1,2)$, $v_2 = (-1,2,-3)$, $v_3 = (1,1,-1)$, and $v_4 = (-2,3,-4)$ of Exercise 3.4.16., namely, $u = v_1 - v_2 + 2v_3 - 2v_4$. Use your answer to that exercise to write u as a linear combination of only three of the v's.
18. The vectors $w_1 = (1,2,-1,1)$, $w_2 = (-1,-3,2,-4)$, $w_3 = (1,0,1,-5)$, and $w_4 = (2,5,-3,4)$ are linearly dependent in $\mathbf{R}^4$. Write one of the vectors as a linear combination of the rest.
19. The vector $z = (2,2,0,-3)$ is a linear combination of the vectors $w_1 = (1,2,-1,1)$,

$w_2 = (-1, -3, 2, -4)$, $w_3 = (1, 0, 1, -5)$, and $w_4 = (2, 5, -3, 4)$ of Exercise 3.4.18., namely, $z = 2w_1 - w_2 + w_3 - w_4$. Use your answer to that exercise to write z as a linear combination of only three of the w's.

20. Suppose v_1 and v_2 are linearly independent in $\mathbf{R}^4$. Show that if v_3 is a vector in $\mathbf{R}^4$ that is not a linear combination of v_1 and v_2, then $\{v_1, v_2, v_3\}$ is a linearly independent set in $\mathbf{R}^4$.

21. Let A be an $m \times n$ matrix and let $v_1, v_2, \ldots, v_k$ be vectors in $\mathbf{R}^n$. Show that if Av_1, $Av_2, \ldots, Av_k$ are linearly independent vectors in $\mathbf{R}^m$, then $v_1, v_2, \ldots, v_k$ are linearly independent in $\mathbf{R}^n$.

22.(a) Let A be an invertible $m \times m$ matrix and let $v_1, v_2, \ldots, v_k$ be vectors in $\mathbf{R}^m$. Show that if $v_1, v_2, \ldots, v_k$ are linearly independent vectors in $\mathbf{R}^m$, then $Av_1, Av_2, \ldots, Av_k$ are also linearly independent.

(b) Find an example of a non-invertible $m \times m$ matrix B and linearly independent vectors $w_1, w_2, \ldots, w_k$ so that $Bw_1, Bw_2, \ldots, Bw_k$ are linearly dependent.

3.5 Basis and Coordinates

As we saw in the previous sections, we expect to describe subspaces in terms of a set of vectors that span the subspace. We would like to express the subspace as efficiently as possible, not using more vectors than we really need. If one of the vectors in a spanning set can be expressed as a linear combination of the others, that vector is unnecessary. Theorem 3.21 shows that this redundancy is associated with linear dependence.

EXAMPLE 3.25

Consider the matrix in Exercise 3.3.16.

$$A = \begin{pmatrix} 1 & 0 & 1 \\ 3 & 1 & 2 \\ -1 & 1 & -2 \end{pmatrix}$$

In that exercise, you showed that the third column, $(1, 2, -2)$, is a linear combination of the first two columns, $(1, 3, -1)$ and $(0, 1, 1)$. For example, $(1, 2, -2) = (1, 3, -1) - (0, 1, 1)$. As in the proof above, we can rewrite this equation to show that the columns are linearly dependent:

$$-\begin{pmatrix} 1 \\ 3 \\ -1 \end{pmatrix} + \begin{pmatrix} 0 \\ 1 \\ 1 \end{pmatrix} + \begin{pmatrix} 1 \\ 2 \\ -2 \end{pmatrix} = \begin{pmatrix} 0 \\ 0 \\ 0 \end{pmatrix}$$

We also saw in that exercise that $(-1, 0, 4)$ is in the range of A because it is a linear combination of the columns of A

$$\begin{pmatrix} -1 \\ 0 \\ 4 \end{pmatrix} = \begin{pmatrix} 1 \\ 3 \\ -1 \end{pmatrix} + \begin{pmatrix} 0 \\ 1 \\ 1 \end{pmatrix} - 2 \begin{pmatrix} 1 \\ 2 \\ -2 \end{pmatrix}$$

On the other hand, because the third column of A is a linear combination of the first two, we can replace it in the expression above to express $(-1, 0, 4)$ more efficiently as a linear combination of only two of the columns of A:

$$\begin{pmatrix} -1 \\ 0 \\ 4 \end{pmatrix} = \begin{pmatrix} 1 \\ 3 \\ -1 \end{pmatrix} + \begin{pmatrix} 0 \\ 1 \\ 1 \end{pmatrix} - 2(\begin{pmatrix} 1 \\ 3 \\ -1 \end{pmatrix} - \begin{pmatrix} 0 \\ 1 \\ 1 \end{pmatrix}) = -\begin{pmatrix} 1 \\ 3 \\ -1 \end{pmatrix} + 3\begin{pmatrix} 0 \\ 1 \\ 1 \end{pmatrix}$$

The following definition gives language to talk about the situation in which we have a non-redundant spanning set.

DEFINITION *If* $\mathcal{W}$ *is a vector space, we say* $\{w_1, w_2, \cdots, w_n\}$ *is a* basis *for* $\mathcal{W}$ *if* $w_1, w_2, \cdots, w_n$ *span* $\mathcal{W}$ *and the set* $\{w_1, w_2, \cdots, w_n\}$ *is linearly independent.*

EXAMPLE 3.26

The *standard basis* or *usual basis* for $\mathbf{R}^n$ is the set of vectors $e_1 = (1, 0, \cdots, 0)$, $e_2 = (0, 1, 0, \cdots, 0)$, $\cdots$, $e_n = (0, \cdots, 0, 1)$, that is, e_j is the vector in $\mathbf{R}^n$ that has 1 as its jth component, and 0 for the other components. To see that these vectors are linearly independent, note that the vector equation $c_1 e_1 + c_2 e_2 + \cdots + c_n e_n = 0$ becomes the system $c_1 = 0$, $c_2 = 0$, $\cdots$, $c_n = 0$ when the vectors on the left are added and the components are equated. To see that they span, if $v = (a_1, a_2, \cdots, a_n)$, then clearly, $v = a_1 e_1 + a_2 e_2 + \cdots + a_n e_n$ so v is a linear combination of e_1, e_2, $\cdots$, e_n. Thus every vector in $\mathbf{R}^n$ is a linear combination these vectors and they span $\mathbf{R}^n$. □

EXAMPLE 3.27

The vectors $(1, -1)$ and $(0, 1)$ are a basis for $\mathbf{R}^2$. We saw at the beginning of the section that these vectors are linearly independent. Moreover, they span $\mathbf{R}^2$, for if $v = (a, b)$ then

$$a \begin{pmatrix} 1 \\ -1 \end{pmatrix} + (b + a) \begin{pmatrix} 0 \\ 1 \end{pmatrix} = \begin{pmatrix} a \\ -a + b + a \end{pmatrix} = \begin{pmatrix} a \\ b \end{pmatrix} = v$$

so every vector in $\mathbf{R}^2$ is a linear combination of these vectors. ☐

This example illustrates that there are many bases for every vector space. We will see that some bases are more convenient than others for certain problems and we would like to choose bases that are convenient for our problem that have other nice properties.

To find a basis for a subspace, we first need a way to describe the subspace, to find a spanning set, and ultimately to find a spanning set that is also linearly independent. Finding a basis for the range of a matrix is not usually difficult because a spanning set is already implicit.

EXAMPLE 3.28

Find a basis for the range of the matrix A where

$$A = \begin{pmatrix} 1 & 1 & 2 \\ -1 & 1 & -4 \\ 2 & -1 & 7 \end{pmatrix}$$

SOLUTION We need to find a set of vectors that spans $\mathcal{R}(A)$ and is linearly independent. We know that the columns of A, $v_1 = (1, -1, 2)$, $v_2 = (1, 1, -1)$, and $v_3 = (2, -4, 7)$, are a spanning set for $\mathcal{R}(A)$. However, we saw in Example 3.20 that the columns of A are linearly dependent: indeed, $3v_1 - v_2 - v_3 = 0$. This means that the set $\{v_1, v_2, v_3\}$ is *not* a basis for $\mathcal{R}(A)$. On the other hand, since $v_3 = 3v_1 - v_2$, the subspace spanned by v_1 and v_2 is the same as the subspace spanned by v_1, v_2, and v_3. In other words, v_1 and v_2 span $\mathcal{R}(A)$. Moreover, v_1 and v_2 are linearly independent (since neither is a multiple of the other), so $\{v_1, v_2\}$ is a basis for $\mathcal{R}(A)$. ☐

EXAMPLE 3.29

Find a basis for $\mathcal{N}(B)$ the nullspace of the matrix $B = \begin{pmatrix} 1 & -1 & 2 & 1 \\ 2 & -1 & 1 & 2 \\ 1 & 0 & -1 & 1 \end{pmatrix}$

SOLUTION The nullspace of B is the set of vectors in $\mathbf{R}^4$ that satisfy $BX = 0$, so first we solve this system.

$$\left(\begin{array}{cccc|c} 1 & -1 & 2 & 1 & 0 \\ 2 & -1 & 1 & 2 & 0 \\ 1 & 0 & -1 & 1 & 0 \end{array}\right) \longrightarrow \left(\begin{array}{cccc|c} 1 & -1 & 2 & 1 & 0 \\ 0 & 1 & -3 & 0 & 0 \\ 0 & 1 & -3 & 0 & 0 \end{array}\right)$$

$$\longrightarrow \left(\begin{array}{cccc|c} 1 & 0 & -1 & 1 & 0 \\ 0 & 1 & -3 & 0 & 0 \\ 0 & 0 & 0 & 0 & 0 \end{array}\right)$$

so the system $BX = 0$ is equivalent to the system $w - y + z = 0$ and $x - 3y = 0$, or $w = y - z$ and $x = 3y$. In other words, X is a solution of the system if and only if $X = (y - z, 3y, y, z)$ for some numbers y and z. This means X is in $\mathcal{N}(B)$ if and only if

$$X = \begin{pmatrix} y - z \\ 3y \\ y \\ z \end{pmatrix} = y \begin{pmatrix} 1 \\ 3 \\ 1 \\ 0 \end{pmatrix} + z \begin{pmatrix} -1 \\ 0 \\ 0 \\ 1 \end{pmatrix}$$

which says $\mathcal{N}(B)$ is spanned by the vectors $(1, 3, 1, 0)$ and $(-1, 0, 0, 1)$. Moreover, these vectors are linearly independent, so $\{(1, 3, 1, 0), (-1, 0, 0, 1)\}$ is a basis for $\mathcal{N}(B)$.   ▯

    In finding a basis for the nullspace of a matrix, we must begin by solving the homogeneous system with the given matrix as coefficient matrix. If we solve the system by Gaussian elimination, ending with a row equivalent matrix in reduced row echelon form, the solution of the system can be read off by writing the pivot variables in terms of the free variables. As we saw above, it is easy to choose a basis for the nullspace so that each free variable corresponds to a basis vector.

    The definition of basis says that every vector in the space can be expressed as a linear combination of the basis vectors. In fact, this can be done in only one way, as the following theorem shows.

THEOREM 3.30

If v_1, v_2, $\cdots$, v_k is a basis for $\mathcal{W}$, then each vector in $\mathcal{W}$ can be expressed as a linear combination of the basis vectors in exactly one way.

PROOF   A basis is a spanning set, so every vector in $\mathcal{W}$ can be expressed as a linear combination of the basis vectors. Suppose the vector w can be expressed in two ways:

$$w = a_1 v_1 + a_2 v_2 + \cdots + a_k v_k = b_1 v_1 + b_2 v_2 + \cdots + b_k v_k$$

Then

$$(a_1 - b_1)v_1 + (a_2 - b_2)v_2 + \cdots + (a_k - b_k)v_k = w - w = 0$$

Since the v's are a basis they are linearly independent so the coefficients in this linear combination must be zero. That is, $a_1 = b_1$, $a_2 = b_2$, $\cdots$, and $a_k = b_k$, so the two expressions are really the same expression twice: in other words, there is only one way to express w as a linear combination of these basis vectors.   ∎

This leads to an extension of the idea of coordinates, generalizing the relationship between the coordinates of a vector in $\mathbf{R}^n$ and the coefficients in the expression of the vector as a linear combination of the standard basis vectors.

DEFINITION   *If v_1, v_2, $\cdots$, v_k is a basis for $\mathcal{W}$ and w is a vector in $\mathcal{W}$, the* coordinates of the vector w with respect to this basis *are the coefficients in the expression of w as a linear combination of the v's. That is, if*

$$w = a_1 v_1 + a_2 v_2 + \cdots + a_k v_k$$

the coordinates of w with respect to this basis are a_1, a_2, $\cdots$, and a_k.

Theorem 3.30 says that this definition makes sense, that each point (i.e., each vector) has exactly one set of coordinates with respect to a given basis. One way to interpret independence is to say that the only coordinates of the zero vector are 0, 0, $\cdots$, and 0. Sometimes, if $\mathcal{B} = \{v_1, v_2, \cdots, v_k\}$ is the basis, we write $[w]_{\mathcal{B}} = (a_1, a_2, \cdots, a_k)$ to express the fact that the a's are the coordinates of w with respect to the basis $\mathcal{B}$.

In Figure 3.6, we see the basis $\{u, v\}$ and the *coordinate system* it generates in part of the plane $\mathbf{R}^2$. The point $w = 2u + 3v$ is shown. The figure makes the geometric relationship between the points of the plane and the coefficients needed to write the corresponding vector as a linear combination of u and v. Such a coordinate

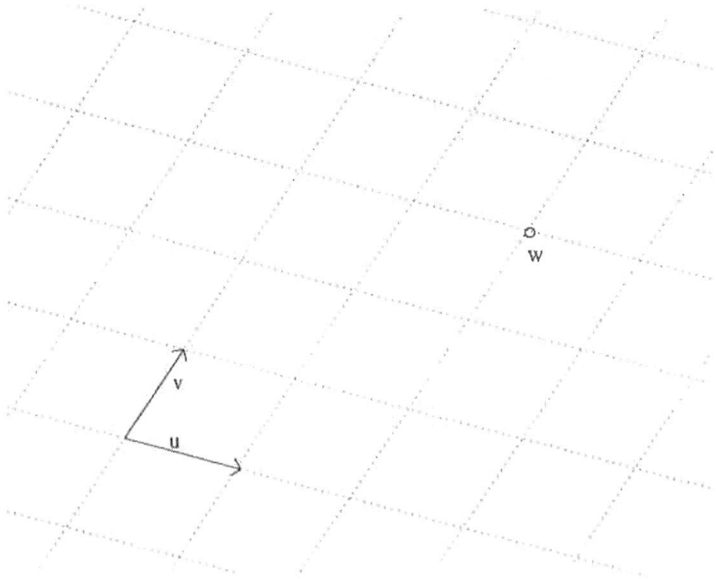

FIGURE 3.6

Coordinate system from the basis $\{u, v\}$

system could be used for many purposes in place of the usual coordinate system, which is based on the standard basis for $\mathbf{R}^2$, $e_1 = (1, 0)$ and $e_2 = (0, 1)$, except that the usual relationship between the coordinates and distance would be incorrect for the u, v–coordinates. We will look at other coordinate systems that do not have this difficulty in later chapters.

EXAMPLE 3.31

The axis of a gyroscope is aligned with the vector $u = (1, 1, 2)$. In order to tailor calculations more closely to this important direction in problems associated with the gyroscope, Rita decides to use the basis u, $v = (1, -1, 0)$, and $w = (1, 1, -1)$ as a coordinate system for space. Find the coordinates of the vector $z = (1, -1, 1)$ with respect to this basis.

SOLUTION The fact that u, v, w form a basis for $\mathbf{R}^3$ means that $z = (1, -1, 1)$ is a linear combination of these vectors. The coordinates of z are just the coefficients in this linear combination: we must therefore solve the system $au + bv + cw = z$

to find a, b, and c. That is, we must solve

$$a \begin{pmatrix} 1 \\ 1 \\ 2 \end{pmatrix} + b \begin{pmatrix} 1 \\ -1 \\ 0 \end{pmatrix} + c \begin{pmatrix} 1 \\ 1 \\ -1 \end{pmatrix} = \begin{pmatrix} 1 \\ -1 \\ 1 \end{pmatrix}$$

Solving the system, we find $a = 1/3$, $b = 1$, and $c = -1/3$. Thus, the coordinates for z with respect to this basis are $[z]_B = (1/3, 1, -1/3)$. $\quad\square$

A basis is a set of vectors that can non-redundantly (because it is a linearly independent set) represent every vector in the space (because it is a spanning set). It is reasonable to investigate what happens if a set of vectors is *not* a basis. Of course, there are sets that are neither linearly independent nor spanning sets. These sets fail to be bases for both reasons; they are not very interesting. More interesting are sets that fail to have one of the properties.

Suppose a set spans a subspace but is linearly dependent. In this case, we know that one of the vectors can be written as a linear combination of the others. This means that we can replace that vector in all linear combinations of vectors in the spanning set by another linear combination not using that vector. An example will make the idea clear.

EXAMPLE 3.32

The vectors $u = (1, -1, 2)$, $v = (2, 0, 1)$, and $w = (0, 2, -3)$ are linearly dependent. The vector $z = u + v - 2w = (3, -5, 9)$ is in the subspace spanned by u, v, and w. Write z as a linear combination of two (or fewer) of u, v, and w.

SOLUTION We see that $2u - v + w = 0$. Since the coefficient of w in this expression is not zero, w can be written as a linear combination of u and v, namely, $w = -2u + v$. Now we can replace w in the expression for z as a linear combination of u, v, and w. To be explicit,

$$z = u + v - 2w = u + v - 2(-2u + v) = u + v + 4u - 2v = 5u - v$$

Thus, $z = 5u - v$ which is a linear combination of just u and v, as we wanted. $\quad\square$

In fact, we can see that in the example above, any linear combination u, v, and w can be replaced by a linear combination of just u and v. In other words, the subspace spanned by u, v, and w is the same as the subspace spanned by u and v. This is a general principle!

> *If a spanning set for a subspace $\mathcal{W}$ is linearly dependent, there is a smaller subset of the given set that also spans $\mathcal{W}$.*

On the other hand, suppose we have a linearly independent set of vectors in a subspace that does not span the subspace. Then, there is a vector in the subspace that is not a linear combination of these vectors. Now, the set consisting of the original linearly independent set and the new vector is still linearly independent! To see why, suppose v_1, v_2, $\cdots$, v_k are linearly independent in $\mathcal{W}$ but that they do not span $\mathcal{W}$. Let u be a vector in $\mathcal{W}$ that is not a linear combination of v_1, v_2, $\cdots$, v_k. Now if we have

$$c_0 u + c_1 v_1 + \cdots + c_k v_k = 0$$

and $c_0 \neq 0$, then we would have

$$u = -\frac{c_1}{c_0} v_1 - \frac{c_2}{c_0} v_2 - \cdots - \frac{c_k}{c_0} v_k$$

contrary to the assumption that u is not a linear combination of the v's. Thus, we must have $c_0 = 0$. Now this gives

$$0u + c_1 v_1 + \cdots + c_k v_k = c_1 v_1 + \cdots + c_k v_k = 0$$

so the independence of v_1, v_2, $\cdots$, v_k implies $c_1 = c_2 = \cdots = c_k = 0$ as well. Thus, the only linear combination of u, v_1, v_2, $\cdots$, and v_k that gives 0 is the trivial linear combination, so these vectors are linearly independent. This argument justifies the following principle. It says that linearly independent sets can be *extended* to larger linearly independent sets.

> **If a set is linearly independent in a subspace, but does not span it, a vector from the subspace can be added to the linearly independent set so that the larger resulting set is still linearly independent.**

An example will illustrate the idea.

EXAMPLE 3.33

The vectors $u = (1, -1, 2)$ and $v = (2, 0, 1)$ are linearly independent in $\mathbf{R}^3$, but do not span $\mathbf{R}^3$. Find a vector x in $\mathbf{R}^3$ so that u, v, and x are linearly independent.

SOLUTION It is not hard to see that the vector $x = (1, 1, 1)$ is not a linear combination of u and v. So the set consisting of u, v, and x is linearly independent. $\quad\square$

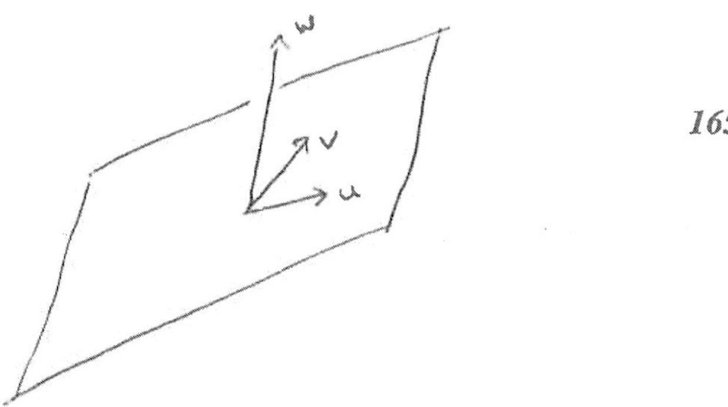

FIGURE 3.7

Extending the independent set $\{u, v\}$

One criticism of the solution given above is that it is not systematic: where did x come from? On the one hand, it can be shown that "almost any" vector will work. This means you can guess a vector x as was done above, and unless you are very unlucky, the first vector you guess will work. Geometrically, this is clear: the span of u and v is a plane in $\mathbf{R}^3$. Since most vectors in $\mathbf{R}^3$ do not lie in this plane, most vectors you choose will not lie in the plane, and any vector not in the plane will work (see Figure 3.7).

However, there is also an easy systematic strategy that is guaranteed to work. In the case above, the vectors e_1, e_2, and e_3 form a basis for $\mathbf{R}^3$. This means that at least one of these vectors is not a linear combination of u and v: indeed, if each of these vectors were a linear combination of u and v, then every vector in $\mathbf{R}^3$ could be written as a linear combination of u and v because every vector in $\mathbf{R}^3$ is a linear combination of e_1, e_2, and e_3. Since it was given that u and v do not span $\mathbf{R}^3$, this is impossible. In other word, at least one of the sets

$$\{u, v, e_1\}, \quad \{u, v, e_2\}, \quad \text{or} \quad \{u, v, e_3\}$$

is guaranteed to be linearly lindependent. (Probably, as in the example above, each of these sets will be independent!)

Exercises 3.5

1. Find a basis for the solution space of the system:

$$\begin{cases} a + 3b - c + 2d &= 0 \\ 2a + 2b + c + 2d &= 0 \\ 4a \quad\;\; + 5c - 7d &= 0 \end{cases}$$

2. Find a basis for the solution space of the system:

$$\begin{cases} u + 3v - w + 2x + y &= 0 \\ u + 2v + 4w + 2x &= 0 \\ 2u + 8v + w + 3x - y &= 0 \end{cases}$$

3. $3x + 2y - z = 0$ is the equation of a plane $\mathcal{P}$ in $\mathbf{R}^3$ that passes through the origin, so $\mathcal{P}$ is a subspace of $\mathbf{R}^3$. Find a basis for $\mathcal{P}$.

4. Let $F = \begin{pmatrix} 1 & -2 & 1 \\ 1 & 1 & -3 \\ -1 & 8 & -9 \end{pmatrix}$

    (a) Find a basis for the row space of F.
    (b) Find a basis for the column space of F.

5. Let $A = \begin{pmatrix} 1 & -1 & 1 & 0 & -2 \\ 2 & 0 & 1 & 1 & -1 \\ 1 & -3 & 2 & -1 & -5 \end{pmatrix}$

    (a) Find a basis for the row space of A.
    (b) Find a basis for the column space of A.
    (c) Is $(1, 1, 1)$ in the column space of A?

6. Let $B = \begin{pmatrix} 1 & 1 & 1 \\ -1 & 1 & -3 \\ 3 & 1 & 5 \\ 2 & -1 & 5 \end{pmatrix}$

    (a) Find a basis for the row space of B.
    (b) Find a basis for the column space of B.
    (c) Is $(1, 3, -1, -4)$ in the column space of B?

7. (a) Show that the vectors $(1, 0, 0)$, $(2, 1, 0)$, and $(1, -2, 1)$ are a basis for $\mathbf{R}^3$.

   (b) What vector in $\mathbf{R}^3$ has coordinates $-3, 2,$ and 1 with respect to this basis?

8. (a) Show that the vectors $(1, 2)$ and $(-1, 1)$ form a basis for $\mathbf{R}^2$.

   (b) Find the coordinates of $(1, 1)$ with respect to this basis.

9. (a) Show that the vectors $(1, 2, 1)$, $(-1, 1, 0)$, and $(2, 0, 0)$ are a basis for $\mathbf{R}^3$.

   (b) Find the coordinates of $(0, 0, 1)$ with respect to this basis.

10. (a) Find the coordinates of $x = (1, 2, 3)$ with respect to Rita's basis $u = (1, 1, 2)$, $v = (1, -1, 0)$, and $w = (1, 1, -1)$ for $\mathbf{R}^3$.

   (b) Find the coordinates of the vector $y = (1, 2, 1)$ with respect to this basis.

11. The vectors $u = (1, 2)$, $v = (1, 0)$, and $w = (2, 1)$ are linearly dependent. Write $z = 2u - v + w = (3, 5)$ as a linear combination of two (or fewer) of these vectors.

12. The vectors $u = (2, 1, -1)$, $v = (1, 1, 0)$, and $w = (1, 2, 1)$ are linearly dependent. Write $z = u + v - 2w = (1, -2, -3)$ as a linear combination of two (or fewer) of these vectors.

13. The vectors $v_1 = (1, -1, 1, 2)$, $v_2 = (3, 0, 1, 4)$, $v_3 = (1, 2, -1, 0)$, and $v_4 = (0, 1, -1, 1)$ are linearly dependent. Write $w = -v_1 + v_2 + v_3 - v_4 = (3, 2, 0, 1)$ as a linear combination of three (or fewer) of these vectors.

14. The vectors $u = (1, 1, 1)$ and $v = (1, -1, -1)$ are linearly independent, but do not span $\mathbf{R}^3$. Find a vector w in $\mathbf{R}^3$ so that u, v, and w are linearly independent.

15. The vectors $v_1 = (1, 1, 1, 0)$, $v_2 = (1, -1, -1, 2)$, and $v_3 = (0, 1, 1, 2)$ are linearly independent, but do not span $\mathbf{R}^4$. Find a vector v_4 in $\mathbf{R}^4$ so that v_1, v_2, v_3, and v_4 are linearly independent.

16. Let $\mathcal{W}$ be the subspace in $\mathbf{R}^4$ consisting of the solution set of the equation $w - 2x + y - 3z = 0$. The vectors $v_1 = (1, 1, 1, 0)$ and $v_2 = (1, -1, 0, 1)$ are in $\mathcal{W}$ and are linearly independent, but they do not span $\mathcal{W}$. Find a vector v_3 in $\mathcal{W}$ so that v_1, v_2, and v_3 are linearly independent.

3.6 Dimension

It is intuitively clear that $\mathbf{R}^3$ is bigger than $\mathbf{R}^2$ in some sense, although, since they each contain infinitely many vectors, it is not clear what "bigger" should mean. The goal of this section is to introduce a precise sense in which $\mathbf{R}^3$ is bigger than $\mathbf{R}^2$: we will use the word "dimension" and it coincides with the use of the word in the colloquial sense as in "$\mathbf{R}^3$ is bigger than $\mathbf{R}^2$ because it is three dimensional and $\mathbf{R}^2$ is only two dimensional." With the concept of dimension, we can make sense of the idea that some vectors spaces are bigger than others or that two vector spaces are the same size.

The first step in the development of this idea is the next result that says a set has to be big to span and a set has to be small to be linearly independent. A basis, which is both spanning and linearly independent, strikes the perfect balance between being big enough to span and small enough to be independent. The consequence we are looking for is that bases span with no redundancy.

THEOREM 3.34

Suppose v_1, v_2, $\cdots$, v_m span the subspace $\mathcal{W}$ and suppose w_1, w_2, $\cdots$, w_n are linearly independent. Then $n \leq m$.

PROOF Suppose that v_1, v_2, $\cdots$, v_m span $\mathcal{W}$ and $n > m$. We will show that in this case, w_1, w_2, $\cdots$, w_n are linearly dependent, which will prove the theorem.

We wish to find scalars $c_1, \cdots, c_n$, not all zero, so that $c_1 w_1 + c_2 w_2 + \cdots + c_n w_n = 0$. Since the v's span $\mathcal{W}$, there are constants $a_j k$ so that for each w_k, we have

$$w_k = a_{1k} v_1 + a_{2k} v_2 + \cdots + a_{mk} v_m$$

Replacing each of the w's by the corresponding linear combination,

$$c_1 (a_{11} v_1 + a_{21} v_2 + \cdots + a_{m1} v_m)$$

$$+ c_2 (a_{12} v_1 + a_{22} v_2 + \cdots + a_{m2} v_m)$$

$$\ddots$$

$$+ c_n ((a_{1n} v_1 + a_{2n} v_2 + \cdots + a_{mn} v_m) = 0$$

and collecting terms, we get

$$(a_{11} c_1 + a_{12} c_2 + \cdots + a_{1n} c_n) v_1$$

$$+(a_{21}c_1 + a_{22}c_2 + \cdots + a_{2n}c_n)v_2$$

$$\ddots$$

$$+(a_{m1}c_1 + a_{m2}c_2 + \cdots + a_{mn}c_n)v_m = 0$$

Of course, one way for this sum to be zero is to choose c's so that each of the coefficients is zero. This would be the case if we can find a non-zero solution of the system

$$\begin{cases} a_{11}c_1 & + & a_{12}c_2 & + & \cdots & + & a_{1n}c_n & = & 0 \\ a_{21}c_1 & + & a_{22}c_2 & + & \cdots & + & a_{2n}c_n & = & 0 \\ & & & & \vdots & & & & \\ a_{m1}c_1 & + & a_{m2}c_2 & + & \cdots & + & a_{mn}c_n & = & 0 \end{cases} \tag{3.6.2}$$

which is a system of m homogeneous equations in n unknowns $(c_1, \cdots, c_n)$. But we know from Theorem 2.16 that any homogeneous system with more unknowns than equations has a non-trivial solution! That is, we can find coefficients c_1, $\cdots$, c_n, not all zero, that solve the system of Equation 3.6.2 and therefore satisfy $c_1w_1 + c_2w_2 + \cdots + c_nw_n = 0$. ∎

Applying Theorem 3.34 to the case of two bases, we get an interesting conclusion.

THEOREM 3.35

If v_1, v_2, $\cdots$, v_m and w_1, w_2, $\cdots$, w_n are both bases for $\mathcal{W}$, then $m = n$.

PROOF   Since v_1, $\cdots$, v_m is a basis, it spans $\mathcal{W}$ and since w_1, $\cdots$, w_n is a basis, it is a linearly independent set in $\mathcal{W}$. Theorem 3.34 therefore implies $m \geq n$.

On the other hand, since v_1, $\cdots$, v_m is a basis, it is a linearly independent set in $\mathcal{W}$ and since w_1, $\cdots$, w_n is a basis, it spans $\mathcal{W}$, so Theorem 3.34 implies $n \geq m$ also. Combining the two inequalities gives $m = n$. ∎

That is, even though there are infinitely many bases for every (non-zero) vector space, every basis for a space consists of the same number of vectors! Since the number of vectors in a basis does not depend on which basis you use, this number is a property of the space itself.

DEFINITION *A vector space $\mathcal{W}$ is called* finite dimensional *if there is a finite set of vectors that forms a basis for $\mathcal{W}$. The* dimension *of $\mathcal{W}$ is the number of vectors in a basis for $\mathcal{W}$.*

The subspace $\{0\}$ has no basis since it contains no linearly independent vectors ($1 \cdot 0 = 0$ so 0 by itself is dependent). By convention, the subspace $\{0\}$ is said to have dimension 0.

EXAMPLE 3.36

Since the standard basis for $\mathbf{R}^n$ has n vectors in it, the dimension of $\mathbf{R}^n$ is n, and we say it is n–dimensional. The definition of dimension in linear algebra agrees with the use of the word in colloquial English: a plane, like $\mathbf{R}^2$, is 2–dimensional and the world we live in, $\mathbf{R}^3$, is 3–dimensional. $\quad\Box$

EXAMPLE 3.37

What is the dimension of the nullspace of the matrix

$$A = \begin{pmatrix} 1 & 2 & -1 & 1 \\ 2 & 1 & -1 & 0 \\ 0 & 3 & -1 & 2 \end{pmatrix}$$

SOLUTION We must find a basis for the nullspace of A, that is, we must solve the system $AX = 0$.

$$\left(\begin{array}{cccc|c} 1 & 2 & -1 & 1 & 0 \\ 2 & 1 & -1 & 0 & 0 \\ 0 & 3 & -1 & 2 & 0 \end{array} \right) \rightarrow \left(\begin{array}{cccc|c} 1 & 2 & -1 & 1 & 0 \\ 0 & -3 & 1 & -2 & 0 \\ 0 & 3 & -1 & 2 & 0 \end{array} \right)$$

$$\rightarrow \left(\begin{array}{cccc|c} 1 & 2 & -1 & 1 & 0 \\ 0 & -3 & 1 & -2 & 0 \\ 0 & 0 & 0 & 0 & 0 \end{array} \right) \rightarrow \left(\begin{array}{cccc|c} 1 & -1 & 0 & -1 & 0 \\ 0 & -3 & 1 & -2 & 0 \\ 0 & 0 & 0 & 0 & 0 \end{array} \right)$$

so if $X = (w, x, y, z)$, we see that $AX = 0$ if and only if $w = x + z$ and $y = 3x + 2z$, and x and z are arbitrary.

Choosing $x = 1$ and $z = 0$, then $x = 0$ and $z = 1$ makes our work easy: $v_1 = (1, 1, 3, 0)$ and $v_2 = (1, 0, 2, 1)$ are a basis for the nullspace of A. To check this, we first note that $Av_1 = 0$ and $Av_2 = 0$ so v_1 and v_2 are in the nullspace. Moreover, if $av_1 + bv_2 = 0$, then from the second component, $a = 0$, and from the fourth component, $b = 0$, so v_1 and v_2 are linearly independent. Finally, if

$X = (w, x, y, z)$ is in the nullspace of A, our calculation above shows that

$$
\begin{pmatrix} w \\ x \\ y \\ z \end{pmatrix} = \begin{pmatrix} x + z \\ x \\ 3x + 2z \\ z \end{pmatrix} = x \begin{pmatrix} 1 \\ 1 \\ 3 \\ 0 \end{pmatrix} + z \begin{pmatrix} 1 \\ 0 \\ 2 \\ 1 \end{pmatrix}
$$

so every vector in the nullspace of A is a linear combination of v_1 and v_2.

Since there are two vectors in this basis for the nullspace of A, there are two vectors in every basis for the nullspace of A, and the dimension of the nullspace of A is 2. $\quad\square$

You may have noticed that the dimension of the nullspace of A in this example was 2 and there were 2 *free variables*, in this case, x and z. This was not a coincidence; the number of free variables is always the dimension of the nullspace in situations like this and the choices of 0's and 1's for the free variables, analogous to the usual basis for $\mathbf{R}^n$, makes the justifications very easy.

THEOREM 3.38

If v_1, v_2, $\cdots$, v_k span the non-zero subspace $\mathcal{W}$, then there is a subset of this set that is a basis. If w_1, w_2, $\cdots$, w_j are linearly independent in the n-dimensional space $\mathcal{W}$, then w_{j+1}, $\cdots$, w_n can be found in $\mathcal{W}$ so that w_1, w_2, $\cdots$, w_j, w_{j+1}, $\cdots$, w_n is a basis for $\mathcal{W}$.

In other words, from a spanning set, vectors can be deleted to obtain a basis, and a linearly independent set can be enlarged or *extended* to a basis.

PROOF Suppose v_1, v_2, $\cdots$, v_k span $\mathcal{W}$. If v_1, v_2, $\cdots$, v_k are linearly independent, then they are a basis and the result is true. If they are linearly dependent, then (Theorem 3.21) one of the vectors is a linear combination of the rest, say v_j is a linear combination of the others. In this case, v_1, $\cdots$, v_{j-1}, v_{j+1}, $\cdots$, v_k span $\mathcal{W}$ since v_j can be replaced in any expression by the linear combination of the others. If v_1, $\cdots$, v_{j-1}, v_{j+1}, $\cdots$, v_k is linearly independent, then they are a basis and the result is true. If not, another vector can be removed and so on. Since there are only k vectors and after each vector is removed, the remaining vectors still span $\mathcal{W}$, this process must stop after at most $k - 1$ steps, and the resulting set is a basis.

Suppose w_1, w_2, $\cdots$, w_j are linearly independent in the n-dimensional space $\mathcal{W}$. If $j = n$ then these vectors are a basis and the result is true adding no vectors. If $j < n$, then the vectors do not span $\mathcal{W}$ and there is a vector w_{j+1} that is not in the subspace spanned by $w_1, \cdots, w_j$. Since none of the vectors in the set $w_1, \cdots,$ w_j, w_{j+1} is a linear combination of the others, this set must be linearly independent (Theorem 3.21). If $j + 1 = n$, these vectors are a basis and the result is true adding one vector. If $j + 1 < n$, then these vectors do not span the space and we may find another vector to add. Continuing this process for $n - j$ steps we get a basis for $\mathcal{W}$. ∎

COROLLARY 3.39

If $\mathcal{V}$ is a vector space with $\dim(\mathcal{V}) = n$ and $v_1, v_2, \cdots, v_n$ is a linearly independent set, then $v_1, v_2, \cdots, v_n$ also spans $\mathcal{V}$, so these vectors form a basis for $\mathcal{V}$. Similarly, if $\mathcal{V}$ is a vector space with $\dim(\mathcal{V}) = n$ and $v_1, v_2, \cdots, v_n$, span $\mathcal{V}$, then $v_1, v_2, \cdots,$ v_n is also a linearly independent set, so these vectors form a basis for $\mathcal{V}$.

Thus, if you already know the dimension of the space you are working with, it is less difficult to show that a particular set is a basis. Usually, it is easier to show that a set is linearly independent than to show it spans a space.

THEOREM 3.40

If $\mathcal{W}_1$ is a subspace of the vector space $\mathcal{W}_2$ and $\dim(\mathcal{W}_1) = \dim(\mathcal{W}_2)$, then $\mathcal{W}_1 = \mathcal{W}_2$.

PROOF Let $k = \dim(\mathcal{W}_1) = \dim(\mathcal{W}_2)$. Let $w_1, w_2, \cdots, w_k$ be a basis for $\mathcal{W}_1$ (which is possible since $\dim(\mathcal{W}_1) = k$). Then the vectors $w_1, w_2, \cdots, w_k$ form a linearly independent set in $\mathcal{W}_1$, hence in $\mathcal{W}_2$. On the other hand, since $k = \dim(\mathcal{W}_2)$, any k linearly independent vectors in $\mathcal{W}_2$ form a basis for $\mathcal{W}_2$. Thus, $w_1, w_2, \cdots, w_k$ are a basis for $\mathcal{W}_2$ and they span $\mathcal{W}_2$. This means that every vector in $\mathcal{W}_2$ is a linear combination of $w_1, w_2, \cdots, w_k$. Since every vector that is a linear combination of these vectors is clearly in $\mathcal{W}_1$, it follows that $\mathcal{W}_2 \subset \mathcal{W}_1$. Since the other containment was part of the hypothesis, equality holds. ∎

To close the section, we will name some of the subspaces associated with a matrix and see some of their relationships. In Section 3.2, we defined the *range of* A to be the subspace of $\mathbf{R}^m$

$$\mathcal{R}(A) = \{b \in \mathbf{R}^m : AX = b \text{ for some } X \text{ in } \mathbf{R}^n\}$$

The dimension of the range of A will be called the *rank of A*. We also defined the column space to be the subspace of $\mathbf{R}^m$ spanned by the columns of A and the row space of A to be the subspace of $\mathbf{R}^n$ spanned by the rows of A. The dimensions of the column space and the row space of A are called the *column rank* and *row rank* of

A. Since we showed the column space and the range are the same (Theorem 3.13), the column rank and the rank are the same. In fact, the row rank and the column rank are equal as well, so we can just use the word "rank" for all these dimensions without distinction! Although we will not prove this fact until Section 4.4, we record it here for reference and begin to use the result. This result will be a consequence of a geometric version of the rank–nullity theorem that will be proved there.

THEOREM 3.41

If A is an $m \times n$ matrix, then the dimension of the range of A is the same as the dimension of the range of A'.

EXAMPLE 3.42

Find bases for the row space and the column space of the matrix

$$B = \begin{pmatrix} 1 & 0 & 2 & 2 \\ 2 & -1 & 3 & 1 \\ -1 & 2 & 0 & 4 \end{pmatrix}$$

Are $(2, 1, -1)$ and $(1, 1, 1)$ in the range of B?

SOLUTION According to Theorem 3.38, since the columns of B span the column space of B, we can find a basis for the column space by selecting the largest possible linearly independent subset of the columns. Let's take the first column, $(1, 2, -1)$. Does it span the column space? Clearly not, because the second column, $(0, -1, 2)$ is not a linear combination of it (i.e., a multiple of it). Thus, the set $(1, 2, -1)$ and $(0, -1, 2)$ is a linearly independent set of columns. Do these two vectors span the column space? Let's check to see if $(2, 3, 0)$ is a linear combination of $(1, 2, -1)$ and $(0, -1, 2)$: if

$$\alpha \begin{pmatrix} 1 \\ 2 \\ -1 \end{pmatrix} + \beta \begin{pmatrix} 0 \\ -1 \\ 2 \end{pmatrix} = \begin{pmatrix} 2 \\ 3 \\ 0 \end{pmatrix}$$

we see immediately that α must equal 2 (an advantage of using a column vector with a 0 component!). The equation $2\alpha - \beta = 3$ from the second components, using $\alpha = 2$ from above, shows that β must be 1. Since this also works in the third component $(-\alpha + 2\beta = 0)$, we see that the third column is a linear combination of the first two columns: $(2, 3, 0) = 2(1, 2, -1) + (0, -1, 2)$. Thus, the first three columns of B are linearly dependent and we do not need to include the third column.

In the same way, we can see that the fourth column is a linear combination of the first two columns:

$$\begin{pmatrix} 2 \\ 1 \\ 4 \end{pmatrix} = 2 \begin{pmatrix} 1 \\ 2 \\ -1 \end{pmatrix} + 3 \begin{pmatrix} 0 \\ -1 \\ 2 \end{pmatrix}$$

This means we do not need to include the fourth column, either. We conclude that the first two columns of B, $\{(1, 2, -1), (0, -1, 2)\}$, are a basis for the column space of B. In particular, this means that the rank of B is 2.

Knowing that the rank of B is two makes the job of finding a basis for the row space of B much easier; we know that we must choose two linearly independent rows of B. Clearly, the first two rows of B are linearly independent, so they must be a basis for the row space. We have found the row space of B is

$$\text{span}\{(1\ 0\ 2\ 2),\ (2\ -1\ 3\ 1)\}$$

In particular, this means that the third row of B is a linear combination of the first two and it is easy to check that it is

$$(-1\ 2\ 0\ 4) = 3(1\ 0\ 2\ 2) - 2(2\ -1\ 3\ 1)$$

To find out if $(2, 1, -1)$ is in the range of B, we need to see if there is a vector X in $\mathbf{R}^4$ so that $BX = (2, 1, -1)$, that is, we need to solve a system of 3 equations in 4 unknowns. On the other hand, since $(1, 2, -1)$ and $(0, -1, 2)$ are a basis for the column space of B, which is the same as the range of B, it is enough to see if $(2, 1, -1)$ is a linear combination of $(1, 2, -1)$ and $(0, -1, 2)$, a system of 3 equations in only 2 unknowns!

$$a \begin{pmatrix} 1 \\ 2 \\ -1 \end{pmatrix} + b \begin{pmatrix} 0 \\ -1 \\ 2 \end{pmatrix} = \begin{pmatrix} 2 \\ 1 \\ -1 \end{pmatrix}$$

From the first component, we get $a = 2$. From the second component, we then get $b = 3$. But these values do not work in the third component, $2(-1) + 3(2) = 4 \neq -1$, so $(2, 1, -1)$ is not a linear combination of $(1, 2, -1)$ and $(0, -1, 2)$ so it is not in the range of B.

In the same way, we check $(1, 1, 1)$

$$c \begin{pmatrix} 1 \\ 2 \\ -1 \end{pmatrix} + d \begin{pmatrix} 0 \\ -1 \\ 2 \end{pmatrix} = \begin{pmatrix} 1 \\ 1 \\ 1 \end{pmatrix}$$

and find that $c = 1$ and $d = 1$ work, so $(1, 1, 1)$ is in the range of B. $\quad\square$

MATLAB finds the rank of the matrix A with the command `rank(A)`. The ease with which this can be done leads to some easy solutions of problems such as that above.

EXAMPLE 3.43

Let's do Example 3.42 again using MATLAB. Recall that we were asked to find bases for the row space and the column space of the matrix

$$B = \begin{pmatrix} 1 & 0 & 2 & 2 \\ 2 & -1 & 3 & 1 \\ -1 & 2 & 0 & 4 \end{pmatrix}$$

We were also asked to decide if $(2, 1, -1)$ and $(1, 1, 1)$ are in the range of B?

First, we enter the matrix B

```
> B=[1 0 2 2; 2 -1 3 1; -1 2 0 4]

B =
     1      0      2      2
     2     -1      3      1
    -1      2      0      4
```

and recall that

```
> B(:,1:3)

ans =
     1      0      2
     2     -1      3
    -1      2      0
```

gives the submatrix of B consisting of all the rows "`:`" and the columns 1, 2, and 3 "`1:3`". Now the rank calculations

```
> rank(B(:,1))

ans =
     1
```

```
> rank(B(:,1:2))

ans =
      2

> rank(B(:,1:3))

ans =
      2

> rank(B)

ans =
      2
```

which parallel the computations we did by hand, show that the second column of B is not a linear combination of the first column (otherwise the rank of the matrix consisting of the first two columns of B would also be 1), but that the third column is a linear combination of the first two columns (the rank of the matrix consisting of the first three columns is the same as the rank of the matrix consisting of the first two columns). Finally, since the rank of matrix consisting of the first two columns of B is 2 and the rank of B is two shows that the first two columns form a basis for the column space of B.

We can easily decide if the vectors $(2, 1, -1)$ or $(1, 1, 1)$ are in the range of B as well.

```
> B1=[B [2;1;-1]]

B1 =
      1      0      2      2      2
      2     -1      3      1      1
     -1      2      0      4     -1

> rank(B1)

ans =
      3

> B2=[B [1;1;1]]
```

```
B2 =
       1        0        2        2        1
       2       -1        3        1        1
      -1        2        0        4        1

> rank(B2)

ans =
       2
```

Since the rank of B is 2, but the rank of B together with $(2, 1, -1)$ is 3, we see that the vector $(2, 1, -1)$ is *not* a linear combination of the columns of B and since the rank of B together with $(1, 1, 1)$ is still 2, we see that the vector $(1, 1, 1)$ *is* a linear combination of the columns of B.   ☐

EXAMPLE 3.44

Find a basis for the range of the matrix D where

$$D = \begin{pmatrix} 1 & -1 & 2 \\ 3 & 1 & 0 \\ 2 & 1 & -1 \end{pmatrix}$$

Decide if $(1, 2, -1)$ is in the range of D.

SOLUTION We'll use MATLAB.

```
> D=[1 -1 2; 3 1 0; 2 1 -1]

D =
       1       -1        2
       3        1        0
       2        1       -1

> rank(D)

ans =
       3
```

Since the dimension of the range of D is 3 and D has only three columns, they are linearly independent. Since the columns of D span the range of D, the three columns of D, $(1, 3, 2)$, $(-1, 1, 1)$, and $(2, 0, -1)$ form a basis for the range of D. However, the range of D is a subspace of $\mathbf{R}^3$, which is also 3–dimensional, so Theorem 3.40 implies $\mathcal{R}(D) = \mathbf{R}^3$. Clearly, since $(1, 2, -1)$ is in $\mathbf{R}^3$, we have $(1, 2, -1)$ is in the range of D. ▯

A word of caution is in order, though. Recall that in finding the rank of the matrix B, we needed to decide if the third and fourth columns of the matrix were linear combinations of the first two, or to put it differently, whether a linear combination of the first two columns minus the third or fourth column is zero. In general, the determination of rank involves a decision whether some number is zero. With exact hand calculation, that is not difficult. On the other hand, since a computer calculation (such as those done by MATLAB) involves small errors in rounding numbers off, the number in question is rarely exactly zero. In practice, then, the computer must decide if the number being tested is zero except for some small round-off error or is small but not actually zero. This cannot be done with complete accuracy all the time and we must be aware that sometimes the rank of a matrix as computed may not be the rank we would get if we were to do all the calculations exactly. In many situations in engineering and science, however, the numbers in the matrix at the beginning are measurements and are likely not exactly correct anyway, so the roundoff errors made by the computer are not the only errors in the problem, and probably not even the most important ones. It turns out that some problems are more sensitive to small errors than others, that is, for some problems a small error in the input leads to small error in the output, but for others, a small error in the input leads to a relatively large error in the output. A problem of the latter sort is said to be *ill–conditioned*. MATLAB is designed to look for this sort of problem and sometimes issues a warning

```
Warning: Matrix is close to singular or badly scaled.
         Results may be inaccurate. RCOND = 1.536259e-17
```

but it will not issue such a warning for the `rank` command. (RCOND is the reciprocal of the condition number and should be "near 1" in problems that are well conditioned.)

We have seen that the *rank* of a matrix is its *column rank* which is the number of linearly independent columns of the matrix, so the rank of an $m \times n$ matrix is no more than n the number of columns of the matrix. On the other hand, the rank of a matrix is also its *row rank*, the number of linearly independent rows of the matrix, so the rank of an $m \times n$ matrix is no more than m the number of rows of the matrix. In other words, the rank of a matrix cannot be more than the smaller of m and n. An $m \times n$ matrix whose rank is the smaller of m and n is said to have *full rank*.

In general, we would hope to use full rank matrices in most applications and most algorithms for matrix computations work best when the matrices are full rank.

It is very easy to determine the rank of a row echelon or reduced row echelon matrix (see page 64) as the columns containing the leading entries are obviously linearly independent and the non-zero rows are clearly linearly independent. (That the number of non-zero rows is the same as the number of columns containing leading entries makes it clear that the dimension of the row space and the dimension of the column space are the same for a row echelon matrix. This observation can be used to give a proof of Theorem 3.41, but we will not do so.)

EXAMPLE 3.45

Find bases for the column space and the row space of the matrix

$$E = \begin{pmatrix} \textcircled{1} & 1 & -1 & 0 & -1 & -3 \\ 0 & \textcircled{2} & -1 & 3 & -2 & 1 \\ 0 & 0 & 0 & 0 & \textcircled{1} & 1 \\ 0 & 0 & 0 & 0 & 0 & 0 \end{pmatrix}$$

SOLUTION We will use the rows and columns containing the leading entries, which have been circled. Clearly the first, second, and fifth columns of E are linearly independent: if

$$a \begin{pmatrix} 1 \\ 0 \\ 0 \\ 0 \end{pmatrix} + b \begin{pmatrix} 1 \\ 2 \\ 0 \\ 0 \end{pmatrix} + c \begin{pmatrix} -1 \\ -2 \\ 1 \\ 0 \end{pmatrix} = \begin{pmatrix} 0 \\ 0 \\ 0 \\ 0 \end{pmatrix}$$

then $c = 0$ from the third equation, so $b = 0$ from the second equation, and $a = 0$ from the first equation.

On the other hand, the third, fourth, and sixth columns are linear combinations of these. To find r, s, and t so that

$$r \begin{pmatrix} 1 \\ 0 \\ 0 \\ 0 \end{pmatrix} + s \begin{pmatrix} 1 \\ 2 \\ 0 \\ 0 \end{pmatrix} + t \begin{pmatrix} -1 \\ -2 \\ 1 \\ 0 \end{pmatrix} = \begin{pmatrix} -3 \\ 1 \\ 1 \\ 0 \end{pmatrix}$$

we must solve an upper triangular system: clearly, from the third equation, $t = 1$, so from the second equation, $s = 3/2$, and from the first equation $r = -7/2$ so the sixth column is a linear combination of the first, second, and fifth.

Similarly, the third column satisfies

$$\begin{pmatrix} -1 \\ -1 \\ 0 \\ 0 \end{pmatrix} = -\frac{1}{2}\begin{pmatrix} 1 \\ 0 \\ 0 \\ 0 \end{pmatrix} + -\frac{1}{2}\begin{pmatrix} 1 \\ 2 \\ 0 \\ 0 \end{pmatrix} + 0\begin{pmatrix} -1 \\ -2 \\ 1 \\ 0 \end{pmatrix}$$

and the fourth column satisfies

$$\begin{pmatrix} 0 \\ 3 \\ 0 \\ 0 \end{pmatrix} = -\frac{3}{2}\begin{pmatrix} 1 \\ 0 \\ 0 \\ 0 \end{pmatrix} + \frac{3}{2}\begin{pmatrix} 1 \\ 2 \\ 0 \\ 0 \end{pmatrix} + 0\begin{pmatrix} -1 \\ -2 \\ 1 \\ 0 \end{pmatrix}$$

Thus, the first, second, and fifth column of E form a basis for the column space and the rank of E is 3.

In the same way, the first three rows of E are linearly independent, and the last row is a linear combination of the first three:

$$a\begin{pmatrix} 1 \\ 1 \\ -1 \\ 0 \\ -1 \\ -3 \end{pmatrix} + b\begin{pmatrix} 0 \\ 2 \\ -1 \\ 3 \\ -2 \\ 1 \end{pmatrix} + c\begin{pmatrix} 0 \\ 0 \\ 0 \\ 0 \\ 1 \\ 1 \end{pmatrix} = \begin{pmatrix} 0 \\ 0 \\ 0 \\ 0 \\ 0 \\ 0 \end{pmatrix}$$

so $a = 0$ from the first equation, $b = 0$ from the second, and $c = 0$ from the third, so the first three rows of E are independent. the last row is the trivial linear combination of the first three:

$$\begin{pmatrix} 0 \\ 0 \\ 0 \\ 0 \\ 0 \\ 0 \end{pmatrix} = 0\begin{pmatrix} 1 \\ 1 \\ -1 \\ 0 \\ -1 \\ -3 \end{pmatrix} + 0\begin{pmatrix} 0 \\ 2 \\ -1 \\ 3 \\ -2 \\ 1 \end{pmatrix} + 0\begin{pmatrix} 0 \\ 0 \\ 0 \\ 0 \\ 1 \\ 1 \end{pmatrix}$$

This strategy will work with any matrix in row echelon form. $\square$

EXAMPLE 3.46

Find bases for the range and the nullspace of the matrix C where

$$C = \begin{pmatrix} 1 & 2 & -1 & 1 \\ -1 & -1 & 3 & 0 \\ -2 & -1 & 8 & 1 \\ 1 & 3 & 1 & 2 \end{pmatrix}$$

SOLUTION Let's find the nullspace of C by solving the system $CX = 0$.

$$\left(\begin{array}{cccc|c} 1 & 2 & -1 & 1 & 0 \\ -1 & -1 & 3 & 0 & 0 \\ -2 & -1 & 8 & 1 & 0 \\ 1 & 3 & 1 & 2 & 0 \end{array} \right) \longrightarrow \left(\begin{array}{cccc|c} 1 & 2 & -1 & 1 & 0 \\ 0 & 1 & 2 & 1 & 0 \\ 0 & 3 & 6 & 3 & 0 \\ 0 & 1 & 2 & 1 & 0 \end{array} \right)$$

$$\longrightarrow \left(\begin{array}{cccc|c} 1 & 0 & -5 & -1 & 0 \\ 0 & 1 & 2 & 1 & 0 \\ 0 & 0 & 0 & 0 & 0 \\ 0 & 0 & 0 & 0 & 0 \end{array} \right)$$

Now letting $X = (w, x, y, z)$, this says X is in the nullspace of C if and only if $w = 5y + z$ and $x = -2y - z$, so that

$$X = \begin{pmatrix} 5y + z \\ -2y - z \\ y \\ z \end{pmatrix} = y \begin{pmatrix} 5 \\ -2 \\ 1 \\ 0 \end{pmatrix} + z \begin{pmatrix} 1 \\ -1 \\ 0 \\ 1 \end{pmatrix}$$

Thus, $(5, -2, 1, 0)$ and $(1, -1, 0, 1)$ span the nullspace of C and since they are linearly independent, they form a basis for the nullspace of C. Notice that each of these vectors corresponds to a free variable.

If we block C by columns, the fact that $X_1 = (5, -2, 1, 0)$ is in the nullspace of C gives

$$0 = CX_1 = \begin{pmatrix} C_1 & C_2 & C_3 & C_4 \end{pmatrix} \begin{pmatrix} 5 \\ -2 \\ 1 \\ 0 \end{pmatrix} = 5C_1 - 2C_2 + C_3 + 0C_4$$

or that $C_3 = -5C_1 + 2C_2$. Similarly, the fact that $X_2 = (1, -1, 0, 1)$ is in the nullspace of C gives

$$0 = CX_2 = \begin{pmatrix} C_1 & C_2 & C_3 & C_4 \end{pmatrix} \begin{pmatrix} 1 \\ -1 \\ 0 \\ 1 \end{pmatrix} = C_1 - C_2 + 0C_3 + C_4$$

or that $C_4 = -C_1 + C_2$. Now, since $\mathcal{R}(C)$ is spanned by C_1, C_2, C_3, and C_4, these two statements say that C_3 and C_4 are both linear combinations of C_1 and C_2, so that $\{C_1, C_2\}$ is a spanning set for $\mathcal{R}(C)$. Since they are also linearly independent, this means is $\{C_1, C_2\}$ is a basis for $\mathcal{R}(C)$. Notice that the first two columns corresponded to the pivot variables. ☐

The following result is another way to think of finding the nullspace of a matrix by row operations. To solve $CX = 0$ by row operations, one replaces this system by equivalent, easier systems. In the case E is the reduced row echelon form of C, the easier system is $EX = 0$, so the equivalence of the systems says they have the same solution space. Looking at it from a matrix factorization point of view is also easy. Given our computational strategies, it is not surprising that there is an intimate connection between the nullspace of a matrix and that of its reduced row echelon form.

PROPOSITION 3.47

Suppose C is an $m \times n$ matrix, L is an invertible $m \times m$ matrix, and $C = LE$. Then the nullspace of C and the nullspace of E are the same.

PROOF   We need to show that if $C = LE$ then $CX = 0$ if and only if $EX = 0$ because the first condition says that X is in the nullspace of C and the latter condition says X is in the nullspace of E.

Clearly, if $EX = 0$, then $CX = (LE)X = L(EX) = L0 = 0$. Conversely, if $CX = 0$, then $L(EX) = 0$ which means $L^{-1}L(EX) = L^{-1}0 = 0$. That is, $EX = 0$. ∎

There are also connections between the range of a matrix and that of its reduced row echelon form, but they are more subtle than for the nullspace. The following result, whose proof is Exercise 3.6.20., is very useful in the case when E is a row echelon matrix.

THEOREM 3.48

Suppose C is an $m \times n$ matrix, L is an invertible $m \times m$ matrix, and $C = LE$. Then the rank of C and the rank of E are the same.

EXAMPLE 3.49

Use Theorem 3.48 to find bases for the matrix B of Example 3.42.

SOLUTION We can use this result to find the rank of B if we find a matrix L and an echelon matrix E so that $B = LE$. It is not difficult to see that

$$B = \begin{pmatrix} 1 & 0 & 2 & 2 \\ 2 & -1 & 3 & 1 \\ -1 & 2 & 0 & 4 \end{pmatrix} = \begin{pmatrix} 1 & 0 & 0 \\ 2 & 1 & 0 \\ -1 & -2 & 1 \end{pmatrix} \begin{pmatrix} 1 & 0 & 2 & 2 \\ 0 & -1 & -1 & -3 \\ 0 & 0 & 0 & 0 \end{pmatrix}$$

Clearly L is invertible, so the result of the theorem is that the rank of B is the same as the rank of

$$E = \begin{pmatrix} 1 & 0 & 2 & 2 \\ 0 & -1 & -1 & -3 \\ 0 & 0 & 0 & 0 \end{pmatrix}$$

which is clearly 2. Of course, as we found in Section 2.5, the E found here is the result of row reducing B and L is a record of the row operations that we did. Ordinarily, we would not actually write L but rather do row operations on B until we reached E, then claim that, as the rank of E is 2, the rank of B is 2 also. $\quad\Box$

Exercises 3.6

For each of the following sets of vectors, decide if they span $\mathbf{R}^3$. If they do, show that they do. If they do not, find a vector that is not in the subspace spanned by the vectors.

1. $\{(1, -1, 0), (3, 1, 0)\}$
2. $\{(1, 0, 0), (-2, 1, 0), (1, 1, -1)\}$
3. $\{(2, 1, -1), (1, 0, -1), (1, 1, 0), (0, 1, 1)\}$
4. $\{(1, 1, 2), (1, 2, 3), (1, 0, 4), (1, -3, 1)\}$

5. Find a basis for the subspace of $\mathbf{R}^4$ spanned by $(1, -1, 5, -5)$, $(1, 1, -1, 1)$, and $(2, 1, 1, -1)$. (See Exercise 3.3.6.) What is the dimension of this subspace?

6. Find a basis for the subspace of $\mathbf{R}^4$ spanned by $(1, -1, 1, 2), (1, 1, -1, 1),$ and $(2, 1, 1, -1)$. (See Exercise 3.3.7.) What is the dimension of this subspace?
7. Find a basis for the subspace of $\mathbf{R}^4$ spanned by $(1, -3, 0, 2), (1, 0, 1, 1), (0, 1, 1, -1),$ and $(2, 1, 1, 3)$. (See Exercise 3.3.8.) What is the dimension of this subspace?
8. What is the dimension of the solution space of the system:

$$\begin{cases} u + 3v - w + 2x + y &= 0 \\ u + 2v + 4w + 2x &= 0 \\ 2u + 8v + w + 3x - y &= 0 \end{cases}$$

(See Exercise 3.5.2.)

For each of the following matrices, find bases for the range and the nullspace and find the dimensions of these subspaces.

9. $\begin{pmatrix} 1 & -1 \\ 2 & 1 \end{pmatrix}$

10. $\begin{pmatrix} 2 & -1 \\ 0 & 0 \\ -4 & 2 \end{pmatrix}$

11. $\begin{pmatrix} 1 & -1 & 1 \\ 2 & -2 & 2 \\ -1 & 1 & -1 \end{pmatrix}$

12. $\begin{pmatrix} 2 & -1 & 0 \\ 1 & 0 & 1 \\ 0 & 1 & 2 \end{pmatrix}$

13. $\begin{pmatrix} 1 & -1 & 1 & 0 \\ 2 & 1 & 0 & 1 \\ 1 & 1 & 3 & 1 \\ 0 & -1 & 4 & 0 \end{pmatrix}$

14. $\begin{pmatrix} 1 & -1 & 1 & 0 \\ 2 & 1 & 0 & 1 \\ 1 & 2 & -1 & 1 \\ 3 & 3 & -1 & 2 \end{pmatrix}$

15. $\begin{pmatrix} 2 & -1 & 0 & 1 \\ 4 & -2 & 0 & 2 \\ -2 & 1 & 0 & -1 \\ 2 & -1 & 0 & 1 \end{pmatrix}$

16. $\begin{pmatrix} 1 & -1 & 2 & 1 \\ 2 & -3 & 4 & -1 \\ -1 & 2 & 0 & 1 \\ -1 & 2 & -1 & 2 \end{pmatrix}$

17. The vectors $v_1 = (1, 1, 1)$ and $v_2 = (0, 1, 1)$ are linearly independent in $\mathbf{R}^3$.
    (a) Find a third vector v_3 so that $\{v_1, v_2, v_3\}$ is linearly independent.
    (b) Since $\mathbf{R}^3$ is three dimensional, $\{v_1, v_2, v_3\}$ must span the space. Express each of the standard basis vectors e_1, e_2, e_3 as a linear combination of your basis vectors, v_1, v_2, v_3.
18. Suppose $\mathcal{U}$ is a three dimensional subspace of $\mathbf{R}^4$ and $\{u_1, u_2, u_3\}$ is a basis for $\mathcal{U}$. Show that $v_1 = u_1 + u_3, v_2 = u_2 - u_3,$ and $v_3 = u_1 + u_2 + u_3$ also forms a basis for $\mathcal{U}$.
19. Show that if $w_1, w_2, \cdots, w_n$ is a basis for $\mathbf{R}^n$ and T is an invertible $n \times n$ matrix, then $Tw_1, Tw_2, \cdots, Tw_n$ is also a basis for $\mathbf{R}^n$.
20. Use the result of Exercise 3.4.22. to prove that if C is an $m \times n$ matrix, L is an invertible $m \times m$ matrix, and $C = LE$, then the (column) rank of C is the same as the (column) rank of E. (See Exercise 3.6.19.)
21. Let V be a finite dimensional vector space.
    For each of the situations in (a)–(e), choose the *best* (one) answer from the list below.

(a) The vectors v_1, v_2, v_3, and v_4 are linearly independent in V, so
(b) The vectors v_1, v_2, v_3, and v_4 are linearly dependent in V, so
(c) The vectors v_1, v_2, v_3, and v_4 span V, so
(d) The vectors v_1, v_2, v_3, and v_4 do not span V, so
(e) The vectors v_1, v_2, v_3, and v_4 are linearly independent and span V, so
      (i) dimension $V \geq 4$.
      (ii) dimension $V \leq 4$.
      (iii) dimension $V = 4$.
      (iv) we cannot deduce anything about dimension V.

22. The matrix C is

$$C = \begin{pmatrix} 1 & -1 & 1 \\ 1 & 2 & -1 \\ 1 & -4 & 3 \end{pmatrix}$$

and the matrix D is a 3×5 matrix with rows D_1, D_2, and D_3.
(a) Write D as a block matrix, blocked in rows, and find CD as a block matrix, blocked in rows.
(b) Suppose, now, that the rank of D is 2 and $D_2 = -2D_1$. Find a basis for the row space of D.
(c) Suppose, as in part (b), that the rank of D is 2 and $D_2 = -2D_1$. Find a basis for the row space of CD.

3.7 Rank–Nullity Theorem

In this section, we will examine the relationship between two of the important subspaces associated with a matrix. Recall from Section 3.2 that the *nullspace* of the $m \times n$ matrix A is the subspace of $\mathbf{R}^n$

$$\mathcal{N}(A) = \{X \in \mathbf{R}^n : AX = 0\}$$

The dimension of the nullspace of A is frequently called the *nullity* of A. Also, we defined the *range of A* to be the subspace of $\mathbf{R}^m$

$$\mathcal{R}(A) = \{b \in \mathbf{R}^m : AX = b \text{ for some } X \text{ in } \mathbf{R}^n\}$$

and defined the *rank of A* to be the dimension of the range of A. Even though for an $m \times n$ matrix A the nullspace is a subspace of $\mathbf{R}^n$ and range of A is a subspace of $\mathbf{R}^m$, Example 3.19 gave a specific instance of a connection between the range and the nullspace of a matrix. More generally, their dimensions are related: that is the content of the Rank–Nullity Theorem proved below. The proof will consist of finding bases for the subspaces involved and relating them to a basis for $\mathbf{R}^n$.

THEOREM 3.50 Rank–Nullity Theorem

If A is an $m \times n$ matrix, then

$$dim\mathcal{R}(A) + dim\mathcal{N}(A) = n$$

EXAMPLE 3.51

In Example 3.42, we found that $(1, 2, -1)$ and $(0, -1, 2)$ form a basis for the range of B where

$$B = \begin{pmatrix} 1 & 0 & 2 & 2 \\ 2 & -1 & 3 & 1 \\ -1 & 2 & 0 & 4 \end{pmatrix}$$

In particular, this means the rank of B is 2. Let's illustrate the Rank–Nullity Theorem by finding a basis for the nullspace of B as well. To do so, we must solve the system

$$\begin{pmatrix} 1 & 0 & 2 & 2 \\ 2 & -1 & 3 & 1 \\ -1 & 2 & 0 & 4 \end{pmatrix} \begin{pmatrix} w \\ x \\ y \\ z \end{pmatrix} = \begin{pmatrix} 0 \\ 0 \\ 0 \end{pmatrix}$$

Row reducing the augmented matrix for this system (or just row reducing B) by replacing the second row by the second row plus -2 times the first row and replacing the third row by the third row plus the first row gives

$$\left(\begin{array}{cccc|c} 1 & 0 & 2 & 2 & 0 \\ 2 & -1 & 3 & 1 & 0 \\ -1 & 2 & 0 & 4 & 0 \end{array} \right) \longrightarrow \left(\begin{array}{cccc|c} 1 & 0 & 2 & 2 & 0 \\ 0 & -1 & -1 & -3 & 0 \\ 0 & 2 & 2 & 6 & 0 \end{array} \right)$$

Now, replacing the third row by the third row plus 2 times the second, we get

$$\longrightarrow \left(\begin{array}{cccc|c} 1 & 0 & 2 & 2 & 0 \\ 0 & -1 & -1 & -3 & 0 \\ 0 & 0 & 0 & 0 & 0 \end{array} \right)$$

which says the solution of the system is $w = -2y - 2z$ and $x = -y - 3z$ where y and z are arbitrary. In other words, choosing $y = 1$ and $z = 0$, then $y = 0$ and

$z = 1$, we find $u_1 = (-2, -1, 1, 0)$ and $u_2 = (-2, -3, 0, 1)$ are a basis for $\mathcal{N}(B)$. In this case, B is a 3×4 matrix, and

$$\operatorname{rank}(B) + \operatorname{nullity}(B) = 2 + 2 = 4 = n$$

as the Rank–Nullity Theorem predicts. The proof of the theorem is organized around calculations like these. Specifically, after finding a basis for the nullspace, in this case, $(-2, -1, 1, 0)$ and $(-2, -3, 0, 1)$, we extend this linearly independent set to a basis for $\mathbf{R}^4$, for example, it is easy to see that

$$\left\{ \begin{pmatrix} -2 \\ -1 \\ 1 \\ 0 \end{pmatrix}, \begin{pmatrix} -2 \\ -3 \\ 0 \\ 1 \end{pmatrix}, \begin{pmatrix} 1 \\ 0 \\ 0 \\ 0 \end{pmatrix}, \begin{pmatrix} 0 \\ 1 \\ 0 \\ 0 \end{pmatrix} \right\}$$

is a basis for $\mathbf{R}^4$. The proof then shows that the images under multiplication by B of the vectors we added to form the basis actually form a basis for the range of B, in this case, that would be

$$Be_1 = \begin{pmatrix} 1 & 0 & 2 & 2 \\ 2 & -1 & 3 & 1 \\ -1 & 2 & 0 & 4 \end{pmatrix} \begin{pmatrix} 1 \\ 0 \\ 0 \\ 0 \end{pmatrix} = \begin{pmatrix} 1 \\ 2 \\ -1 \end{pmatrix}$$

and

$$Be_2 = \begin{pmatrix} 1 & 0 & 2 & 2 \\ 2 & -1 & 3 & 1 \\ -1 & 2 & 0 & 4 \end{pmatrix} \begin{pmatrix} 0 \\ 1 \\ 0 \\ 0 \end{pmatrix} = \begin{pmatrix} 0 \\ -1 \\ 2 \end{pmatrix}$$

and we found in Example 3.42 that these vectors are indeed a basis for the range of B. $\quad \Box$

Another way to view the Rank–Nullity Theorem is suggested by Example 3.46. The Rank–Nullity Theorem says the dimension of the nullspace of a matrix plus the dimension of the range of the matrix is the number of columns of the matrix. In Example 3.46, we saw that the dimension of nullspace is the number of free variables in the row equivalent reduced row echelon matrix and the dimension of the range is the number of pivot variables. Since every column corresponds to either a free variable or a pivot, but not both, this sum is the number of columns.

We are now ready to prove the Rank–Nullity Theorem.

PROOF Supposing A is an $m \times n$ matrix and the nullity of A is j, let $u_1, u_2, \cdots, u_j$ be a basis for the nullspace of A, which is a subspace of $\mathbf{R}^n$. Since this is a linearly independent set, it can be extended to a basis for $\mathbf{R}^n$, that is, we can find $v_1, v_2, \cdots,$ v_k so that $\{v_1, \cdots, v_k, u_1, \cdots, u_j\}$ is a basis for $\mathbf{R}^n$. Because the dimension of $\mathbf{R}^n$ is n, this means that $k + j = n$. Since the nullity of A is j, we need to prove that the rank of A is k. Let $w_1 = Av_1$, $w_2 = Av_2$, $\cdots$, $w_k = Av_k$. Clearly, the w's are in the range of A; we want to show that they are actually a basis for the range!

Suppose z is in $\mathcal{R}(A)$, say, $z = Ax$ for some x in $\mathbf{R}^n$. Since the u's and v's form a basis for $\mathbf{R}^n$, there are scalars so that

$$x = a_1 v_1 + a_2 v_2 + \cdots + a_k v_k + b_1 u_1 + b_2 u_2 + \cdots + b_j u_j$$

This means that

$$z = Ax = a_1 Av_1 + a_2 Av_2 + \cdots + a_k Av_k + b_1 Au_1 + b_2 Au_2 + \cdots + b_j Au_j$$

Remembering that the u's are in the nullspace of A and the definition of the w's, this is

$$z = a_1 w_1 + a_2 w_2 + \cdots + a_k w_k$$

so z is a linear combination of the w's. Since this is true for all vectors z in $\mathcal{R}(A)$, the w's span the range.

We claim that the w's are linearly independent, also. Suppose that there are scalars so that

$$c_1 w_1 + c_2 w_2 + \cdots + c_k w_k = 0$$

Then, we get

$$0 = c_1 Av_1 + c_2 Av_2 + \cdots + c_k Av_k = A(c_1 v_1 + c_2 v_2 + \cdots + c_k v_k)$$

This says that the vector $c_1 v_1 + c_2 v_2 + \cdots + c_k v_k$ is in the nullspace of A. Since the u's are a basis for the nullspace of A, there are scalars so that

$$c_1 v_1 + c_2 v_2 + \cdots + c_k v_k = d_1 u_1 + d_2 u_2 + \cdots + d_j u_j$$

or

$$c_1 v_1 + c_2 v_2 + \cdots + c_k v_k + (-d_1)u_1 + (-d_2)u_2 + \cdots + (-d_j)u_j = 0$$

Now the u's and v's were constructed to be a basis for $\mathbf{R}^n$, so they are linearly independent, that is, the only way this equation could be true is for $c_1 = c_2 = \cdots = c_k = 0$ and $-d_1 = -d_2 = \cdots = -d_j = 0$. Thus, we began with a linear combination of the w's being zero and have discovered that this implies their coefficients are all zero, in other words, the w's are linearly independent.

We have shown that $w_1, w_2, \cdots, w_k$ is a basis for the range of A, which means that the rank of A is k. Since $k + j = n$, we have proved the result.   ∎

Sometimes it is difficult to remember for an $m \times n$ matrix whether the rank–nullity theorem says the rank plus the nullity adds to m or n. Remembering the idea of the proof — that a special basis for the space containing the nullspace is constructed — will help remind you that it is the number of columns of A since the nullspace in contained in $\mathbf{R}^n$.

If A is an $m \times n$ matrix, the system $AX = b$ is a system of m equations in n unknowns. If the rank of A is k, then according to the Rank–Nulliuty Theorem, the dimension of the nullspace of A plus the rank adds up to n. Put a little differently, this says that the dimension of the nullspace is n minus the rank. In other words, the number of variables that can be solved for is the rank and the number of free variables plus the rank is the total number of variables.

EXAMPLE 3.52

Suppose A is a 6×9 matrix with rank 4. What are the dimensions of the nullspace of A and the range and nullspace of A'?

SOLUTION The Rank–Nullity Theorem says that

$$\dim(\mathcal{R}(A)) + \dim(\mathcal{N}(A)) = 9$$

which means, since $\dim(\mathcal{R}(A)) = 4$, that the dimension of the nullspace of A is $9 - 4 = 5$. We also know that the rank of A is the same as the rank of A', so $\dim(\mathcal{R}(A')) = 4$. Since A' is a 9×6 matrix, the Rank–Nullity Theorem also says

$$\dim(\mathcal{R}(A')) + \dim(\mathcal{N}(A')) = 6$$

so $\dim(\mathcal{N}(A')) = 2$.   □

We can use the Rank–Nullity Theorem to save effort in finding bases for the important subspaces for a matrix.

EXAMPLE 3.53

Find bases for the range and the nullspace of the matrix C where

$$C = \begin{pmatrix} 1 & 2 & 0 & -1 \\ -1 & -2 & -2 & 3 \\ 1 & 2 & -2 & 1 \end{pmatrix}$$

SOLUTION As in the proof of the Rank–Nullity Theorem, we will first find a basis for the nullspace of C. Solving $CX = 0$ by row reduction gives

$$\left(\begin{array}{cccc|c} 1 & 2 & 0 & -1 & 0 \\ -1 & -2 & -2 & 3 & 0 \\ 1 & 2 & -2 & 1 & 0 \end{array} \right) \longrightarrow \left(\begin{array}{cccc|c} 1 & 2 & 0 & -1 & 0 \\ 0 & 0 & -2 & 2 & 0 \\ 0 & 0 & -2 & 2 & 0 \end{array} \right)$$

$$\longrightarrow \left(\begin{array}{cccc|c} 1 & 2 & 0 & -1 & 0 \\ 0 & 0 & -2 & 2 & 0 \\ 0 & 0 & 0 & 0 & 0 \end{array} \right) \longrightarrow \left(\begin{array}{cccc|c} 1 & 2 & 0 & -1 & 0 \\ 0 & 0 & 1 & -1 & 0 \\ 0 & 0 & 0 & 0 & 0 \end{array} \right)$$

In other words, the solution to the system is $w = -2x + z$ and $y = z$ where x and z are arbitrary. Thus, for example, the vectors $u_1 = (-2, 1, 0, 0)$ and $u_2 = (1, 0, 1, 1)$ form a basis for $\mathcal{N}(C)$, the nullspace of C. Since C is a 3×4 matrix, the Rank–Nullity Theorem says the rank of C is

$$\text{rank}(C) = 4 - \text{nullity}(C) = 4 - 2 = 2$$

Thus, two linearly independent columns of C will form a basis for the range of C! Clearly, since the second column is twice the first, the first two columns of C will not do, but we can see that $(1, -1, 1)$ and $(0, -2, -2)$ are linearly independent. Since there are two of them and, by the calculation above, since the rank of C is 2, these must span the range of C. That this is the case is easily checked:

$$\begin{pmatrix} 2 \\ -2 \\ 2 \end{pmatrix} = 2 \begin{pmatrix} 1 \\ -1 \\ 1 \end{pmatrix} + 0 \begin{pmatrix} 0 \\ -2 \\ -2 \end{pmatrix}$$

and

$$\begin{pmatrix} -1 \\ 3 \\ 1 \end{pmatrix} = (-1) \begin{pmatrix} 1 \\ -1 \\ 1 \end{pmatrix} + (-1) \begin{pmatrix} 0 \\ -2 \\ -2 \end{pmatrix}$$

In the proof of the theorem, this answer corresponds to choosing u_1, u_2, e_1, and e_3 for the basis for $\mathbf{R}^4$ and observing that therefore the first and third columns of C, that is, Ce_1 and Ce_3 are a basis for the range of C.

Examining the row echelon form for C that we found above in which the leading entries were the $(1, 1)$ entry and the $(2, 3)$ entry, we see that this calculation connects the proof of the Rank–Nullity Theorem and the ideas of Theorems 3.41 and 3.48: the number of columns of the row echelon form that contain leading entries (the rank of the matrix) plus the number of entries that do not (the number of free variables in the solution of the homogeneous system, that is, the nullity of the matrix) is the total number of columns of the matrix (n). Moreover, Corollary 3.56 below shows that because the first and third columns of the row echelon matrix are a basis for its column space (they contain the leading entries), the first and third columns of C will be a basis for the column space of C. This is precisely the structure of the proof of Theorem 3.50! $\square$

LEMMA 3.54

If B is an invertible $k \times k$ matrix and v_1, v_2, $\cdots$, v_j are vectors in $\mathbf{R}^k$, then Bv_1, Bv_2, $\cdots$, Bv_j are linearly independent if and only if v_1, v_2, $\cdots$, v_j are linearly independent.

PROOF   (See Exercises 3.4.21., 3.4.22., and 3.6.19.) If the vectors v_1, v_2, $\cdots$, v_j are linearly dependent, say

$$a_1 v_1 + a_2 v_2 + \cdots + a_j v_j = 0$$

and not all the a's are zero, then

$$a_1(Bv_1) + a_2(Bv_2) + \cdots + a_j(Bv_j) = B(a_1 v_1 + a_2 v_2 + \cdots + a_j v_j) = B0 = 0$$

so the vectors Bv_1, Bv_2, $\cdots$, Bv_j are dependent.

Conversely, if the vectors v_1, v_2, $\cdots$, v_j are linearly independent and b_1, b_2, $\cdots$, b_j are scalars so that

$$b_1(Bv_1) + b_2(Bv_2) + \cdots + b_j(Bv_j) = 0$$

then

$$B(b_1 v_1 + b_2 v_2 + \cdots + b_j v_j) = 0$$

and, since B is invertible,

$$b_1 v_1 + b_2 v_2 + \cdots + b_j v_j = B^{-1}B(b_1 v_1 + b_2 v_2 + \cdots + b_j v_j) = B^{-1}0 = 0$$

so all the coefficients, b_1, b_2, $\cdots$, b_j, must be zero and the vectors Bv_1, Bv_2, $\cdots$, Bv_j are linearly independent.   ∎

THEOREM 3.55

If A is an $m \times n$ matrix, S is an $m \times m$ invertible matrix and T is an $n \times n$ invertible matrix, then

$$\text{rank}(SA) = \text{rank}(A) = \text{rank}(AT)$$

PROOF The rank of SA is the dimension of the column space of SA and the rank of A is the dimension of the column space of A. The rank of A is j if we can find j linearly independent columns of A and there are not $j + 1$ independent columns of A. Now the columns of SA are just S times the columns of A:

$$SA = S \begin{pmatrix} A_1 & A_2 & \cdots & A_n \end{pmatrix} = \begin{pmatrix} SA_1 & SA_2 & \cdots & SA_n \end{pmatrix}$$

Thus, because S is invertible, Lemma 3.54 shows that a set of columns of A is linearly independent if and only if the corresponding columns of SA are linearly independent. This means that the ranks of A and SA are the same (and that the corresponding columns form bases for their ranges).

Suppose $v_1, v_2, \cdots, v_j$ are a basis of the nullspace of AT. Then, since T is invertible, Lemma 3.54 implies $Tv_1, Tv_2, \cdots, Tv_j$ are linearly independent vectors. Moreover, each of these vectors is in the nullspace of A since $A(Tv_i) = (AT)v_i = 0$. On the other hand, these vectors span the nullspace of A. Indeed, if w is in the nullspace of A, then $(AT)(T^{-1}w) = Aw = 0$, so $T^{-1}w$ is in the nullspace of AT. This means there are scalars $c_1, c_2, \cdots, c_j$ so that $T^{-1}w = c_1v_1 + c_2v_2 + \cdots + c_jv_j$ and

$$w = T(T^{-1}w) = T(c_1v_1 + c_2v_2 + \cdots + c_jv_j) = c_1(Tv_1) + c_2(Tv_2) + \cdots + c_j(Tv_j)$$

Thus, w is a linear combination of these vectors. This means that $Tv_1, Tv_2, \cdots, Tv_j$ are a basis for the nullspace of A.

In other words, the dimensions of the nullspaces of A and AT are the same. Since both A and AT are $m \times n$ matrices, by the rank–nullity theorem,

$$\text{rank}(A) = n - \text{nullity}(A) = n - \text{nullity}(AT) = \text{rank}(AT)$$

∎

COROLLARY 3.56

Suppose B and E are two $m \times n$ matrices and L is an invertible matrix so that $B = LE$, then the dependence relations among the columns of B and dependence relations among the columns of E are the same. In particular, if a set of columns of E form a basis for the column space of E, the same set of columns of B form a basis for the column space of B.

PROOF   This is essentially the first paragraph of the proof of the theorem above, where A is E and S is L.   ▌

THEOREM 3.57

Suppose B and E are two $m \times n$ matrices and L is an invertible matrix so that $B = LE$, then the row space of B is the same as the row space of E.

PROOF   Block B and E into rows and consider $B = LE$ as a block matrix multiplication.

$$\begin{pmatrix} B_1 \\ B_2 \\ \cdots \\ B_m \end{pmatrix} = \begin{pmatrix} l_{11} & l_{12} & \cdots & l_{1m} \\ l_{21} & l_{22} & \cdots & l_{2m} \\ & & \vdots & \vdots \\ l_{m1} & l_{m2} & \cdots & l_{mm} \end{pmatrix} \begin{pmatrix} E_1 \\ E_2 \\ \cdots \\ E_m \end{pmatrix}$$

In particular, the ith row of B is computed as

$$B_i = l_{i1}E_1 + l_{i2}E_2 + \cdots + l_{im}E_m$$

so every row of B is a linear combination of rows of E. This means the span of the rows of B is a subspace of the span of the rows of E. Conversely, if $U = L^{-1}$, then $E = UB$ and the same argument shows every row of E is a linear combination of the rows of B, which means the row space of E is a subspace of the row space of B. These two containments mean that the row space of B and the row space of E are the same.   ▌

We have used row operations to find solve systems of equations and to find determinants. Corollary 3.56 and Theorem 3.57 are the keys to applying the row operations to find bases of row or column spaces of matrices. (Of course, since finding the nullspace of a matrix is synonymous with solving a homogeneous system, we have already been using row operations to find nullspaces.) In the application of these theorems, the matrix E will usually be a matrix in row echelon or reduced row echelon form because they arise in the calculations naturally and their rows and columns are very easy to check for independence.

EXAMPLE 3.58

Find bases for the nullspace, row space, and the column space of

$$B = \begin{pmatrix} 1 & -1 & 2 & 1 & 4 \\ -1 & 1 & 1 & 0 & -5 \\ 2 & -2 & 7 & 2 & 8 \\ 1 & -1 & 5 & 1 & 4 \end{pmatrix}$$

SOLUTION First, we row reduce B to get the row echelon matrix E that satisfies $B = LE$ for L invertible. (We will not need, and do not compute, L.)

$$
\begin{pmatrix}
1 & -1 & 2 & 1 & 4 \\
-1 & 1 & 1 & 0 & -5 \\
2 & -2 & 7 & 2 & 8 \\
1 & -1 & 5 & 1 & 4
\end{pmatrix}
\rightarrow
\begin{pmatrix}
1 & -1 & 2 & 1 & 4 \\
0 & 0 & 3 & 1 & -1 \\
0 & 0 & 3 & 0 & 0 \\
0 & 0 & 3 & 0 & 0
\end{pmatrix}
\rightarrow
$$

$$
\begin{pmatrix}
1 & -1 & 2 & 1 & 4 \\
0 & 0 & 3 & 0 & 0 \\
0 & 0 & 3 & 1 & -1 \\
0 & 0 & 3 & 0 & 0
\end{pmatrix}
\rightarrow
\begin{pmatrix}
1 & -1 & 2 & 1 & 4 \\
0 & 0 & 1 & 0 & 0 \\
0 & 0 & 3 & 1 & -1 \\
0 & 0 & 3 & 0 & 0
\end{pmatrix}
\rightarrow
$$

$$
\begin{pmatrix}
1 & -1 & 2 & 1 & 4 \\
0 & 0 & 1 & 0 & 0 \\
0 & 0 & 0 & 1 & -1 \\
0 & 0 & 0 & 0 & 0
\end{pmatrix}
$$

This calculation implies $B = LE$ where

$$
E =
\begin{pmatrix}
\textcircled{1} & -1 & 2 & 1 & 4 \\
0 & 0 & \textcircled{1} & 0 & 0 \\
0 & 0 & 0 & \textcircled{1} & -1 \\
0 & 0 & 0 & 0 & 0
\end{pmatrix}
$$

Since L is invertible, $BX = 0$ if and only if $EX = L^{-1}BX = 0$, so the nullspace of B is the same as the nullspace of E. The three equations necessary for finding the nullspace of E are $u - v + 2w + x + 4y = 0$, $w = 0$, and $x - y = 0$, so we find v and y are arbitrary and $u = v - 5y$, $w = 0$, and $x = y$, so $(1, 1, 0, 0, 0)$ and $(-5, 0, 0, 1, 1)$ are a basis for the nullspace of E, hence, for the nullspace of B.

Now, it is easy to see that the first, second, and third rows of E are linearly independent and span the row space of E; since Theorem 3.57 says the row space of E is the same as the row space of B, these vectors, $(1, -1, 2, 1, 4)$, $(0, 0, 1, 0, 0)$ and $(0, 0, 0, 1, -1)$, form a basis for the row space of B.

It is also easy to see that the first, third, and fourth columns of E are linearly independent and span the column space of E. Indeed, denoting the columns of E by E_1, E_2, and so on, we see $E_2 = -E_1 + 0E_3 + 0E_4$ and $E_5 = 5E_1 + 0E_3 - E_4$. Corollary 3.56 says that B_1, B_3, and B_4 form a basis for the column space of A and $B_2 = -B_1 + 0B_3 + 0B_4$ and $B_5 = 5B_1 + 0B_3 - B_4$. In other words, the vectors $(1, -1, 2, 1)$, $(2, 1, 7, 5)$, and $(1, 0, 2, 1)$ form a basis for the column space of B.   ▯

Theorem 3.57 and Corollary 3.56 can also be used to prove Theorem 3.41. If $A = LE$ where L is invertible and E is in echelon form, then the row rank of E is the same as the row rank of A and column rank of E is the same as the column rank of A. Now, because E is in echelon form, the row rank of E is clearly the number of non-zero rows of E. On the other hand, since the columns of E that contain a leading entry of some row of E are easily seen to form a basis for the column space of E, the column rank of E is be the number of leading entries of the rows of E, that is, it is the number of non-zero rows of E. Thus, the column rank of E and the row rank of E are each equal to the number of non-zero rows of E and are equal to each other.

Exercises 3.7

For each of the following matrices, find a basis for the range (column space), find a basis for the range of the adjoint, find a basis for the nullspace, find a basis for the nullspace of the adjoint, and find the dimensions of all four subspaces.

1. $\begin{pmatrix} -2 & 1 \\ 4 & -2 \end{pmatrix}$

2. $\begin{pmatrix} 2 & 1 \\ 3 & -2 \end{pmatrix}$

3. $\begin{pmatrix} -1 & 1 & 0 \\ 1 & -2 & 2 \end{pmatrix}$

4. $\begin{pmatrix} 1 & 2 & -1 \\ 2 & 1 & 1 \\ 1 & 5 & -4 \end{pmatrix}$

5. $\begin{pmatrix} 1 & 1 & 0 & -1 \\ 2 & -1 & 1 & 0 \\ 0 & 3 & -1 & -2 \end{pmatrix}$

6. $\begin{pmatrix} 0 & 1 & 1 & -1 \\ 1 & -1 & 1 & 2 \\ 2 & -2 & 1 & -1 \end{pmatrix}$

7. $\begin{pmatrix} 2 & 1 & -1 & -1 \\ 1 & -1 & -1 & 0 \\ 1 & 1 & -1 & 1 \\ 1 & 4 & 0 & 0 \end{pmatrix}$

8. $\begin{pmatrix} 1 & 1 & 0 & 1 & -1 \\ 0 & 2 & -1 & -1 & 1 \\ 1 & 3 & -1 & -2 & 1 \\ 0 & 2 & -1 & 1 & 0 \end{pmatrix}$

9. Suppose A is a 5×9 matrix and suppose the dimension of the nullspace of A is 4. Use Theorems 3.41 and 3.50 to find the dimension of the range of A, the dimension of the

range of A', and the dimension of the nullspace of A'.

10. Suppose C is a 7×12 matrix with rank 4. Use Theorems 3.41 and 3.50 to find the dimension of the range of C, the dimension of the nullspace of C, the dimension of the range of C', and the dimension of the nullspace of C'.

11. Suppose D is a 11×8 matrix and suppose that the nullspace of D is 3 dimensional. Use Theorems 3.41 and 3.50 to find the dimension of the range of D, the dimension of the nullspace of D, the dimension of the range of D', and the dimension of the nullspace of D'.

12. Theorem 3.57 says that if $B = LE$ where L is invertible then the row space of B is the same as the row space of E. Corollary 3.56 says that in this situation, the column space of B has the same dimension as the column space of E. Give an example of B and E so that $B = LE$ for some invertible matrix L such that the column space of B and the column space of E are not the same subspace.

13. John and Mary are taking linear algebra. One of the problems in their homework assignment was to find the nullspace $\mathcal{N}(A)$ of a certain 4×5 matrix. John's answer was

$$\mathcal{N}(A) = \text{span}\left\{ \begin{pmatrix} -2 \\ -2 \\ 0 \\ 2 \\ -6 \end{pmatrix}, \begin{pmatrix} 1 \\ 5 \\ 4 \\ -3 \\ 11 \end{pmatrix}, \begin{pmatrix} 3 \\ 5 \\ 2 \\ -4 \\ 13 \end{pmatrix}, \begin{pmatrix} 0 \\ -2 \\ -2 \\ 1 \\ -4 \end{pmatrix} \right\}$$

Mary's answer was

$$\mathcal{N}(A) = \text{span}\left\{ \begin{pmatrix} 1 \\ 1 \\ 0 \\ -1 \\ 3 \end{pmatrix}, \begin{pmatrix} -2 \\ 0 \\ 2 \\ 1 \\ -2 \end{pmatrix}, \begin{pmatrix} -1 \\ 3 \\ 4 \\ -1 \\ 5 \end{pmatrix} \right\}$$

(a) Are their answers consistent with each other? (Explain your answer!)

(b) What is the dimension of $\mathcal{N}(A)$?

14. Tom and Janice are taking linear algebra. One of the problems in their homework assignment was to find a basis for the column space of a certain 5×6 matrix B. Tom's and Janice's answers were

$$\text{Tom:} \quad \left\{ \begin{pmatrix} 2 \\ 1 \\ 2 \\ -3 \\ 2 \end{pmatrix}, \begin{pmatrix} 1 \\ 1 \\ 0 \\ -1 \\ -1 \end{pmatrix}, \begin{pmatrix} 3 \\ 2 \\ 2 \\ -4 \\ 1 \end{pmatrix}, \begin{pmatrix} 3 \\ -3 \\ 0 \\ 3 \\ 1 \end{pmatrix} \right\}$$

and

$$\text{Janice:} \quad \left\{ \begin{pmatrix} 1 \\ -1 \\ 1 \\ 0 \\ 2 \end{pmatrix}, \begin{pmatrix} 2 \\ 0 \\ 1 \\ -1 \\ 1 \end{pmatrix}, \begin{pmatrix} 0 \\ 1 \\ 1 \\ -2 \\ 1 \end{pmatrix} \right\}$$

    (a) How can you tell immediately that they do not agree?

    (b) Assuming Janice is correct, what change should Tom make in his answer?

15. Suppose E is a $m \times n$ matrix and F is an $n \times p$ matrix. In this exercise, you will prove that the rank of EF is no more than the smaller of the rank of E and the rank of F.

    (a) Show that the nullspace of F is contained in the nullspace of EF, that is, show that every vector in the nullspace of F is in the nullspace of EF.

    (b) Use part (a) and the rank–nullity theorem (Theorem 3.50) to show that the rank of EF is less than or equal to the rank of F.

    (c) Show that the range of EF is contained in the range of E and conclude that the rank of EF is less than or equal to the rank of E.

3.8   Reprise: Theoretical Foundations of Solution of Systems

In this section, we will try to summarize the most important ideas of the preceding material as it relates to the solution of systems of equations. While the motivation for the development of subspace, linear independence, dimension, and range and nullspace of a matrix is rooted in the study of linear systems, this motivation sometimes gets lost in the jumble of ideas. The results presented here make specific connections between the solution of systems of equations and the ideas of range and nullspace of a matrix and their dimensions. When solving systems of equations, it is important to keep these ideas in mind: they allow you to interpret the arithmetic you do and to plan your attack on the problem.

The Relation between Homogeneous and Non-homogeneous Systems:

Although the language used in the statement of the theorem below is the language introduced in Chapter 2, you should recognize that it is the motivation for the development of nullspace: the solution set of the homogeneous system $AX = 0$ is the nullspace of A and first part of the theorem reminds us that the nullspace is a subspace! Thus, the theorem explains how vectors in the nullspace of the coefficient matrix lead to solutions of the system, as long as the system is consistent in the first place. In particular, although the solution set of the system $AX = b$ is not a subspace if $b \neq 0$, the solution set will have the same "dimension" as the nullspace of A: second part of the theorem says that the solution space of the system $AX = b$

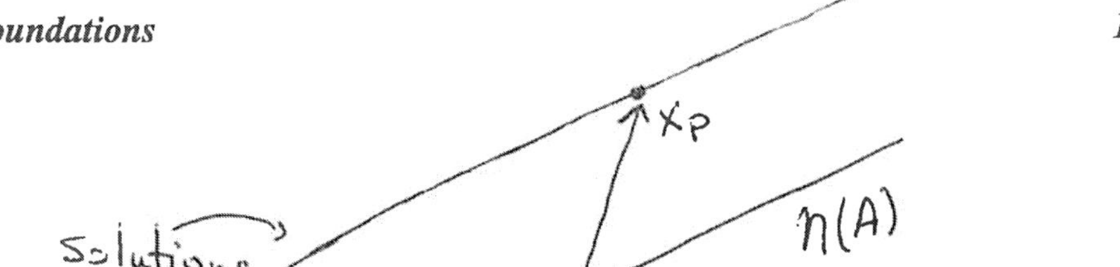

FIGURE 3.8

The geometry of solution of systems

is just a translate (by X_p) of the nullspace of A. In particular, a consistent system will have a unique solution exactly when the nullspace of the coefficient matrix is the trivial subspace $\{0\}$.

As noted in Section 2.2, in the solution of linear differential equations, one usually attempts to find one solution of the equation in question and *all* solutions of the associated homogeneous equation. This technique is a generalization of Theorem 2.2 which is restated below.

THEOREM 3.59

Let A be an $m \times n$ matrix.

1. *The solution set of the homogeneous system*

$$AX = 0$$

   is a subspace, that is, if X_1 and X_2 are both solutions of the homogeneous system and α is a number, then αX_1 and $X_1 + X_2$ are also solutions.

2. *If X_3 and X_4 are solutions of the system*

$$AX = b,$$

   then $X_3 - X_4$ is a solution of the homogeneous system $AX = 0$.

3. *If X_p is a solution of the non-homogeneous system*

$$AX = b,$$

   then every solution of the non-homogeneous system can be written as $X = X_p + X_0$, where X_0 is a solution of the homogeneous system $AX = 0$.

General Theory of Solutions:

The basic question to be answered is "How many solutions does the system $AX = b$ have?" Part of the answer, deduced from the theorem above, is that the system has either no solutions, exactly one solution, or infinitely many solutions. (This is because all subspaces except the zero subspace contain infinitely many vectors.) In addition to the theorems below, the Duality Theorem 4.22, an extension of the Rank–Nullity Theorem 3.50, is useful in considering these issues.

THEOREM 3.60

Let A be an $m \times n$ matrix.

1. *If b is in the range of A, the system $AX = b$ has one or more solutions. If b is not in the range of A, then the system $AX = b$ has no solutions.*

2. *If 0 is the only vector in $\mathcal{N}(A)$, the nullspace of A, then either $AX = b$ has no solutions or $AX = b$ has exactly one solution. If there are non-zero vectors in $\mathcal{N}(A)$, the nullspace of A, then either $AX = b$ has no solutions or $AX = b$ has infinitely many solutions.*

COROLLARY 3.61

Let A be an $m \times n$ matrix.
The system $AX = b$ has exactly *one solution for* every *b if and only if*
$m = n$ and $rank(A) = n$.

PROOF   Suppose every system $AX = b$ has exactly one solution. Since every vector b in $\mathbf{R}^m$ leads to a consistent system, the range of A must be $\mathbf{R}^m$, that is, the rank of A is m. The Rank–Nullity Theorem (Theorem 3.50) says that the dimension of the nullspace of A is $n - m$. On the other hand, since $AX = 0$ has exactly one solution, the dimension of the nullspace must be 0: in other words, $n = m$.

If $n = m$ and the rank of A is n, then the dimension of the range, which is a subspace of $\mathbf{R}^m$, is m, so by Theorem 3.40, the range is equal to $\mathbf{R}^m$, that is, every vector in $\mathbf{R}^m$ is in the range. Moreover, the dimension of the null space is $n - n = 0$, so the nullspace consists only of the zero subspace and there is exactly one solution for every system $AX = b$.   ∎

EXAMPLE 3.62

Suppose A is a 6×9 matrix and the dimension of the nullspace of A is 3. Are there vectors b in $\mathbf{R}^6$ for which $AX = b$ is inconsistent? If $AX = b$ is consistent, how many solutions does the system have?

SOLUTION The Rank–Nullity Theorem says that the dimension of the range of A is $9 - 3 = 6$. Thus, $\mathcal{R}(A)$ is a 6–dimensional subspace of $\mathbf{R}^6$, and it must be equal to $\mathbf{R}^6$ by Theorem 3.40. In particular, this means every vector in $\mathbf{R}^6$ is in the range and $AX = b$ is consistent for every b in $\mathbf{R}^6$. Now the nullspace of A is 3 dimensional, which means the nullspace of A contains infinitely many vectors. Thus, whenever X_p satisfies $AX_p = b$, then so does $X_p + X_h$ for infinitely many different vectors X_h from the nullspace of A. In other words, every system has infinitely many solutions. $\square$

Invertibility:

The special case of square matrices is especially important. The results are summarized in the following theorem about invertibility of matrices. Equation solving is connected to invertibility by the following observation: If A is invertible, the unique solution to the system $AX = b$ is $X = A^{-1}b$, although this is not usually the most efficient way to compute the solution.

All the information in the following theorem has already been discussed, however, it is helpful to think about it together to integrate the pieces of your knowledge. The proof of the equivalences is simply a matter of finding the earlier results to quote.

THEOREM 3.63

Let A be an $n \times n$ matrix. The following statements about A are equivalent, that is, if any one of them is true all the rest are true, and if any one is false, all the rest are false.

(a)   *A is invertible, that is, there is a matrix C so that $CA = AC = I$.*

(b)   *A is left invertible, that is, there is a matrix C so that $CA = I$.*

(c)   *A is right invertible, that is, there is a matrix C so that $AC = I$.*

(d)   *The only solution of the homogeneous system $AX = 0$ is $X = 0$.*

(e)   *The system $AX = b$ is consistent for every b in $\mathbf{R}^n$.*

(f)   *$\mathcal{R}(A)$, the range of A, is $\mathbf{R}^n$.*

(g)   *$\mathcal{N}(A)$, the nullspace of A, is zero.*

(h)   *$Rank(A) = n$.*

(i)   *The dimension of $\mathcal{N}(A)$, the nullspace of A, is zero.*

(j)   *$Det(A)$, the determinant of A, is not zero.*

PROOF   We will prove the result by proving chains of implications that form loops so that each statement implies the rest.

Recall that Corollary 2.32 says that a matrix is invertible if and only if its determinant is non-zero; in other words, *(a)* is equivalent to *(j)*.

Clearly, if C is an inverse of A, then it is a left inverse and a right inverse; that is, *(a)* implies *(b)* and *(a)* implies *(c)*.

On the other hand, if $AC = I$, then

$$\det(A)\det(C) = \det(AC) = \det(I) = 1$$

and similarly, if $CA = I$

$$\det(C)\det(A) = \det(CA) = \det(I) = 1$$

so if either *(b)* or *(c)* holds, then $\det(A) \neq 0$ and *(b)* implies *(j)* and *(c)* implies *(j)*.

To see that *(b)* implies *(d)*, suppose X_0 is a solution of $AX_0 = 0$. Then, for C a left inverse of A, we have $CAX_0 = C0$ which says $X_0 = IX_0 = 0$.

To see that *(c)* implies *(e)*, suppose b is a vector in $\mathbf{R}^n$ and suppose C a right inverse of A. Letting $X = Cb$, we see that $AX = ACb = Ib = b$. In other words, the system $AX = b$ is solvable for every right side b.

Now, the definition of the range is that it is the set of right hand sides for which $AX = b$ is solvable; since every system is solvable, the range of A is all of $\mathbf{R}^n$. In other words, *(e)* implies *(f)*.

Similarly, by definition, the nullspace is the set of vectors for which $AX = 0$. According to *(d)*, this consists of the single vector 0, so that *(d)* implies *(g)*.

The rank of A is just the dimension of the range, and since the range is a subset of $\mathbf{R}^n$, *(f)* and *(h)* are equivalent. Similarly, the nullspace is the zero subspace if and only if its dimension is 0, so *(g)* and *(i)* are equivalent.

The Rank–Nullity Theorem (Theorem 3.50) says that the rank of A and the nullity of A add up to the dimension of $\mathbf{R}^n$ which is n. Thus, *(h)* and *(i)* are equivalent.

If every system $AX = b$ is consistent, then we can find vectors $C_1, C_2, \cdots, C_n$ in $\mathbf{R}^n$ so that $AC_j = e_j$, the jth standard basis vector. Creating the matrix C that has these vectors as columns, we get

$$AC = A \left(C_1\ C_2\ \cdots\ C_n \right) = \left(AC_1\ AC_2\ \cdots\ AC_n \right)$$
$$= \left(e_1\ e_2\ \cdots\ e_n \right) = I$$

In other words, C is a right inverse for A and *(f)* implies *(c)*.   ∎

EXAMPLE 3.64

Suppose A is a 4×4 matrix and $\det(A) = 2.76$. Is the system $AX = b$ solvable for every b in $\mathbf{R}^4$? If $AX = b$ is consistent, how many solutions does it have?

SOLUTION We are given that $\det(A) \neq 0$, so *(j)* of Theorem 3.63 is true. This means that *all* the statements of the Theorem are true. In particular, *(e)* says that every system $AX = b$ is consistent. Moreover, since the only solution of $AX = 0$ is $X = 0$ (statement *(d)*). This means that whenever $AX = b$ is consistent, the solution is unique.   ⬚

Let's look again at the statics example with which we began the chapter (page 127). We are interested in solution of systems of the form $AX = b$ where

$$A = \begin{pmatrix} -3 & 3 & 0 & 0 \\ 3 & -8 & 5 & 0 \\ 0 & 5 & -5.4 & .4 \\ 0 & 0 & .4 & -.4 \end{pmatrix}$$

Now an easy calculation shows that $(1, 1, 1, 1)$ is a vector in the nullspace of A, indeed, it is a basis for the nullspace of A. It follows that the rank of A is $4 - 1 = 3$ and that $AX = b$ is solvable if and only if the vector b is in the 3–dimensional subspace $\mathcal{R}(A)$ of $\mathbf{R}^{(}4)$. In fact, a basis for $\mathcal{R}(A)$ is $\{(-3, 3, 0, 0), (3, -8, 5, 0), (0, 0, .4, -.4)\}$. In particular, this says that there is *no static equilibrium position for some sets of applied forces!* Moreover, it says that whenever there is an equilibrium position for some set of applied forces, there are infinitely many such positions. Indeed, we see explicitly that if (d_1, d_2, d_3, d_4) is a static equilibrium position for some set of applied forces then, because $(1, 1, 1, 1)$ is a basis for the nullspace of A, so is

$$(d_1,\ d_2,\ d_3,\ d_4) + c(1,\ 1,\ 1,\ 1) = (d_1 + c,\ d_2 + c,\ d_3 + c,\ d_4 + c)$$

This makes physical sense because it says that the other equilibrium positions are all translates of the whole picture to the right or left by c units.

Easy calculations show that $(2, 1, 0, -5)$ is not in the range of A so there is no static equilibrium position for this set of applied forces. Other calculations show that $(2, 1, 0, -3)$ is in the range of A so there are equilibrium positions for this set of applied forces. Indeed, it turns out that the equilibrium positions for this set of applied forces are $(-23/30 + c, -1/10 + c, 1/2 + c, 8 + c)$ where c is an arbitrary translation.

Now it is easier to understand the meaning of the range of A if we identify it in another way. It turns out that the subspace $\mathcal{W} = \{(b_1, b_2, b_3, b_4) : b_1 + b_2 + b_3 + b_4 = 0\}$ is the same as the range of A! To see this, notice that $\mathcal{W}$ is the nullspace of the 1×4 matrix $F = \begin{pmatrix} 1 & 1 & 1 & 1 \end{pmatrix}$. Since the rank of F is 1, the dimension of the nullspace of F is $4 - 1 = 3$. In other words, $\mathcal{W} = \mathcal{N}(F)$ is also a 3–dimensional subspace of $\mathbf{R}^4$. Moreover, each of the vectors $(-3, 3, 0, 0)$, $(3, -8, 5, 0)$, and $(0, 0, .4, -.4)$ is in $\mathcal{W}$. Since these vectors span the range of A, we see that $\mathcal{R}(A)$ is a subspace of $\mathcal{W}$ and since their dimensions are the same, we have $\mathcal{R}(A) = \mathcal{W}$. Thus, a set of applied forces leads to a position for which there is static equilibrium if and only if the sum of the applied forces is zero! Now this makes physical sense as well: if the sum of the forces were not zero, then the whole system would be dragged off to the left or the right forever, and not be in equilibrium at rest. The material of this chapter has given us ways to discuss and understand the consequences of the systems of equations we wrote down to describe the mechanical system.

Exercises 3.8

1. For each of the situations (a)–(f) below, decide which of the statements in the box can correctly complete the sentence. *Include all correct responses.*

   (a) If A is a 7×12 matrix whose rank is 5, then . . .
   (b) If A is a 7×12 matrix whose rank is 7, then . . .
   (c) If A is a 7×12 matrix whose rank is 9, then . . .

   (d) If A is an 11×7 matrix whose rank is 5, then . . .
   (e) If A is an 11×7 matrix whose rank is 7, then . . .
   (f) If A is an 11×7 matrix whose rank is 9, then . . .

   > (i) $AX = b$ is solvable for every vector b.
   > (ii) there are some vectors b for which $AX = b$ is not solvable.
   > (iii) for some vectors b, the system $AX = b$ has exactly one solution.
   > (iv) for some vectors b, the system $AX = b$ has infinitely many solutions.
   > (v) the given information is contradictory, no such system is possible.

2.

$$A = \begin{pmatrix} 3 & -1 & 0 \\ 5 & 3 & -1 \\ -2 & -4 & 1 \end{pmatrix}$$

   The vectors $v = (1, 3, 14))$ forms basis for the nullspace of A.
   (a) What is the nullity of A? the rank of A? the rank of A'? the nullity of A'?
   (b) Find bases for the range of A, the range of A', and the nullspace of A'.
   (c) Is the equation $AX = (-1, 2, -3)$ solvable?
   (d) Is the equation $AX = (7, 8, -1)$ solvable?
   (e) Is the equation $A'X = (2, 1, -1)$ solvable?

3.

$$B = \begin{pmatrix} 2 & 1 & 3 \\ 1 & -2 & 2 \\ 2 & -3 & 1 \end{pmatrix}$$

   The nullspace of B consists of the zero vector.
   (a) What is the nullity of B? the rank of B? the rank of B'? the nullity of B'?
   (b) Find bases for the range of B, the range of B', and the nullspace of B'.
   (c) Is the equation $BX = (2, 1, -1)$ solvable?
   (d) Is the equation $BX = (5, -1, 3)$ solvable?
   (e) Is the equation $B'X = (1, 7, 4)$ solvable?

4.
$$C = \begin{pmatrix} 1 & 0 & 3 & -1 \\ 0 & 1 & 1 & 2 \\ 2 & -1 & 5 & -4 \end{pmatrix}$$

The vectors $v_1 = (-3, -1, 1, 0)$ and $v_2 = (1, -2, 0, 1)$ are a basis for the nullspace of C.

(a) What is the nullity of C? the rank of C? the rank of C'? the nullity of C'?
(b) Find bases for the range of C, the range of C', and the nullspace of C'.
(c) Is the equation $CX = (2, 3, 1)$ solvable?
(d) Is the equation $CX = (-1, 1, 3)$ solvable?
(e) Is the equation $C'X = (1, -1, 2, -3)$ solvable?

5.
$$D = \begin{pmatrix} 1 & 2 & 0 & 1 \\ 0 & 1 & 1 & 0 \\ -1 & 1 & 3 & -1 \end{pmatrix}$$

(a) Find a basis for the nullspace of D.
(b) What is the nullity of D? the rank of D? the rank of D'? the nullity of D'?
(c) Find bases for the range of D, the range of D', and the nullspace of D'.
(d) Is the equation $DX = (2, 3, 1)$ solvable?
(e) Is the equation $DX = (1, 1, 2)$ solvable?
(f) Is the equation $D'X = (1, 0, 0, -1)$ solvable?
(g) Is the equation $D'X = (1, 2, 1, -1)$ solvable?

6. For each real number t, let $F(t)$ be the matrix
$$F(t) = \begin{pmatrix} 0 & 0 & 0 & 1 \\ t & 2 & 1 & 1 \\ 1 & -2 & t & -1 \\ 0 & 3 & 1 & 2 \end{pmatrix}$$

(a) Find $\det\big(F(t)\big)$. (It is a function of t.)
(b) For which value (or values) of t are the columns of $F(t)$ linearly dependent?

7. For each real number t, let $G(t)$ be the matrix
$$G(t) = \begin{pmatrix} 1 & 3 & -4 & 4 \\ 0 & -1 & 1 & -1 \\ -3 & 0 & t & 7 \\ t & -1 & 4 & -5 \end{pmatrix}$$

(a) For which value (or values) of t are the columns of $G(t)$ linearly dependent?
(b) For which values of t is G invertible?

8. For each of the following, decide if the statement is *always true* or *always false* or *sometimes true, sometimes false* when the given condition is true.

(a) **Given:** The vectors v_1, v_2, v_3, v_4, v_5, v_6 span $\mathbf{R}^6$.
  Statement: The vectors v_1, v_2, v_3, v_4, v_5, v_6 are linearly independent.

      always true   always false   sometimes true, sometimes false

(b) **Given:** B is a 6×6 matrix with $\det(B) \neq 0$.
  Statement: The equation $BX = b$ has infinitely many solutions.

      always true   always false   sometimes true, sometimes false

(c) **Given:** B is a 6×6 matrix, b is in $\mathbf{R}^6$, $BX = 0$ has infinitely many solutions.
  Statement: The equation $BX = b$ has infinitely many solutions.

      always true   always false   sometimes true, sometimes false

(d) **Given:** B is a 6×6 matrix and the columns of B are linearly independent.
  Statement: The equation $BX = b$ has exactly one solution.

      always true   always false   sometimes true, sometimes false

(e) **Given:** E is a 6×8 matrix, b is in $\mathbf{R}^6$, $\mathcal{N}(E)$, the nullspace of E is 2–dimensional.
  Statement: The equation $EX = b$ has infinitely many solutions.

      always true   always false   sometimes true, sometimes false

4

Inner Products and the Geometry of $\mathbf{R}^n$

4.1 Coordinatizing a Plane in Space

Many physical and engineering problems can be most efficiently approached by choosing a special coordinate system. For example, in computer aided manufacturing, many machine tools have their own internal coordinate system and using that coordinate system in the description of the process will make programming the machine tool much easier. Since a problem may involve several different sets of coordinates, changing between them is essential. Moreover, since we are describing physical situations, the coordinates must accurately describe the sizes and shapes of the objects. In this chapter we will address the problem of creating coordinate systems and keeping track of sizes and shapes. Size and shape will be preserved if we make changes that preserve length and angle; so we must be able to describe length and angle in the coordinate systems we use.

If a machine tool works in a single plane, it seems reasonable that its internal coordinate system would be described in the usual x, y–coordinate system in the plane: that is, it is described by a square grid in which each square has unit length. Physically, however, this plane could be anywhere, that is, it need not have a simple description in some other coordinate system. The following problem is suggestive of the sort thinking that needs to go into thinking about describing positions in this plane, given their description in some other system.

> **Problem** The equation $2x - 3y + 2z = 11$ describes a plane in $\mathbf{R}^3$. Set up a unit length, square grid on this plane and use it to find the other corners of the two squares in the plane that have vertices at $(3, -1, 1)$ and $(4, 1, 3)$.

4.2 Inner Products

Geometric intuition gained from living in a three dimensional Euclidean world is a big help in understanding facts about the vector space $\mathbf{R}^3$. We can extend this thinking by analogy to other settings. Indeed, it is very profitable to think of all vector spaces as looking a lot like $\mathbf{R}^2$ or $\mathbf{R}^3$ except perhaps "bigger". The most important notion in this analogy is the idea of orthogonality or perpendicularity. We will show that the analogue of the Pythagorean Theorem holds, as well as other results from high school Euclidean geometry. This means that we can draw pictures that look "just like" what is really going on in $\mathbf{R}^n$.

The ultimate goal of this chapter is the proof of the *Duality Theorem* (4.22). This is an extension of the Rank–Nullity Theorem (3.50) that tells the geometric relationship of the important subspaces associated with a matrix in addition to giving their sizes as the Rank–Nullity Theorem does.

The most important ideas from geometry are related to the concepts of length and angle. These geometric ideas are introduced into linear algebra by defining inner products on vector spaces.

DEFINITION *If V is a real vector space, an* inner product *on V is a scalar valued function, $\langle \cdot, \cdot \rangle$, of pairs of vectors from V such that*

(i) $\langle v, v \rangle > 0$ *for* $v \neq 0$ *and* $\langle 0, 0 \rangle = 0$.

(ii) $\langle u, v \rangle = \langle v, u \rangle$.

(iii) $\langle u, \alpha v + \beta w \rangle = \alpha \langle u, v \rangle + \beta \langle u, w \rangle$.

A vector space that also has an inner product defined on it is called an *inner product space*. The best known example of an inner product space is Euclidean space $\mathbf{R}^3$ with the "dot" product. Notice that property *(ii)* says that, in the case of real scalars, the inner product is symmetric: $\langle u, v \rangle = \langle v, u \rangle$. Properties (ii) and (iii) together imply

$$\langle \alpha v + \beta w, u \rangle = \alpha \langle v, u \rangle + \beta \langle w, u \rangle \tag{4.2.1}$$

for real inner product spaces. An inner product is called a *scalar product* by some authors in contrast with the *cross product* of two vectors in $\mathbf{R}^3$ which is a vector rather than a number.

EXAMPLE 4.1

If $V = \mathbf{R}^3$, the standard inner product is

$$\langle u, v \rangle = u_1 v_1 + u_2 v_2 + u_3 v_3$$

More generally, for $V = \mathbf{R}^n$, the *Euclidean inner product*, (or *usual inner product* or *standard inner product*) is

$$\langle u, v \rangle = u_1 v_1 + u_2 v_2 + \cdots + u_n v_n$$

$\Box$

EXAMPLE 4.2

If V is the vector space of continuous real–valued functions defined on the interval $[0, 1]$, then we can define an inner product by

$$\langle f, g \rangle = \int_0^1 f(t) g(t) \, dt$$

This inner product is important in electrical engineering and signal processing; for example, if f represents power output, then RMS power output is $\sqrt{\langle f, f \rangle}$. $\Box$

If V is a complex vector space, an *inner product on* V is a scalar valued function, $\langle \cdot, \cdot \rangle$, of pairs of vectors from V such that

(i) $\langle v, v \rangle > 0$ for $v \neq 0$ and $\langle 0, 0 \rangle = 0$.

(ii) $\langle u, v \rangle = \overline{\langle v, u \rangle}$.

(iii) $\langle u, \alpha v + \beta w \rangle = \alpha \langle u, v \rangle + \beta \langle u, w \rangle$.

The *Euclidean inner product* (or *usual inner product* or *standard inner product*) on $V = \mathbf{C}^n$ is

$$\langle u, v \rangle = \overline{u_1} v_1 + \overline{u_2} v_2 + \cdots + \overline{u_n} v_n$$

Note that (ii) and (iii) above imply

$$\langle \alpha v + \beta w, u \rangle = \overline{\alpha} \langle v, u \rangle + \overline{\beta} \langle w, u \rangle \tag{4.2.2}$$

for complex inner product spaces, so the inner product is linear in one variable and conjugate linear in the other. The conjugate is there so that $\langle v, v \rangle > 0$ for v

non–zero. A choice must be made about which variable should be conjugate linear: some books (most math books) make the second variable conjugate linear, other books (most engineering books) make the first variable conjugate linear. We follow the engineers because there is a convenient interpretation of the inner product in terms of matrix multiplication that is more natural for this choice.

For an $m \times n$ matrix A, recall that the *adjoint of A*, denoted A' is the conjugate transpose of A. Using this with column vectors and interpreting a 1×1 matrix as being synonymous with the number it contains, we find that for the Euclidean inner product on $\mathbf{R}^n$ or $\mathbf{C}^n$,

$$\langle u, v \rangle = u'v$$

In particular, if your calculator or computer software (such as MATLAB) can multiply matrices and take adjoints, this formulation of the inner product will probably be the most convenient way to compute it.

EXAMPLE 4.3

Let $u = (1.2, -3.6, 4.73, -.64)$ and $v = (3.5, 2.93, .86, -2.17)$ in $\mathbf{R}^4$. Find $\langle u, v \rangle$.

SOLUTION

$$\langle u, v \rangle = (1.2)(3.5) + (-3.6)(2.93) + (4.73)(.86) + (-.64)(-2.17)$$
$$= 4.2 - 10.548 + 4.0678 + 1.3888 = -.8914$$

Using MATLAB, we get

```
> u=[1.2;-3.6;4.73;-.64]

u =
    1.2000
   -3.6000
    4.7300
   -0.6400

> v=[3.5;2.93;.86;-2.17]

v =
    3.5000
    2.9300
    0.8600
   -2.1700
```

```
> innprod = u'*v

innprod =
   -0.8914
```

□

The properties in the definition mean that expanding inner products of linear combinations works much like multiplying polynomials in high school algebra.

EXAMPLE 4.4

Suppose $\mathcal{V}$ is a subspace of $\mathbf{R}^n$ and x and y are vectors such that $\langle x, x \rangle = 3$, $\langle y, y \rangle = 5$, and $\langle x, y \rangle = -2$. Find $\langle 2x + 3y, x - 5y \rangle$.

SOLUTION We use part (iii) of the definition and Equation (4.2.1) to expand the inner product $\langle 2x + 3y, 4x - 5y \rangle$ into a sum of four terms which can be computed from the information given. This expansion is analogous to the expansion of the product of two binomials in high school algebra.

First, we take $u = 2x + 3y$, $v = x$, $w = y$, $\alpha = 4$, and $\beta = -5$ in (iii) of the definition of inner product to get

$$\langle 2x + 3y, 4x - 5y \rangle = 4\langle 2x + 3y, x \rangle + (-5)\langle 2x + 3y, y \rangle$$

Now, each of the terms on the right can be expanded using Equation (4.2.1) with $v = x$, $w = y$, $\alpha = 2$, $\beta = 3$ in each of the two terms and $u = x$ in the first and $u = y$ in the second term to get

$$= 4\left(2\langle x, x \rangle + 3\langle y, x \rangle\right) + (-5)\left(2\langle x, y \rangle + 3\langle y, y \rangle\right)$$

$$= 8\langle x, x \rangle + 12\langle y, x \rangle - 10\langle x, y \rangle - 15\langle y, y \rangle$$

We know $\langle x, x \rangle = 3$, $\langle y, y \rangle = 5$, and $\langle x, y \rangle = -2$, but we have not been given $\langle y, x \rangle$. However, property (ii) of the definition of inner product says that $\langle y, x \rangle = \langle x, y \rangle = -2$. Finally, we conclude

$$\langle 2x + 3y, 4x - 5y \rangle = 8\langle x, x \rangle + 12\langle y, x \rangle - 10\langle x, y \rangle - 15\langle y, y \rangle$$

$$= 8(3) + 12(-2) - 10(-2) - 15(5) = -55$$

□

Inner products give rise to the geometric concepts of length and angle; these are theorems in $\mathbf{R}^2$ and $\mathbf{R}^3$ where length and angle already have meaning and they are helpful analogies in other spaces.

DEFINITION *The* norm (or length) of the vector v *is*

$$\|v\| = \sqrt{\langle v, v \rangle}$$

We associate "length" with the norm since this agrees with the usual notion of length of a vector in $\mathbf{R}^2$ or $\mathbf{R}^3$. We want to develop some of the properties of the norm to show that the norm is consistent with our notion of length. For example, it seems reasonable that if a vector is multiplied by 5 that its length should be multiplied by 5 as well. This is indeed the case; from the properties of inner products,

$$\|\alpha v\|^2 = \langle \alpha v, \alpha v \rangle = \alpha\alpha\langle v, v \rangle = \alpha^2\|v\|^2$$

so, taking square roots,

$$\|\alpha v\| = |\alpha|\|v\| \tag{4.2.3}$$

If p and q are points in $\mathbf{R}^n$, the distance between p and q is just the length of the vector between them, that is, it is $\|p - q\|$. For example, if $p = (p_1, p_2)$ and $q = (q_1, q_2)$ are points in $\mathbf{R}^2$, then the distance from p to q is just

$$\|p - q\| = \|(p_1 - q_1, p_2 - q_2)\| = \sqrt{(p_1 - q_1)^2 + (p_2 - q_2)^2}$$

which is the familiar distance formula for the plane.

The *Cauchy–Schwarz Inequality*, proved at the end of the section, states that $|\langle u, v \rangle| \leq \|u\|\|v\|$. This means that in $\mathbf{R}^n$

$$-1 \leq \frac{\langle u, v \rangle}{\|u\|\|v\|} \leq 1$$

This permits us to make the following definition of the "angle between two vectors". We can make this definition in any real inner product space and the definition of orthogonal applies to any inner product space.

DEFINITION *If u and v are vectors in $\mathbf{R}^n$, the* angle between u and v *is an angle θ*

$$\cos(\theta) = \frac{\langle u, v \rangle}{\|u\|\|v\|}$$

In particular, u and v are called perpendicular *or* orthogonal *if* $\langle u, v \rangle = 0$.

This leads to language that is suggestive of analogies with our geometric experience. Moreover, these analogies can be made precise to give versions of theorems from Euclidean geometry that are true and have easy proofs in any inner product space, for example, the Pythagorean Theorem, the Parallelogram Law, and the perpendicularity of the diagonals of a rhombus.

EXAMPLE 4.4, continued. Suppose, as above, $\mathcal{V}$ is a subspace of $\mathbf{R}^n$ and x and y are vectors such that $\langle x, x \rangle = 3$, $\langle y, y \rangle = 5$, and $\langle x, y \rangle = -2$. Find $\|2x + 3y\|$. Find the angle between the vectors $2x + 3y$ and $x - 5y$.

SOLUTION Again, we use part (iii) of the definition and Equation (4.2.1) to expand the inner product $\|2x + 3y\|^2 = \langle 2x + 3y, 2x + 3y \rangle$ into a sum of four terms which can be computed from the information given.

$$\|2x + 3y\|^2 = \langle 2x + 3y, 2x + 3y \rangle$$

$$= 2\langle 2x + 3y, x \rangle + 3\langle 2x + 3y, y \rangle$$

$$= 4\langle x, x \rangle + 6\langle y, x \rangle + 6\langle x, y \rangle + 9\langle y, y \rangle$$

$$= 4(3) + 6(-2) + 6(-2) + 9(5) = 33$$

so $\|2x + 3y\| = \sqrt{33}$.

To find the angle between the vectors, we need also to find the length of $x - 5y$, which we do in the same way.

$$\|x - 5y\|^2 = \langle x - 5y, x - 5y \rangle$$

$$= \langle x, x \rangle - 5\langle y, x \rangle - 5\langle x, y \rangle + 25\langle y, y \rangle$$

$$= 3 - 5(-2) - 5(-2) + 25(5) = 118$$

so $\|x - 5y\| = \sqrt{118}$. Now, if θ is the angle between $2x + 3y$ and $x - 5y$, then

$$\cos(\theta) = \frac{\langle 2x + 3y, x - 5y \rangle}{\|2x + 3y\|\|x - 5y\|} = \frac{-55}{\sqrt{33}\sqrt{118}} = -0.8814$$

This means θ is about 2.65 radians, or about 152 degrees.   ☐

Physics makes it is clear that, at least in two or three dimensions, vectors that are perpendicular to each other are more convenient for calculations. The generalization to higher dimensional spaces is also correct, and perpendicularity is computationally important because basing numerical calculations on collections of perpendicular vectors leads to greater accuracy as well.

DEFINITION *The set of vectors $\{v_1, v_2, \cdots, v_k\}$ is said to be an* orthogonal set *if*

$$\langle v_i, v_j \rangle = 0 \quad for \ 1 \leq i, j \leq k, \ i \neq j$$

and said to be an orthonormal set *if, in addition,*

$$\langle v_j, v_j \rangle = 1$$

EXAMPLE 4.5

The standard basis vectors for $\mathbf{R}^3$ are an orthonormal basis. The vectors $e_1 = (1, 0, 0)$, $e_2 = (0, 1, 0)$, and $e_3 = (0, 0, 1)$ satisfy $\|e_1\| = \|e_2\| = \|e_3\| = 1$ and $\langle e_1, e_2 \rangle = \langle e_1, e_3 \rangle = \langle e_2, e_3 \rangle = 0$. Many of the facts about the standard basis vectors for $\mathbf{R}^3$ follow from their being an orthonormal basis for the space. $\quad\Box$

EXAMPLE 4.6

The Fourier basis for the space of functions on $[-\pi, \pi]$, that is,

$$\{1\} \cup \{\sin(nx) : n \text{ a positive integer }\} \cup \{\cos(nx) : n \text{ a positive integer }\}$$

is an orthogonal basis. (See Exercise 4.2.6.). An important part of the effectiveness of Fourier series is the orthogonality of the functions in the Fourier basis. $\quad\Box$

EXAMPLE 4.7

Let v_1, v_2, and v_3 be an orthonormal basis for the subspace $\mathcal{V}$ of $\mathbf{R}^6$. Find $\|2v_1 - 3v_2 + v_3\|$.

SOLUTION We use part (iii) of the definition and Equation (4.2.1) to expand the inner product $\|2v_1 - 3v_2 + v_3\|^2 = \langle 2v_1 - 3v_2 + v_3, 2v_1 - 3v_2 + v_3 \rangle$ into a sum of nine terms which can be computed from orthonormality which says $\langle v_1, v_1 \rangle = \langle v_2, v_2 \rangle = \langle v_3, v_3 \rangle = 1$ and $\langle v_1, v_2 \rangle = \langle v_1, v_3 \rangle = \langle v_2, v_3 \rangle = 0$.

Thus,

$$\langle 2v_1 - 3v_2 + v_3, 2v_1 - 3v_2 + v_3 \rangle$$

$$= 2\langle 2v_1 - 3v_2 + v_3, v_1 \rangle + (-3)\langle 2v_1 - 3v_2 + v_3, v_2 \rangle + 1\langle 2v_1 - 3v_2 + v_3, v_3 \rangle$$

$$= 4\langle v_1, v_1 \rangle - 6\langle v_2, v_1 \rangle + 2\langle v_3, v_1 \rangle - 6\langle v_1, v_2 \rangle$$

$$+9\langle v_2, v_2 \rangle - 3\langle v_3, v_2 \rangle + 2\langle v_1, v_3 \rangle - 3\langle v_2, v_3 \rangle + \langle v_3, v_3 \rangle$$

The value of each of these terms is known from the orthonormality, so we get

$$= 4 \cdot 1 - 6 \cdot 0 + 2 \cdot 0 - 6 \cdot 0 + 9 \cdot 1 - 3 \cdot 0 + 2 \cdot 0 - 3 \cdot 0 + 1 = 14$$

so $\|2v_1 - 3v_2 + v_3\| = \sqrt{14}$. Notice that all the terms with different vectors contribute nothing to the sum because of the orthogonality and the norm is just the square root of the sum of the squares of the coefficients in the expansion of the vector. $\Box$

The geometric intuition of a linearly independent set of vectors is that they should be pointing in different directions, with no three vectors lying in the same plane, etc. If a set of vectors is each pointing perpendicular to the rest, we should expect them to be linearly independent. This is precisely the content of the next result.

PROPOSITION 4.8

An orthogonal set of non–zero vectors is linearly independent.

PROOF   Suppose $v_1, v_2, \cdots v_k$ is an orthogonal set of non–zero vectors and $\alpha_1, \cdots \alpha_k$ are scalars such that

$$\alpha_1 v_1 + \alpha_2 v_2 + \cdots + \alpha_k v_k = 0$$

Taking the inner product of both sides of this equation with v_j, we get

$$\langle v_j, \alpha_1 v_1 + \alpha_2 v_2 + \cdots + \alpha_k v_k \rangle = \langle v_j, 0 \rangle$$

or

$$\alpha_1 \langle v_j, v_1 \rangle + \alpha_2 \langle v_j, v_2 \rangle + \cdots + \alpha_k \langle v_j, v_k \rangle = 0$$

Since v's form an orthogonal set, all the terms but one are zero and we have

$$\alpha_j \langle v_j, v_j \rangle = 0$$

Since the vectors are all non–zero, $\langle v_j, v_j \rangle \neq 0$ which means $\alpha_j = 0$. But this is true for all j, so the only linear combination of the v's that gives zero is the trivial one, and the vectors are linearly independent. $\blacksquare$

Of course, we are interested in linearly independent sets as building blocks for bases for our vector spaces. In order for a basis to be used, vectors must be written as a linear combination of the basis vectors. The next theorem shows that finding coordinates with respect to an orthogonal basis is much easier than solving a system of equations as we might expect to have to do.

THEOREM 4.9

Let w_1, w_2, $\cdots$, w_k be an orthogonal set of non–zero vectors. The vector v is in the subspace spanned by the w's if and only if

$$v = \frac{\langle w_1, v \rangle}{\|w_1\|^2} w_1 + \frac{\langle w_2, v \rangle}{\|w_2\|^2} w_2 + \cdots + \frac{\langle w_k, v \rangle}{\|w_k\|^2} w_k$$

PROOF The vector v is in the subspace spanned by the w's if and only if there are constants $\alpha_1, \cdots, \alpha_k$ so that

$$v = \alpha_1 w_1 + \alpha_2 w_2 + \cdots + \alpha_k w_k$$

Taking the inner product of each side of this equality with w_j gives

$$\langle w_j, v \rangle = \langle w_j, \alpha_1 w_1 + \alpha_2 w_2 + \cdots + \alpha_k w_k \rangle$$

so

$$\langle w_j, v \rangle = \alpha_1 \langle w_j, w_1 \rangle + \alpha_2 \langle w_j, w_2 \rangle + \cdots + \alpha_k \langle w_j, w_k \rangle$$
$$= \alpha_j \langle w_j, w_j \rangle$$

because the w's are orthogonal. Since none is zero, this means

$$\alpha_j = \frac{\langle w_j, v \rangle}{\|w_j\|^2}$$

as in the theorem. ∎

The vectors $u = (.8, -.6)$ and $v = (.6, .8)$ are orthogonal and have length 1, that is, they are an orthonormal basis for $\mathbf{R}^2$. As Figure 4.1 illustrates, we can use this basis to get coordinates for the plane. Theorem 4.9 indicates how to find the vector $w = (5, 3)$ in the plane as a linear combination of u and v

$$w = \begin{pmatrix} 5 \\ 3 \end{pmatrix} = \left\langle \begin{pmatrix} .8 \\ -.6 \end{pmatrix}, \begin{pmatrix} 5 \\ 3 \end{pmatrix} \right\rangle \begin{pmatrix} .8 \\ -.6 \end{pmatrix} + \left\langle \begin{pmatrix} .6 \\ .8 \end{pmatrix}, \begin{pmatrix} 5 \\ 3 \end{pmatrix} \right\rangle \begin{pmatrix} .6 \\ .8 \end{pmatrix}$$

$$= 2.2 \begin{pmatrix} .8 \\ -.6 \end{pmatrix} + 5.4 \begin{pmatrix} .6 \\ .8 \end{pmatrix}$$

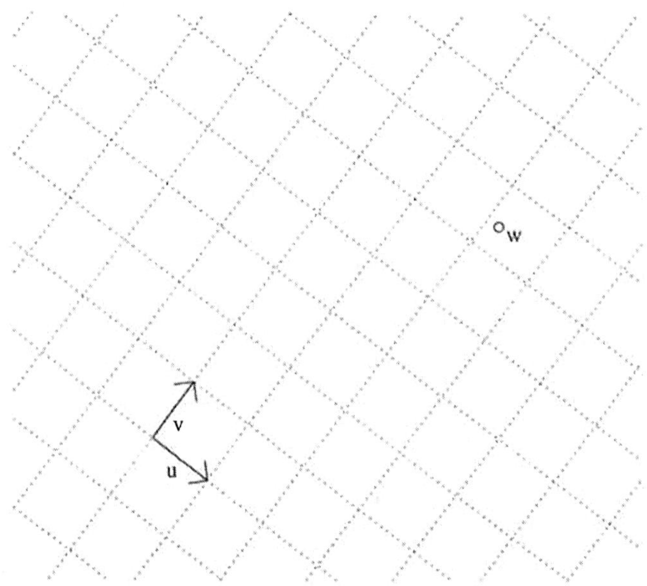

FIGURE 4.1

Coordinate system from the orthonormal basis $\{u, v\}$

That is, the coordinates of w with respect to the coordinate system induced by the basis $\{u, v\}$ are 2.2 and 5.4 since $w = 2.2u + 5.4v$. One of the advantages of orthonormal coordinate systems is that the coordinates can be used in *all* the ways we use the standard coordinate system because the properties we use are that the axes are perpendicular and the units we use to measure distances are 1 along each axis (because $\|u\| = \|v\| = 1$). For example, the distance of the point w from the origin is

$$\|w\| = \sqrt{\left\langle \begin{pmatrix} 5 \\ 3 \end{pmatrix}, \begin{pmatrix} 5 \\ 3 \end{pmatrix} \right\rangle} = \sqrt{5^2 + 3^2} = \sqrt{34}$$

doing the computation in the usual coordinates. Since

$$\|\alpha u + \beta v\|^2 = \langle \alpha u + \beta v, \alpha u + \beta v \rangle$$

$$= \langle \alpha u, \alpha u \rangle + \langle \alpha u, \beta v \rangle + \langle \beta v, \alpha u \rangle + \langle \beta v, \beta v \rangle$$

$$= \alpha^2 \langle u, u \rangle + \alpha\beta \langle u, v \rangle + \beta\alpha \langle v, u \rangle + \beta^2 \langle v, v \rangle$$

$$= \alpha^2 \cdot 1 + \alpha\beta \cdot 0 + \beta\alpha \cdot 0 + \beta^2 \cdot 1 = \alpha^2 + \beta^2$$

This calculation shows that $\|\alpha u + \beta v\| = \sqrt{\alpha^2 + \beta^2}$, in other words, that the length of a vector can be computed as the square root of the sum of the squares of the u, v–coordinates just as in the usual coordinates. Indeed, in this case,

$$\|w\| = \sqrt{2.2^2 + 5.4^2} = \sqrt{4.84 + 29.16} = \sqrt{34}$$

as before.

We conclude the section with the proofs of the Cauchy–Schwarz Inequality and the Triangle Inequality.

THEOREM 4.10   *Cauchy–Schwarz Inequality*

Let u and v be vectors in an inner product space with u non-zero.
If $v = \alpha u$ for some scalar α, then $|\langle u, v \rangle| = \|u\|\|v\|$.
If $v \neq \alpha u$ for any scalar α, then $|\langle u, v \rangle| < \|u\|\|v\|$.

PROOF   We will give the proof for real scalars, and below indicate the changes necessary for complex scalars. Let

$$f(t) = \|tu - v\|^2 = \|u\|^2 t^2 - 2\langle u, v \rangle t + \|v\|^2$$

for t real. Since $f(t)$ is the square of the norm of a vector, $f(t) \geq 0$. Evaluating f at $t = \langle u, v \rangle / \|u\|^2$, we get the inequality

$$\|u\|^2 \left(\frac{\langle u, v \rangle}{\|u\|^2} \right)^2 - 2\langle u, v \rangle \left(\frac{\langle u, v \rangle}{\|u\|^2} \right) + \|v\|^2 \geq 0.$$

Simplifying, we get

$$-\frac{\langle u, v \rangle^2}{\|u\|^2} + \|v\|^2 \geq 0,$$

which is equivalent to $|\langle u, v \rangle| \leq \|u\| \|v\|$. Now equality occurs in this inequality if and only if $f(t) = \|tu - v\|^2 = 0$, which implies $v = \alpha u$ where $\alpha = \langle u, v \rangle / \|u\|^2$.

If we have complex scalars, choose θ so that $\langle u, e^{i\theta} v \rangle \geq 0$ and let $f(t) = \|tu - e^{i\theta} v\|^2$. The Cauchy–Schwartz inequality follows from $f(t) \geq 0$ and equality holds if and only if $v = \alpha u$ where $\alpha = e^{-i\theta} |\langle u, v \rangle| / \|u\|^2$. ∎

(Geometrically, tu describes the line through u and the origin. We have chosen f to be the square of the distance between v and the point on this line corresponding to t. The choice of t in this proof seems less mysterious when we notice that the function f is a quadratic polynomial in t. Now, as is easily seen from elementary calculus, the minimum value of the quadratic polynomial $at^2 + 2bt + c$ is $c - b^2/a$ and it occurs at $t = -b/a$. In our case, $a = \|u\|^2$, $b = -\langle u, v \rangle$, and $c = \|v\|^2$, so our choice of t was the one that minimizes f, corresponding to the point on the line closest to v, and therefore gives the strongest conclusion.)

COROLLARY 4.11   Triangle Inequality

Let u and v be vectors an inner product space with u non-zero.
If $v = \alpha u$ for some non-negative scalar α, then $\|u + v\| = \|u\| + \|v\|$.
If $v \neq \alpha u$ for a non-negative scalar α, then $\|u + v\| < \|u\| + \|v\|$.

PROOF   For real scalars, by the Cauchy–Schwarz inequality,

$$\|u + v\|^2 = \|u\|^2 + 2\langle u, v \rangle + \|v\|^2 \leq \|u\|^2 + 2|\langle u, v \rangle| + \|v\|^2$$

$$\leq \|u\|^2 + 2\|u\|\|v\| + \|v\|^2 = (\|u\| + \|v\|)^2.$$

Thus, for every u and v, $\|u + v\| \leq \|u\| + \|v\|$ and equality holds if and only if $|\langle u, v \rangle| = \|u\|\|v\|$ and $\langle u, v \rangle = |\langle u, v \rangle|$. The first equality implies that $v = \alpha u$ and putting this into the second equality implies α is non-negative.

To modify the proof for complex scalars, we note that

$$\|u + v\|^2 = \|u\|^2 + 2\mathrm{Re}(\langle u, v \rangle) + \|v\|^2 \leq \|u\|^2 + 2|\langle u, v \rangle| + \|v\|^2$$

∎

Exercises 4.2

For each of the following, find $\|v\|$ and $\langle v, w \rangle$.

1. In $\mathbf{R}^3$ with the usual inner product; $v = (1, 2, -1)$ and $w = (1, -1, 1)$.
2. In $\mathbf{R}^3$ with the usual inner product; $v = (2, 1, -1)$ and $w = (1, -1, 1)$.
3. In $\mathbf{R}^4$ with the usual inner product; $v = (1, 1, 2, -2)$ and $w = (2, 0, 1, 1)$.
4. In $\mathbf{R}^4$ with the usual inner product; $v = (3, 0, 1, -1)$ and $w = (1, -2, 1, -1)$.
5. In $\mathbf{C}^3$ with the usual inner product; $v = (1, 1 + i, 2i)$ and $w = (1 - 2i, 2, 2 + i)$.
6. In $\mathbf{C}[-\pi, \pi]$, the set of real-valued continuous functions on the interval $[-\pi, \pi]$, with inner product $\langle f, g \rangle = \int_{-\pi}^{\pi} f(x)g(x)\, dx$; where $v = \sin mx$ and $w = \cos nx$ for positive integers m and n.

In each of the following, find the angle between v and w. (Use the usual inner product.)

7. $v = (3, 2, -1)$ and $w = (1, 0, -2)$.
8. $v = (2, -1, 2)$ and $w = (4, 4, -2)$.
9. $v = (1, -1, 2)$ and $w = (-1, 1, -3)$.
10. $v = (2, 2, -1)$ and $w = (0, -1, -1)$.
11. $v = (1, -1, 2, 0)$ and $w = (3, -1, -1, 5)$.
12. $v = (1, -1, 2, -1)$ and $w = (1, 3, 0, -2)$.
13. $v = (2, 1, 1, 0)$ and $w = (1, 3, 1, -2)$.

14. Let $u = (2, -2, 1)$ and $v = (1, 2, -5)$. Find $\|u\|$, $\|v\|$, and $\|u + v\|$ and observe that $\|u + v\| \leq \|u\| + \|v\|$.
15. Let $u = (1, -2, 1, 3)$ and $v = (2, 1, -2, 1)$. Find $\|u\|$, $\|v\|$, and $\|u + v\|$ and observe that $\|u + v\| \leq \|u\| + \|v\|$.
16. A missile is following the straight line path $f(t) = tu - v$ where $u = (1, -3, 2)$ and $v = (2, 1, 1)$. Write $\|f(t)\|^2$, the square of the distance of the missile from the origin, as a polymomial in t and then find the point at which the missile is closest to the origin.
17. Use the formula $\langle x, y \rangle = x'y$ to show that if A is an $m \times n$ matrix, u is in $\mathbf{R}^n$, and v is in $\mathbf{R}^m$, then

$$\langle Au, v \rangle = \langle u, A'v \rangle$$

This fact is used as the definition of the adjoint in more abstract settings.

18. Prove the Pythagorean Theorem for inner product spaces: If v and w are perpendicular, then
$$\|v\|^2 + \|w\|^2 = \|v + w\|^2$$

19. The Parallelogram Law from Euclidean Geometry is: The sum of the squares of the lengths of the diagonals is equal to the sum of the squares of the lengths of the sides. If u and v are vectors that form the sides of a parallelogram, then the diagonals are $u + v$ and $u - v$. Prove the vector form of the Parallelogram Law
$$\|u + v\|^2 + \|u - v\|^2 = 2\left(\|u\|^2 + \|v\|^2\right)$$

20. If u and v are vectors in $\mathbf{R}^n$ then 0, u, v, and $u + v$ are the vertices of a parallelogram. This parallelogram will be a rhombus exactly when $\|u\| = \|v\|$. Using inner products, show that the diagonals of a rhombus, $u + v$ and $u - v$, are perpendicular.

21. If u and v are vectors in $\mathbf{R}^n$ then 0, u, and v are the vertices of a triangle. This is an equilateral triangle if the lengths of the three sides are equal, that is, if $\|u\| = \|v\| = \|u - v\|$. Using inner products, show that the angle between each pair of sides of an equilateral triangle is $60°$, that is, show that for each pair of sides, $|\cos\theta| = .5$.

22.(a) Use Proposition 4.8 to show that the vectors $w_1 = (1, 3)$ and $w_2 = (-3, 1)$ form a basis for $\mathbf{R}^2$.

   (b) Use Theorem 4.9 to write $v = (2, -1)$ as a linear combination of w_1 and w_2.

23.(a) Use Proposition 4.8 to show that the vectors $w_1 = (1, 1, 0)$, $w_2 = (1, -1, 1)$, and $w_3 = (-1, 1, 2)$ are a basis for $\mathbf{R}^3$.

   (b) Use Theorem 4.9 to write $v = (2, -1, 3)$ as a linear combination of w_1, w_2, and w_3.

24. The vectors u_1, u_2, u_3, u_4 are orthogonal vectors that span the subspace $\mathcal{U}$ of $\mathbf{R}^{11}$. Moreover, $\|u_1\| = 1$, $\|u_2\| = 2$, $\|u_3\| = 3$, and $\|u_4\| = 1$.

   (a) What is the dimension of the subspace $\mathcal{U}$?

   (b) Find $\|v\|$ for $v = 3u_1 - 2u_2 + 4u_3 - u_4$?

25. The vectors u, v, and w are in $\mathbf{R}^n$ and we are given that $\|u\| = 1$, $\|v\| = 2$, $\|w\| = 3$, that $\langle u, v \rangle = -1$, $\langle u, w \rangle = 2$, and that w is perpendicular to v.

   (a) Find $\|u - 3v + 2w\|$.

   (b) Show that u, v, and w are linearly independent.

26.(a) Show that $v_1 = (1, 1, 1)$; $v_2 = (1, 0, -1)$; and $v_3 = (1, -2, 1)$ form an orthogonal set, and quote theorems that imply they form a basis for $\mathbf{R}^3$.

   (b) Write $w = (1, -1, 2)$ as a linear combination of v_1, v_2, and v_3.

27.(a) Show that $v_1 = (2, 2, 1)$; $v_2 = (-1, 2, -2)$; and $v_3 = (2, -1, -2)$ form an orthogonal basis for $\mathbf{R}^3$.

   (b) Write $w = (3, -1, 2)$ as a linear combination of v_1, v_2, and v_3.

28.(a) Show that $v_1 = (1, 1, 1, 1)$; $v_2 = (1, 1, -1, -1)$; $v_3 = (1, -1, 1, -1)$; and $v_4 = (1, -1, -1, 1)$ form an orthogonal basis for $\mathbf{R}^4$.

   (b) Write $w = (2, 1, -1, 2)$ as a linear combination of v_1, v_2, v_3, and v_4.

29.(a) Show that $v_1 = (1, -1, 1)$ and $v_2 = (3, 2, -1)$ are orthogonal vectors in $\mathbf{R}^3$.

   (b) Is $w - (2, 1, -1)$ in the subspace spanned by v_1 and v_2?

   (c) Find a non-zero vector in $\mathbf{R}^3$ that is perpendicular to each of v_1 and v_2.

30.(a) Show that $v_1 = (1, 0, 1, 1)$; $v_2 = (1, 1, -1, 0)$; and $v_3 = (1, -1, 0, -1)$ are orthogonal vectors in $\mathbf{R}^4$.
  (b) Is $w = (6, -1, 2, 1)$ in the subspace spanned by v_1, v_2, and v_3?
  (c) Find a non-zero vector in $\mathbf{R}^4$ that is perpendicular to each of v_1, v_2, and v_3.

4.3   The Gram–Schmidt Algorithm

While the results of the previous section show us that if we have an orthonormal basis, it is easy to write vectors as linear combinations of the basis vectors, it does not help us find an orthonormal basis to start with. In this section, we will address that issue. The Gram–Schmidt algorithm[1] shows how to change an arbitrary basis into an orthonormal basis while retaining some of the characteristics of the original basis.

The geometric meaning of the proof of the Gram–Schmidt algorithm is closely related to Theorem 4.9. While that result discusses the expansion of a vector in terms of an orthogonal basis, the expression in the theorem makes sense even if the vector is not in the subspace spanned by the vectors. What vector is being described in this case? This vector will be called the projection of the given vector onto the subspace. We will use the results of this section to study projections more in Chapter 5 and we will show that the projection is the vector in the subspace that best approximates the vector we started with.

What is happening in the Gram–Schmidt algorithm is that, at each stage, we use the difference between the given vector and its projection onto the subspace generated so far. The point of this construction is that the difference between a vector and its projection is perpendicular to the subspace. That is, if v is a vector and $w_1, w_2, \cdots, w_j$ are orthogonal, non-zero vectors that have been constructed so far, then the projection of v onto the subspace spanned by $w_1, w_2, \cdots, w_j$ is

$$\frac{\langle w_1, v \rangle}{\langle w_1, w_1 \rangle} w_1 + \frac{\langle w_2, v \rangle}{\langle w_2, w_2 \rangle} w_2 + \cdots + \frac{\langle w_j, v \rangle}{\langle w_j, w_j \rangle} w_j$$

as in Theorem 4.9. The difference between this vector and v is perpendicular to the

---

[1]The algorithm is named for actuary J. P. Gram and mathematician E. Schmidt who worked at the end of the nineteenth century.

subspace. Indeed, for each i,

$$\left\langle w_i,\, v - \left(\frac{\langle w_1, v\rangle}{\langle w_1, w_1\rangle} w_1 + \frac{\langle w_2, v\rangle}{\langle w_2, w_2\rangle} w_2 + \cdots + \frac{\langle w_j, v\rangle}{\langle w_j, w_j\rangle} w_j \right) \right\rangle$$

$$(4.3.4)$$

$$= \langle w_i, v\rangle - \frac{\langle w_1, v\rangle}{\langle w_1, w_1\rangle} \langle w_i, w_1\rangle - \frac{\langle w_2, v\rangle}{\langle w_2, w_2\rangle} \langle w_i, w_2\rangle - \cdots - \frac{\langle w_j, v\rangle}{\langle w_j, w_j\rangle} \langle w_i, w_j\rangle$$

$$= \langle w_i, v\rangle - \frac{\langle w_1, v\rangle}{\langle w_1, w_1\rangle} \cdot 0 - \cdots - \frac{\langle w_i, v\rangle}{\langle w_i, w_i\rangle} \langle w_i, w_i\rangle - \cdots - \frac{\langle w_j, v\rangle}{\langle w_j, w_j\rangle} \cdot 0$$

$$= \langle w_i, v\rangle - \langle w_i, v\rangle = 0$$

The first step of this algorithm is familiar to students of physics in the resolution of a force into a force parallel to a given direction and another perpendicular to that direction. For example, as in Theorem 4.9 if we want to express v as a sum of a vector in the direction of w_1 (the projection) and a vector perpendicular to w_1, we compute

$$\frac{\langle w_1, v\rangle}{\langle w_1, w_1\rangle} w_1 \quad \text{and} \quad w_2 = v - \frac{\langle w_1, v\rangle}{\langle w_1, w_1\rangle} w_1$$

where the first vector points in the direction of w_1 and, by the calculation of Equation (4.3.4), the second is perpendicular to it (see Figure 4.2).

Since this difference is perpendicular to the subspace, we take it to be the second vector in the Gram–Schmidt algorithm. To continue, we project the third given vector onto the subspace spanned by the first two and use the difference as the third orthogonal vector. (See Figure 4.3.)

The construction continues recursively: each step in the algorithm breaks a given vector v_{j+1} into the projection of the vector onto the subspace spanned by the previous vectors

$$\frac{\langle w_1, v_{j+1}\rangle}{\langle w_1, w_1\rangle} w_1 + \cdots + \frac{\langle w_j, v_{j+1}\rangle}{\langle w_j, w_j\rangle} w_j$$

as in Theorem 4.9, and a vector perpendicular to that span. We use the part perpendicular to the span as the next orthogonal vector.

The next theorem, the Gram–Schmidt Algorithm, is a careful statement of these ideas.

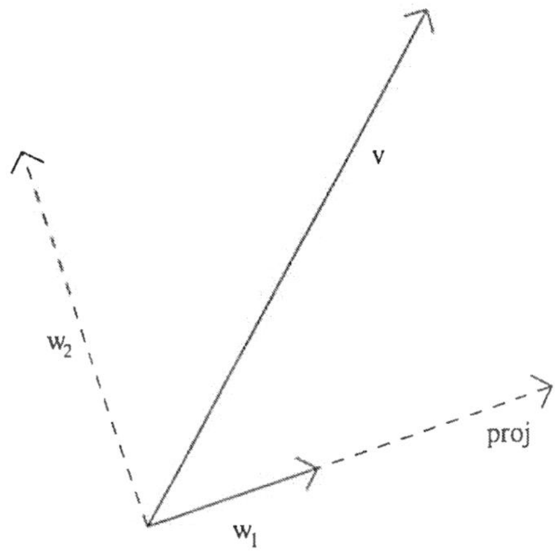

FIGURE 4.2

First projection

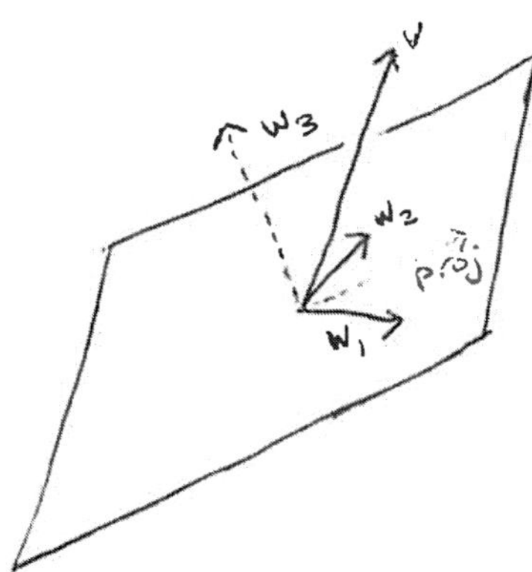

FIGURE 4.3

Second projection

THEOREM 4.12   Gram–Schmidt Algorithm

Suppose v_1, v_2, $\cdots$, v_m are vectors in $\mathbf{R}^n$. Let $w_1 = v_1$ and for $j \geq 1$, let

$$w_{j+1} = v_{j+1} - \frac{\langle w_1, v_{j+1}\rangle}{\langle w_1, w_1\rangle} w_1 - \cdots - \frac{\langle w_j, v_{j+1}\rangle}{\langle w_j, w_j\rangle} w_j \tag{4.3.5}$$

where any term involving w_k when $w_k = 0$ is taken to be zero. Then the set w_1, w_2, $\cdots$, w_n is orthogonal and for each j,

$$span\{v_1, v_2, \cdots, v_j\} = span\{w_1, w_2, \cdots, w_j\}$$

Since orthogonal sets of non–zero vectors are linearly independent, the final condition of the theorem shows that the vectors w_j are all non–zero if and only if the vectors v_j are linearly independent. If the starting vectors v_j are dependent, then some of the w's will be zero. If the formula were taken literally in all cases, the terms corresponding to these vectors would be of the form $\frac{0}{0}0$; these terms are to be replaced by 0 in the formula for computing further vectors. The non-zero vectors in the resulting orthogonal set can be normalized to give an orthonormal set

$$q_j = \frac{1}{\|w_j\|} w_j$$

that forms a basis for the subspace spanned by the v's.

PROOF   We suppose that the w_j's have been defined by equation (4.3.5) and prove the theorem by induction. Clearly, the set $\{w_1\}$ is an orthogonal set and $span\{v_1\}$ is the same as $span\{w_1\}$.

The second step is to let

$$w_2 = v_2 - \frac{\langle w_1, v_2\rangle}{\langle w_1, w_1\rangle} w_1$$

The calculation that w_2 is orthogonal to w_1 is just Equation (4.3.4).

$$\langle w_1, w_2\rangle = \langle w_1, v_2 - \frac{\langle w_1, v_2\rangle}{\langle w_1, w_1\rangle} w_1\rangle = \langle w_1, v_2\rangle - \frac{\langle w_1, v_2\rangle}{\langle w_1, w_1\rangle}\langle w_1, w_1\rangle$$

$$= \langle w_1, v_2\rangle - \langle w_1, v_2\rangle = 0$$

so w_2 is orthogonal w_1. To see that the span of w_1 and w_2 is the same as the span of v_1 and v_2, it is enough to check that each of w_1 and w_2 is a linear combination of v_1 and v_2 and that each of v_1 and v_2 is a linear combination of w_1 and w_2. First,

$$w_1 = v_1 = 1v_1 + 0v_2 \quad \text{and} \quad w_2 = v_2 - \frac{\langle w_1, v_2\rangle}{\langle w_1, w_1\rangle} w_1 = v_2 - \frac{\langle w_1, v_2\rangle}{\langle w_1, w_1\rangle} v_1$$

so each of the w's is a linear combination of the v's. Conversely, we see that

$$v_1 = w_1 = 1w_1 + 0w_2 \quad \text{and} \quad v_2 = \frac{\langle w_1, v_2 \rangle}{\langle w_1, w_1 \rangle} w_1 + w_2$$

so each of the v's is a linear combination of the w's. We continue by induction.

Suppose now that $\{w_1, \cdots, w_j\}$ is an orthogonal set whose span is the same as that of $\{v_1, \cdots, v_j\}$. To see that $\{w_1, \cdots, w_j, w_{j+1}\}$ is orthogonal, we need only check that w_{j+1} is orthogonal to each of $w_1, \cdots, w_j$ since the rest are given to be orthogonal to each other. This is the calculation of Equation (4.3.4).

To see that the spans are the same, notice that w_{j+1} is a linear combination of $w_1, \cdots, w_j$, and v_{j+1}. Since the span of the $w_1, \cdots, w_j$ is the same as the span of $v_1, \cdots, v_j$, this shows that w_{j+1} is a linear combination of $v_1, \cdots, v_j$, and v_{j+1}, so $\mathrm{span}\{w_1, \cdots, w_j, w_{j+1}\}$ is a subspace of $\mathrm{span}\{v_1, \cdots, v_j, v_{j+1}\}$. On the other hand, rewriting the equation defining w_{j+1} gives

$$v_{j+1} = \frac{\langle w_1, v_{j+1} \rangle}{\langle w_1, w_1 \rangle} w_1 + \cdots + \frac{\langle w_j, v_{j+1} \rangle}{\langle w_j, w_j \rangle} w_j \rangle + w_{j+1}$$

so v_{j+1} is in $\mathrm{span}\{w_1, \cdots, w_j, w_{j+1}\}$ and the induction hypothesis then shows that $\mathrm{span}\{v_1, \cdots, v_j, v_{j+1}\}$ is a subspace of $\mathrm{span}\{w_1, \cdots, w_j, w_{j+1}\}$. ∎

EXAMPLE 4.13

Find an orthonormal basis for the subspace M of $\mathbf{R}^4$ spanned by $v_1 = (0, 1, 1, 1)$, $v_2 = (1, 0, 1, 1)$, and $v_3 = (1, 1, 0, 1)$.

SOLUTION Using Gram–Schmidt, we find

$$w_1 = v_1 = \begin{pmatrix} 0 \\ 1 \\ 1 \\ 1 \end{pmatrix}$$

$$w_2 = v_2 - \frac{\langle w_1, v_2 \rangle}{\langle w_1, w_1 \rangle} w_1 = \begin{pmatrix} 1 \\ 0 \\ 1 \\ 1 \end{pmatrix} - \frac{2}{3} \begin{pmatrix} 0 \\ 1 \\ 1 \\ 1 \end{pmatrix} = \begin{pmatrix} 1 \\ -\frac{2}{3} \\ \frac{1}{3} \\ \frac{1}{3} \end{pmatrix}$$

$$w_3 = v_3 - \frac{\langle w_1, v_3 \rangle}{\langle w_1, w_1 \rangle} w_1 - \frac{\langle w_2, v_3 \rangle}{\langle w_2, w_2 \rangle} w_2$$

$$= \begin{pmatrix} 1 \\ 1 \\ 0 \\ 1 \end{pmatrix} - \frac{2}{3} \begin{pmatrix} 0 \\ 1 \\ 1 \\ 1 \end{pmatrix} - \frac{2/3}{5/3} \begin{pmatrix} 1 \\ -\frac{2}{3} \\ \frac{1}{3} \\ \frac{1}{3} \end{pmatrix} = \begin{pmatrix} \frac{3}{5} \\ \frac{3}{5} \\ -\frac{4}{5} \\ \frac{1}{5} \end{pmatrix}$$

Normalizing, we get

$$q_1 = \frac{1}{\|w_1\|} w_1 = \begin{pmatrix} 0 \\ \frac{1}{\sqrt{3}} \\ \frac{1}{\sqrt{3}} \\ \frac{1}{\sqrt{3}} \end{pmatrix}$$

$$q_2 = \frac{1}{\|w_2\|} w_2 = \begin{pmatrix} \frac{3}{\sqrt{15}} \\ -\frac{2}{\sqrt{15}} \\ \frac{1}{\sqrt{15}} \\ \frac{1}{\sqrt{15}} \end{pmatrix}$$

$$q_3 = \frac{1}{\|w_3\|} w_3 = \begin{pmatrix} \frac{3}{\sqrt{35}} \\ \frac{3}{\sqrt{35}} \\ -\frac{4}{\sqrt{35}} \\ \frac{1}{\sqrt{35}} \end{pmatrix}$$

so that q_1, q_2, and q_3 are an orthonormal basis for M.   ▯

If the vectors in the original set are linearly dependent, then a zero vector is produced by the algorithm at the point where one of the vectors is a linear combination of the preceding ones.

EXAMPLE 4.14

Find an orthonormal basis for the subspace M of $\mathbf{R}^3$ spanned by $v_1 = (0, 1, 1)$, $v_2 = (1, 0, 1)$, $v_3 = (1, 2, 3)$, and $v_4 = (1, 1, 0)$.

SOLUTION As usual, we will begin by taking $w_1 = v_1$. So w_2 is defined by

$$w_2 = \begin{pmatrix} 1 \\ 0 \\ 1 \end{pmatrix} - \frac{1}{2} \begin{pmatrix} 0 \\ 1 \\ 1 \end{pmatrix} = \begin{pmatrix} 1 \\ -\frac{1}{2} \\ \frac{1}{2} \end{pmatrix}$$

Now, we can check that w_1 and w_2 are perpendicular. Continuing,

$$w_3 = \begin{pmatrix} 1 \\ 2 \\ 3 \end{pmatrix} - \frac{5}{2}\begin{pmatrix} 0 \\ 1 \\ 1 \end{pmatrix} - \frac{3/2}{3/2}\begin{pmatrix} 1 \\ -\frac{1}{2} \\ \frac{1}{2} \end{pmatrix} = \begin{pmatrix} 0 \\ 0 \\ 0 \end{pmatrix}$$

Since w_3 is zero, since v_3 is a linear combination of w_1 and w_2, we see that v_3 is a linear combination of v_1 and v_2 (whose span is the same as w_1 and w_2). We just drop w_3 from the further computations.

$$w_4 = \begin{pmatrix} 1 \\ 1 \\ 0 \end{pmatrix} - \frac{1}{2}\begin{pmatrix} 0 \\ 1 \\ 1 \end{pmatrix} - \frac{1/2}{3/2}\begin{pmatrix} 1 \\ -\frac{1}{2} \\ \frac{1}{2} \end{pmatrix} - \begin{pmatrix} 0 \\ 0 \\ 0 \end{pmatrix} = \begin{pmatrix} \frac{2}{3} \\ \frac{2}{3} \\ -\frac{2}{3} \end{pmatrix}$$

Now the orthonormal basis may be obtained by normalizing these vectors.

$$q_1 = \frac{1}{\|w_1\|}w_1 = \begin{pmatrix} 0 \\ \frac{1}{\sqrt{2}} \\ \frac{1}{\sqrt{2}} \end{pmatrix}$$

$$q_2 = \frac{1}{\|w_2\|}w_2 = \begin{pmatrix} \frac{2}{\sqrt{6}} \\ -\frac{1}{\sqrt{6}} \\ \frac{1}{\sqrt{6}} \end{pmatrix}$$

$$q_3 = \frac{1}{\|w_4\|}w_4 = \begin{pmatrix} \frac{1}{\sqrt{3}} \\ \frac{1}{\sqrt{3}} \\ -\frac{1}{\sqrt{3}} \end{pmatrix}$$

Notice that these calculations imply that M is a 3–dimensional subspace of $\mathbf{R}^3$, that is, $M = \mathbf{R}^3$. $\quad\Box$

The answer obtained from using the Gram–Schmidt algorithm depends on the order of the vectors in the basis, as well as the vectors themselves. This can be illustrated by finding an orthonormal basis using Gram–Schmidt starting with the vectors in another order in either of the examples above (see Exercises 4.3.8. and 4.3.9.).

EXAMPLE 4.15

Solve the Problem of Section 4.1: The equation $2x - 3y + 2z = 11$ describes a plane in $\mathbf{R}^3$. Set up a unit length, square grid on this plane and use it to find the other corners of the two squares in the plane that have vertices at $(3, -1, 1)$ and $(4, 1, 3)$.

SOLUTION Checking, we see that, indeed, the two points that are given actually lie on the plane: $2(3) - 3(-1) + 2(1) = 11$ and $2(4) - 3(1) + 2(3) = 11$. The first observation is that the plane $2x - 3y + 2z = 11$, that is, the set

$$\mathcal{P} = \{ \begin{pmatrix} x \\ y \\ z \end{pmatrix} \text{ in } \mathbf{R}^3 : 2x - 3y + 2z = 11 \}$$

is *not* a subspace: it does not contain the origin because $2(0) - 3(0) + 2(0) = 0 \neq 11$. Indeed, this plane is of the form $\{ X \in \mathbf{R}^3 : AX = b \}$ where $A = \begin{pmatrix} 2 & -3 & 2 \end{pmatrix}$ and $b = 11$, which was shown not to be a subspace in Exercise 3.2.11. This means that it does not make sense to ask for a basis for this plane, and it is not helpful to know whether points in the plane (i.e. vectors with tail at the origin and tip in the plane) are linearly independent or not, and so on.

On the other hand, we know from Theorem 2.2 that if X_p satisfies the equation $AX = b$, then every solution of the equation can be written as $X_p + X_0$ where X_0 is a solution of $AX = 0$. Since we are especially interested in the points $(3, -1, 1)$ and $(4, 1, 3)$, we should take one of them to be X_p, say $X_p = (3, -1, 1)$. (In effect, X_p will become the "origin" of the coordinate system we are setting up on the plane; choosing $(4, 1, 3)$ will give a different system, but not better or worse.) Now, the set $\{ X : AX = 0 \}$ *is* a subspace, indeed, it is the nullspace of the matrix A. This set has a basis, and since every point in the plane $\mathcal{P}$ is the sum of X_p and a point in this subspace, if we "coordinatize" the nullspace, we can transport the coordinate system to the plane $\mathcal{P}$.

Now, if we are going to do the coordinatization in such a way that lengths are to be preserved, we want to choose an orthonormal coordinate system for $\{ X : AX = 0 \}$. To do so, we first find a basis, then use Gram–Schmidt. Since $(4, 1, 3) = (3, -1, 1) + (1, 2, 2) = X_p + (1, 2, 2)$, the vector $(1, 2, 2)$ is in $\{ X : AX = 0 \}$ and it is an especially interesting vector in this subspace. Let's use it for one of the basis vectors. Since the rank of A is 1 and the dimension of $\mathbf{R}^3$ is 3, the dimension of the nullspace of A is $3 - 1 = 2$. The equation $2x - 3y + 2z = 0$ can be solved for x as $x = 1.5y - z$. Choosing $y = 0$ and $z = 1$ gives $(-1, 0, 1)$ as a second vector in $\mathcal{N}(A)$. We can easily see that $v_1 = (1, 2, 2)$ and $v_2 = (-1, 0, 1)$ are linearly independent, so they form a basis for $\mathcal{N}(A)$.

Taking $w_1 = v_1 = (1, 2, 2)$ and

$$w_2 = v_2 - \frac{\langle w_1, v_2 \rangle}{\langle w_1, w_1 \rangle} w_1 = \begin{pmatrix} -1 \\ 0 \\ 1 \end{pmatrix} - \frac{1}{9} \begin{pmatrix} 1 \\ 2 \\ 2 \end{pmatrix} = \begin{pmatrix} -\frac{10}{9} \\ -\frac{2}{9} \\ \frac{7}{9} \end{pmatrix}$$

Normalizing, we get

$$q_1 = \frac{1}{\|w_1\|} w_1 = \begin{pmatrix} \frac{1}{3} \\ \frac{2}{3} \\ \frac{2}{3} \end{pmatrix} \approx \begin{pmatrix} .3333 \\ .6667 \\ .6667 \end{pmatrix}$$

and

$$q_2 = \frac{1}{\|w_2\|} w_2 = \begin{pmatrix} -\frac{10}{3\sqrt{17}} \\ -\frac{2}{3\sqrt{17}} \\ \frac{7}{3\sqrt{17}} \end{pmatrix} \approx \begin{pmatrix} -.8085 \\ -.1617 \\ .5659 \end{pmatrix}$$

The vectors q_1 and q_2 are an orthonormal basis for $\mathcal{N}(A)$ and for any pair of real numbers s and t, the vector $sq_1 + tq_2$ is in $\mathcal{N}(A)$ and

$$\|sq_1 + tq_2\| = \left\| \begin{pmatrix} s \\ t \end{pmatrix} \right\| = \sqrt{s^2 + t^2}$$

Moreover the angle between the vectors $s_1q_1 + t_1q_2$ and $s_2q_1 + t_2q_2$ is the same as the between the corresponding vectors in the s, t–plane

$$\langle s_1q_1 + t_1q_2, s_2q_1 + t_2q_2 \rangle = s_1s_2 + t_1t_2 = \left\langle \begin{pmatrix} s_1 \\ t_1 \end{pmatrix}, \begin{pmatrix} s_2 \\ t_2 \end{pmatrix} \right\rangle$$

This means that any calculations of distance or angle we do for vectors in $\mathcal{N}(A)$ can be carried out with the coordinates s and t instead.

Now every point in the plane $\mathcal{P}$ can be written $X_p + X_0$ where X_0 is in $\mathcal{N}(A)$, that is, every point in $\mathcal{P}$ can be written as $X_p + sq_1 + tq_2$. Thus the correspondence

$$\begin{pmatrix} s \\ t \end{pmatrix} \longleftrightarrow X_p + sq_1 + tq_2 \approx \begin{pmatrix} 3 \\ -1 \\ 1 \end{pmatrix} + s \begin{pmatrix} .3333 \\ .6667 \\ .6667 \end{pmatrix} + t \begin{pmatrix} -.8085 \\ -.1617 \\ .5659 \end{pmatrix}$$

puts a unit length, square grid (corresponding to the usual grid in the s, t–plane) on the plane $\mathcal{P}$ which is the plane with equation $2x - 3y + 2z = 11$. (See Figure 4.4.)

FIGURE 4.4

The coordinatized plane

Now the points $(3, -1, 1) = X_p + 0q_1 + 0q_2$ and $(4, 1, 3) = X_p + 3q_1 + 0q_2$ are two vertices of a square. The length of this side of the square is

$$\left\| \begin{pmatrix} 3 \\ -1 \\ 1 \end{pmatrix} - \begin{pmatrix} 4 \\ 1 \\ 3 \end{pmatrix} \right\| = \|(X_p + 0q_1 + 0q_2) - (X_p + 3q_1 + 0q_2)\| = \|3q_1\| = 3$$

Another way to see this is to look at the corresponding points in the s, t–plane: $(0, 0)$ and $(3, 0)$. Of course in the s, t–plane, the other vertices of the square are either $(0, 3)$ and $(3, 3)$ or they are $(0, -3)$ and $(3, -3)$.

This means that the other vertices of the square in $\mathcal{P}$ are

$$X_p + 0q_1 + 3q_2 \approx \begin{pmatrix} 0.5746 \\ -1.4851 \\ 2.6977 \end{pmatrix}$$

and

$$X_p + 3q_1 + 3q_2 \approx \begin{pmatrix} 1.5746 \\ 0.5149 \\ 4.6977 \end{pmatrix}$$

or they are

$$X_p + 0q_1 - 3q_2 \approx \begin{pmatrix} 5.4254 \\ -0.5149 \\ -0.6977 \end{pmatrix}$$

and

$$X_p + 3q_1 - 3q_2 \approx \begin{pmatrix} 6.4254 \\ 1.4851 \\ 1.3023 \end{pmatrix}$$

$\square$

Exercises 4.3

1. The vectors $v_1 = (1, 1)$ and $v_2 = (2, 1)$ form a basis for $\mathbf{R}^2$.
   (a) Make a sketch showing $w_1 = v_1$, v_2, the projection
$$\frac{\langle w_1, v_2 \rangle}{\langle w_1, w_1 \rangle} w_1$$
   and
$$w_2 = v_2 - \frac{\langle w_1, v_2 \rangle}{\langle w_1, w_1 \rangle} w_1$$
   (b) Use the Gram–Schmidt orthogonalization process starting with the vectors v_1 and v_2 to create an orthonormal basis for $\mathbf{R}^2$.

2. The vectors $v_1 = (3, 1)$ and $v_2 = (1, 2)$ form a basis for $\mathbf{R}^2$.
   (a) Make a sketch showing $w_1 = v_1$, v_2, the projection
$$\frac{\langle w_1, v_2 \rangle}{\langle w_1, w_1 \rangle} w_1$$
   and
$$w_2 = v_2 - \frac{\langle w_1, v_2 \rangle}{\langle w_1, w_1 \rangle} w_1$$
   (b) Use the Gram–Schmidt orthogonalization process starting with the vectors v_1 and v_2 to create an orthonormal basis for $\mathbf{R}^2$.

3. The vectors $v_1 = (-1, 1)$, $v_2 = (1, 2)$, and $v_3 = (2, -1)$ form a spanning set for $\mathbf{R}^2$.
   (a) Explain how you can tell without any calculation that these vectors are linearly dependent.
   (b) Noting that v_1 and v_2 are linearly independent, predict what must happen in computing w_3.
   (c) Use the Gram–Schmidt orthogonalization process starting with the vectors v_1, v_2, and v_3 to create an orthonormal basis for $\mathbf{R}^2$. Compare your answer with your prediction in (b) above.

4. The vectors $v_1 = (1, 1, -1)$; $v_2 = (2, 1, 2)$; and $v_3 = (2, -1, -1)$ are a basis for $\mathbf{R}^3$. Use the Gram–Schmidt orthogonalization process to create an orthonormal basis for $\mathbf{R}^3$.

5. The vectors $v_1 = (1, 1, -1, 1)$; $v_2 = (1, 0, 1, 2)$; $v_3 = (1, -2, -1, 0)$; and $v_4 = (0, 2, 1, -1)$ are a basis for $\mathbf{R}^4$. Use the Gram–Schmidt orthogonalization process to create an orthonormal basis for $\mathbf{R}^4$.

6. The vectors $v_1 = (1, 0, -1)$ and $v_2 = (2, 1, -1)$ span the subspace $\mathcal{U}$ in $\mathbf{R}^3$. Use the Gram–Schmidt orthogonalization process to create an orthonormal basis for $\mathcal{U}$.

7. The vectors $v_1 = (1, 0, -1, 1)$; $v_2 = (2, 1, 1, -1)$; and $v_3 = (1, -1, -1, 0)$ span the subspace $\mathcal{W}$ in $\mathbf{R}^4$. Use the Gram–Schmidt orthogonalization process to create an orthonormal basis for $\mathcal{W}$.

8. As in Example 4.13, find an orthonormal basis for the subspace M of $\mathbf{R}^4$ spanned by $v_1 = (0, 1, 1, 1)$, $v_2 = (1, 0, 1, 1)$, and $v_3 = (1, 1, 0, 1)$ by using the Gram–Schmidt algorithm with the given vectors in the order v_2, v_3, v_1.

9. As in Example 4.14, find an orthonormal basis for the subspace M of $\mathbf{R}^3$ spanned by $v_1 = (0, 1, 1)$, $v_2 = (1, 0, 1)$, $v_3 = (1, 2, 3)$, and $v_4 = (1, 1, 0)$ by using the Gram–Schmidt algorithm with the given vectors in the order v_2, v_3, v_1, v_4.

10. Let $\mathcal{W}$ be the hyperplane (i.e. 3-dimensional subspace) in $\mathbf{R}^4$ with equation $2a + b - c + 2d = 0$. Find an orthonormal basis for $\mathcal{W}$.

11. Let M be the subspace of $\mathbf{R}^4$ spanned by $v_1 = (1, 1, 0, 0)$; $v_2 = (0, 1, 1, 0)$; $v_3 = (0, 0, 1, 1)$; and $v_4 = (1, 0, 0, 1)$.

    (a) Use the Gram-Schmidt orthogonalization process to find an orthonormal basis for M.

    (b) Extend the basis you found in part (a) to an orthonormal basis for all of $\mathbf{R}^4$.

    (c) Is $w = (1, -1, 0, 1)$ in M? (Justify your answer.)

12. Let $C = \begin{pmatrix} 1 & 0 & 3 & -1 \\ 0 & 1 & 1 & 2 \\ 2 & -1 & 5 & -4 \end{pmatrix}$ (Compare with problem 3.8.4.).

    (a) Find an orthonormal basis for the nullspace of C.

    (b) Find an orthonormal basis for the range of C'.

    (c) Find the angles between the vectors you found in parts (a) and (b).

    (d) Find an orthonormal basis for the nullspace of C'.

    (e) Find an orthonormal basis for the range of C.

    (f) What do you notice about your answers to parts (d) and (e).

13. The equation $3x - y + 2z = 6$ describes a plane in $\mathbf{R}^3$ and the points $(1, -1, 1)$ and $(2, 0, 0)$ lie on this plane. Set up a unit length, square grid on this plane by writing every point on the plane as $X_p + sq_1 + tq_2$ where q_1 and q_2 are orthogonal vectors of length one with $X_p = (1, -1, 1)$ and $(2, 0, 0) = X_p + sq_1$ for some s.

4.4   Orthogonal Complements and the Duality Theorem

Just as the orthogonality of vectors in Euclidean space is important, perpendicularity of subspaces is a useful concept.

DEFINITION *Let M be a subspace of $\mathbf{R}^n$ or $\mathbf{C}^n$. The* orthogonal complement of M, written $M^\perp$, is the set

$$M^\perp = \{v : \langle w, v \rangle = 0 \text{ for all } w \in M\}$$

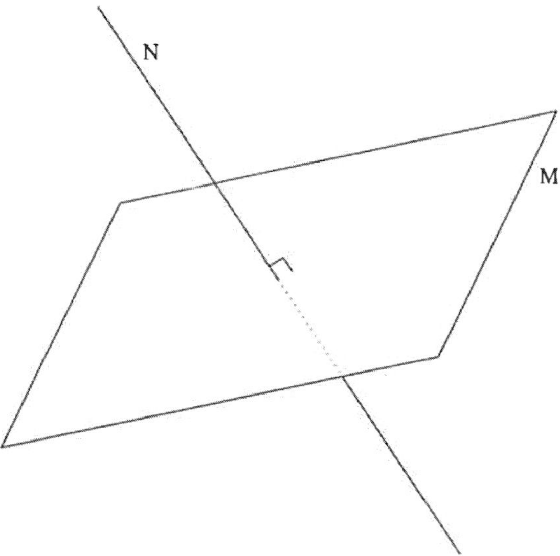

FIGURE 4.5

Orthogonal complements: $M = N^{\perp}$ **and** $N = M^{\perp}$

PROPOSITION 4.16

If M is a subspace of $\mathbf{R}^n$ or $\mathbf{C}^n$, then $M^{\perp}$ is also a subspace.

PROOF We need to show that $M^{\perp}$ is closed under addition and scalar multiplication. Suppose v_1 and v_2 are in $M^{\perp}$ and suppose a is a number. If w is in M, then $\langle w, av_1 \rangle = a\langle w, v_1 \rangle = a0 = 0$. Since this is true for all w in M, av_1 is in $M^{\perp}$ and $M^{\perp}$ is closed under scalar multiplication. Similarly, $\langle w, v_1 + v_2 \rangle = \langle w, v_1 \rangle + \langle w, v_2 \rangle = 0 + 0 = 0$, so $v_1 + v_2$ is in $M^{\perp}$. ∎

For example, the orthogonal complement of a plane through the origin in $\mathbf{R}^3$ is a line through the origin and the orthogonal complement of a line through the origin is a plane through the origin (see Figure 4.5).

Finding the orthogonal complement from the definition looks like a formidable task: we must check $\langle w, v \rangle = 0$ for *every* vector v in $\mathbf{R}^n$ and there are infinitely many such vectors for most subspaces! For example, if we wish to find the orthogonal complement of the subspace of $\mathbf{R}^3$ spanned by $(1, 1, 2)$ and $(2, -1, -3)$, then we should try $v = (a, b, c)$ and check

$$0 = \left\langle \begin{pmatrix} 1 \\ 1 \\ 2 \end{pmatrix}, \begin{pmatrix} a \\ b \\ c \end{pmatrix} \right\rangle = a + b + 2c$$

since $(1, 1, 2)$ is in the subspace;

$$0 = \left\langle \begin{pmatrix} 2 \\ -1 \\ -3 \end{pmatrix}, \begin{pmatrix} a \\ b \\ c \end{pmatrix} \right\rangle = 2a - b - 3c$$

since $(2, -1, -3)$ is in the subspace;

$$0 = \left\langle \begin{pmatrix} 3 \\ 0 \\ -1 \end{pmatrix}, \begin{pmatrix} a \\ b \\ c \end{pmatrix} \right\rangle = 3a - c$$

since $(1, 1, 2) + (2, -1, -3) = (3, 0, -1)$ is in the subspace;

$$0 = \left\langle \begin{pmatrix} -4 \\ 5 \\ 13 \end{pmatrix}, \begin{pmatrix} a \\ b \\ c \end{pmatrix} \right\rangle = -4a + 5b + 13c$$

since $2(1, 1, 2) - 3(2, -1, -3) = (-4, 5, 13)$ is in the subspace;

$$0 = \left\langle \begin{pmatrix} 7 \\ -5 \\ -14 \end{pmatrix}, \begin{pmatrix} a \\ b \\ c \end{pmatrix} \right\rangle = 7a - 5b - 14c$$

since $-(1, 1, 2) + 4(2, -1, -3) = (7, -5, -14)$ is in the subspace; and so on! It appears that we need to solve a system of infinitely many equations in three unknowns to find a description of the orthogonal complement!

Therefore, the following characterization of orthogonal complement is convenient to use when a basis is given for a subspace: it permits finding the orthogonal complement by solving only a few equations. It is a good example of the usefulness of theory in matrix analysis. Its proof is Exercise 4.4.12.

THEOREM 4.17

Suppose M is the subspace of $\mathbf{R}^n$ spanned by $\{w_1, w_2, \cdots, w_k\}$. Then

$$M^\perp = \{v : \langle w_j, v \rangle = 0 \, for \, j = 1, 2, \cdots, k\}$$

EXAMPLE 4.18

Let M be the subspace of $\mathbf{R}^4$ spanned by $w_1 = (1, 2, 2, -1)$ and $w_2 = (1, 1, 0, 1)$. Find a basis for $M^\perp$.

SOLUTION We use Theorem 4.17: it says a vector v is in $M^\perp$ if and only if $\langle w_1, v \rangle = 0$ and $\langle w_2, v \rangle = 0$. If $v = (a, b, c, d)$, then

$$\langle w_1, v \rangle = \left\langle \begin{pmatrix} 1 \\ 2 \\ 2 \\ -1 \end{pmatrix}, \begin{pmatrix} a \\ b \\ c \\ d \end{pmatrix} \right\rangle = a + 2b + 2c - d$$

and

$$\langle w_2, v \rangle = \left\langle \begin{pmatrix} 1 \\ 1 \\ 0 \\ 1 \end{pmatrix}, \begin{pmatrix} a \\ b \\ c \\ d \end{pmatrix} \right\rangle = a + b + 0c + d$$

so the two equations above are

$$\begin{cases} \langle w_1, v \rangle &=& a + 2b + 2c - d &=& 0 \\ \langle w_2, v \rangle &=& a + b + d &=& 0 \end{cases}$$

Solving the system gives $a = 2c - 3d$, $b = -2c + 2d$, and c and d are arbitrary. Choosing $c = 1$, $d = 0$ gives the vector $(2, -2, 1, 0)$ and choosing $c = 0$, $d = 1$ gives the vector $(-3, 2, 0, 1)$ as vectors forming a basis for the solution space of the system. Thus, the vectors $(2, -2, 1, 0)$ and $(-3, 2, 0, 1)$ are a basis for $M^\perp$.

It is easy to check that these vectors are in $M^\perp$ because they are each perpendicular to w_1 and w_2. Aside from the calculation above, we do not have other checks that this is correct. Theorem 4.19 below gives some other checks, for example, that the dimension of $M^\perp$ is 2 in this case, that give us confidence that we have not made any errors. □

The following theorem gives the important basic facts about orthogonal complements. Properties *(i)* and *(ii)* justify the use of the word "complement" for $M^\perp$. Although property *(iv)* seems quite natural, it is rather subtle. In particular, this property fails for infinite dimensional spaces. In property *(iv)*, the set $(M^\perp)^\perp$ can be described as "the set of vectors that are orthogonal to every vector that is orthogonal to the vectors in M". Clearly, every vector in M is such a vector: it is a like asking "Who's buried in Grant's tomb?". Of course, Grant is! The subtlety comes in that General Grant's wife, Julia, is *also* buried in Grant's tomb! After the fairly easy containment $M \subset (M^\perp)^\perp$ (Grant is buried in Grant's tomb), what must be done in the proof of *(iv)* is to show that *no other vectors besides those in M can be in* $(M^\perp)^\perp$ (there is no room for Mrs. Grant!). The proof must be subtle because it is not true in infinite dimensions which means the proof must use the finite dimensionality somehow; in infinite dimensional spaces, there *is* sometimes room for Mrs. Grant! Figure 4.5 on page 233 illustrates parts (i), (iii), and (iv) of the theorem.

THEOREM 4.19

Let M be a subspace of $\mathbf{R}^n$.

(i) $M \cap M^\perp = (0)$.

(ii)   *If v is a vector in* $\mathbf{R}^n$, *then there are vectors w in M and u in* $M^\perp$ *so that* $v = w + u$.

(iii) $\dim(M) + \dim(M^\perp) = \dim(\mathbf{R}^n)$.

(iv) $(M^\perp)^\perp = M$.

PROOF   *(i)*. Let v be in $M \cap M^\perp$. Since v is in M and v is in $M^\perp$, $\langle v, v \rangle = 0$. The only vector orthogonal to itself is 0 so $v = 0$.

(ii). Let $w_1, w_2, \cdots, w_m$ be an orthogonal basis for M. Since these vectors are linearly independent in $\mathbf{R}^n$, we can find vectors $v_1, v_2, \cdots, v_k$ so that $w_1, w_2, \cdots, w_m, v_1, v_2, \cdots, v_k$ is a basis for $\mathbf{R}^n$. Orthogonalize this basis with the Gram–Schmidt algorithm to get an orthogonal basis for $\mathbf{R}^n$. Since $w_1, w_2, \cdots, w_m$ was an orthogonal linearly independent set to start with, the new basis is $w_1, w_2, \cdots, w_m, u_1, u_2, \cdots, u_k$. The fact that this is an orthogonal basis implies that each of the u's is orthogonal to all the w's. The w's span M, so by Theorem 4.17, each u_j is in $M^\perp$. Every vector in $\mathbf{R}^n$ is a linear combination of the w's and the u's, that is, if v is in $\mathbf{R}^n$, there are numbers $a_1, \cdots, a_m$, and $b_1, \cdots, b_k$ so that

$$v = a_1 w_1 + a_2 w_2 + \cdots + a_m w_m + b_1 u_1 + \cdots + b_k u_k$$

Taking $w = a_1 w_1 + a_2 w_2 + \cdots + a_m w_m$ in M and $u = b_1 u_1 + \cdots + b_k u_k$ in $M^\perp$, we have $v = w + u$.

(iii). Let $w_1, w_2, \cdots, w_m, u_1, u_2, \cdots, u_k$ be the basis for $\mathbf{R}^n$ introduced in *(ii)*, so that $\dim(M) = m$ and $\dim(\mathbf{R}^n) = m + k$. We need to show that $\dim(M^\perp) = k$. Since $u_1, u_2, \cdots, u_k$ are linearly independent vectors in $M^\perp$, we only need to show they span $M^\perp$. So, suppose u is a vector in $M^\perp$. Since $M^\perp$ is a subspace of $\mathbf{R}^n$, u is in $\mathbf{R}^n$ and there are numbers so that

$$u = c_1 w_1 + c_2 w_2 + \cdots + c_m w_m + d_1 u_1 + \cdots + d_k u_k$$

Let $z = c_1 w_1 + c_2 w_2 + \cdots + c_m w_m$. Since the w_j's form a basis for M, clearly, z is in M. On the other hand,

$$z = u - d_1 u_1 - \cdots - d_k u_k$$

and each of the vectors on the right is in $M^\perp$, so z is in $M^\perp$. By *(i)*, the only vector in both M and $M^\perp$ is 0, so $z = 0$. Thus

$$u = c_1 w_1 + c_2 w_2 + \cdots + c_m w_m + d_1 u_1 + \cdots + d_k u_k = 0 + d_1 u_1 + \cdots + d_k u_k$$

so u is a linear combination of $u_1, \cdots, u_k$. Since this is true for every u in $M^\perp$, these vectors span $M^\perp$ and are a basis for it. This means the dimension of $M^\perp$ is k and the conclusion follows.

(iv). If v is a vector in M and w is a vector in $M^\perp$, then $\langle w, v \rangle = 0$. Since this is true for every vector w in $M^\perp$, it follows that v is in $(M^\perp)^\perp$. Thus, $M \subset (M^\perp)^\perp$.

We find from *(ii)* that

$$\dim(M) + \dim(M^\perp) = \dim(\mathbf{R}^n)$$

and, applying this to $M^\perp$, that

$$\dim(M^\perp) + \dim\left((M^\perp)^\perp\right) = \dim(\mathbf{R}^n)$$

It follows from this that

$$\dim(M) = \dim\left((M^\perp)^\perp\right)$$

and the containment proved above together with Lemma 3.40 show that $M = (M^\perp)^\perp$. ∎

EXAMPLE 4.20

Let M be the subspace of $\mathbf{R}^4$ spanned by $w_1 = (1, -1, 1, 3)$, $w_2 = (2, -1, 0, 1)$, and $w_3 = (-1, 0, 1, 2)$. Find a basis for $M^\perp$. Find vectors w in M and u in $M^\perp$ so that $w + u = (1, 0, 1, 0)$.

SOLUTION According to Theorem 4.17, a vector v is in $M^\perp$ if and only if $\langle w_j, v \rangle = 0$ for $j = 1, 2, 3$. Letting $v = (a, b, c, d)$, these equations are

$$\begin{cases} \langle w_1, v \rangle & = & a - b + c + 3d & = & 0 \\ \langle w_2, v \rangle & = & 2a - b + d & = & 0 \\ \langle w_3, v \rangle & = & -a + c + 2d & = & 0 \end{cases}$$

Solving this system

$$\left(\begin{array}{cccc|c} 1 & -1 & 1 & 3 & 0 \\ 2 & -1 & 0 & 1 & 0 \\ -1 & 0 & 1 & 2 & 0 \end{array}\right) \longrightarrow \cdots \longrightarrow \left(\begin{array}{cccc|c} 1 & 0 & -1 & -2 & 0 \\ 0 & 1 & -2 & -5 & 0 \\ 0 & 0 & 0 & 0 & 0 \end{array}\right)$$

gives $a = c + 2d$ and $b = 2c + 5d$ with c and d arbitrary. Thus $u_1 = (1, 2, 1, 0)$ and $u_2 = (2, 5, 0, 1)$ form a basis for $M^\perp$. This solves the first part of the problem.

The second part of the problem is possible because of part (ii) of Theorem 4.19. Notice that the dimension of $M^\perp$ is 2 and that, by part (iii) of the theorem, $\dim(M) + 2 = 4$ so the dimension of M is also 2. In particular this means that w_1, w_2, and w_3 are linearly dependent, so one is a linear combination of the others: indeed, $w_3 = w_1 - w_2$. Thus, w_1 and w_2 form a basis for M.

The essence of the proof of (ii) is that because the only vector that M and $M^\perp$ have in common is 0 (part (i) of the Theorem), that a basis for M and a basis for $M^\perp$ can be combined to form a basis for $\mathbf{R}^4$. (The point of using orthogonal bases in the proof was to actually get bases for M and $M^\perp$ instead of M and something else.) In our case, this should mean that $\{w_1, w_2, u_1, u_2\}$ form a basis for $\mathbf{R}^4$. This is indeed the case: it is not difficult to check that they are linearly independent:

```
> w1=[1; -1; 1; 3]

w1 =

        1
       -1
        1
        3

> w2=[2; -1; 0; 1]

w2 =

        2
       -1
        0
        1

> u1=[1; 2; 1; 0]

u1 =

        1
        2
        1
        0
```

```
> u2=[2; 5; 0; 1]

u2 =
     2
     5
     0
     1

> b=[w1 w2 u1 u2]

b =
     1      2      1      2
    -1     -1      2      5
     1      0      1      0
     3      1      0      1

> rank(b)

ans =
     4
```

So the column vectors of b, which are the vectors we are interested in, are linearly independent.

Since these vectors are a basis for $\mathbf{R}^4$, every vector in $\mathbf{R}^4$ can be written as a linear combination of them, including $v = (1, 0, 1, 0)$. Then, as in the proof of (ii), the vector w can be taken to be the resulting linear combination of w_1 and w_2 and u to be the resulting linear combination of u_1 and u_2. Thus, we want to find constants c_1, c_2, d_1, and d_2 so that $c_1 w_1 + c_2 w_2 + d_1 u_1 + d_2 u_2 = v$. This system is easily solved using MATLAB.

```
> x=b\[1;0;1;0]

x =
          0
     0.3333
     1.0000
    -0.3333
```

This calculation says

$$v = 0w_1 + \frac{1}{3}w_2 + 1u_1 - \frac{1}{3}u_2$$

Recalling that $\{w_1, w_2\}$ are a basis for M and $\{u_1, u_2\}$ are a basis for $M^\perp$, we should take $w = 0w_1 + \frac{1}{3}w_2 = (\frac{2}{3}, -\frac{1}{3}, 0, \frac{1}{3})$ and $u = 1u_1 - \frac{1}{3}u_2 = (\frac{1}{3}, \frac{1}{3}, 1, -\frac{1}{3})$.

It is easily checked that with these vectors, $w + u = v$, that w is in M because it is a linear combination of w_1, w_2, and w_3, and that u is in $M^\perp$ because it is orthogonal to w_1, w_2, and w_3.

Moreover, this is the only possible answer: if $w_0 + u_0 = v$ where w_0 is in M and u_0 is $M^\perp$ is also a splitting, then we get $w + u = v = w_0 + u_0$ which would mean $w - w_0 = u_0 - u$. The vector represented by this equality is in M because it is $w - w_0$ and also in $M^\perp$ because it is $u_0 - u$. But the only vector in M and $M^\perp$, by (i) of Theorem 4.19, is 0, so $w - w_0 = u_0 - u = 0$ which says $w_0 = w$ and $u_0 = u$.   ⬚

Before we prove the Duality Theorem, we need an easy formula relating the inner product to matrix multiplication; it is this formula that motivates the approach to inner products in this book. The formula was Exercise 4.2.17.

LEMMA 4.21

If A is an $m \times n$ matrix, u is in $\mathbf{R}^n$ (or $\mathbf{C}^n$), and v is in $\mathbf{R}^m$ (or $\mathbf{C}^m$),

$$\langle Au, v \rangle = \langle u, A'v \rangle$$

PROOF

$$\langle Au, v \rangle = (Au)'v = (u'A')v = u'(A'v) = \langle u, A'v \rangle$$

∎

We are ready to state and prove the Duality Theorem. It is a modern version of *Fredholm's Alternative* (see Exercise 4.4.13.). Fredholm proved his alternative in connection with work he was doing at the beginning of the twentieth century on partial differential equations. If A is a matrix, $\mathcal{R}(A)$ will denote the range of A and $\mathcal{N}(A)$ will denote the nullspace of A.

THEOREM 4.22   ***Duality Theorem***

If A is an $m \times n$ matrix, then $\mathrm{rank}(A) = \mathrm{rank}(A')$ *and*

$$\mathcal{R}(A') = \mathcal{N}(A)^{\perp} \qquad \mathcal{N}(A') = \mathcal{R}(A)^{\perp}$$
$$\mathcal{R}(A) = \mathcal{N}(A')^{\perp} \qquad \mathcal{N}(A) = \mathcal{R}(A')^{\perp}$$

PROOF   Suppose v is in $\mathcal{N}(A')$, that is, $A'v = 0$. If w is in $\mathcal{R}(A)$, say $w = Au$ for some u in $\mathbf{R}^n$, then using Lemma 4.21

$$\langle v, w \rangle = \langle v, Au \rangle = \langle A'v, u \rangle = \langle 0, u \rangle = 0$$

Since this is true for all w in $\mathcal{R}(A)$, it follows that v is in $\mathcal{R}(A)^{\perp}$, that is, that $\mathcal{N}(A') \subset \mathcal{R}(A)^{\perp}$. On the other hand, if v is in $\mathcal{R}(A)^{\perp}$, then for any u in $\mathbf{R}^n$, since Au is in the range of A,

$$\langle A'v, u \rangle = \langle v, Au \rangle = 0$$

That is, $A'v$ is a vector in $\mathbf{R}^n$ that is perpendicular to *every* vector in $\mathbf{R}^n$. The only such vector is 0, so $A'v = 0$ and v is in $\mathcal{N}(A')$. Thus, $\mathcal{R}(A)^{\perp} \subset \mathcal{N}(A')$ and together with the above containment, this gives $\mathcal{N}(A') = \mathcal{R}(A)^{\perp}$.

Now taking orthogonal complements of both sides and using *(iv)* of Theorem 4.19, this gives

$$\mathcal{N}(A')^{\perp} = (\mathcal{R}(A)^{\perp})^{\perp} = \mathcal{R}(A)$$

The other equalities follow from these by replacing A by A' and using $(A')' = A$.

To prove the first assertion, we recall the Rank–Nullity Theorem says that

$$\dim \mathcal{R}(A) + \dim \mathcal{N}(A) = n$$

and *(iii)* of Theorem 4.19 says that

$$\dim(\mathcal{N}(A)^{\perp}) + \dim \mathcal{N}(A) = n$$

also, so we can conclude

$$\dim \mathcal{R}(A) = \dim(\mathcal{N}(A)^{\perp})$$

Since we have seen $\mathcal{R}(A') = \mathcal{N}(A)^{\perp}$, this means

$$\dim \mathcal{R}(A) = \dim \mathcal{R}(A')$$

∎

EXAMPLE 4.23

In Exercise 3.7.5., you found bases for the range of C, the range of C', the nullspace of C and the nullspace of C' for the matrix

$$C = \begin{pmatrix} 1 & 1 & 0 & -1 \\ 2 & -1 & 1 & 0 \\ 0 & 3 & -1 & -2 \end{pmatrix}$$

One answer to this problem is that

$$\mathcal{R}(C) = \mathrm{span}\left\{ \begin{pmatrix} 1 \\ 2 \\ 0 \end{pmatrix}, \begin{pmatrix} 1 \\ -1 \\ 3 \end{pmatrix} \right\}$$

$$\mathcal{R}(C') = \mathrm{span}\left\{ \begin{pmatrix} 1 \\ 1 \\ 0 \\ -1 \end{pmatrix}, \begin{pmatrix} 2 \\ -1 \\ 1 \\ 0 \end{pmatrix} \right\}$$

$$\mathcal{N}(C) = \mathrm{span}\left\{ \begin{pmatrix} -1 \\ 1 \\ 3 \\ 0 \end{pmatrix}, \begin{pmatrix} 1 \\ 0 \\ -2 \\ 1 \end{pmatrix} \right\}$$

$$\mathcal{N}(C') = \mathrm{span}\left\{ \begin{pmatrix} -2 \\ 1 \\ 1 \end{pmatrix} \right\}$$

The theorem above implies the vector $(-2, 1, 1)$ that spans $\mathcal{N}(C')$ is perpendicular to all the vectors in the range of C, in particular to $(1, 2, 0)$ and $(1, -1, 3)$ and this is easily checked

$$\left\langle \begin{pmatrix} -2 \\ 1 \\ 1 \end{pmatrix}, \begin{pmatrix} 1 \\ 2 \\ 0 \end{pmatrix} \right\rangle = -2 + 2 + 0 = 0$$

and

$$\left\langle \begin{pmatrix} -2 \\ 1 \\ 1 \end{pmatrix}, \begin{pmatrix} 1 \\ -1 \\ 3 \end{pmatrix} \right\rangle = -2 - 1 + 3 = 0$$

Similarly, the theorem implies the vectors in the range of C' are perpendicular to the vectors in the nullspace of C, for example,

$$\left\langle \begin{pmatrix} 1 \\ 1 \\ 0 \\ -1 \end{pmatrix}, \begin{pmatrix} -1 \\ 1 \\ 3 \\ 0 \end{pmatrix} \right\rangle = -1 + 1 + 0 + 0 = 0$$

$$\left\langle \begin{pmatrix} 1 \\ 1 \\ 0 \\ -1 \end{pmatrix}, \begin{pmatrix} 1 \\ 0 \\ -2 \\ 1 \end{pmatrix} \right\rangle = 1 + 0 + 0 - 1 = 0$$

$$\left\langle \begin{pmatrix} 2 \\ -1 \\ 1 \\ 0 \end{pmatrix}, \begin{pmatrix} -1 \\ 1 \\ 3 \\ 0 \end{pmatrix} \right\rangle = -2 - 1 + 3 + 0 = 0$$

and

$$\left\langle \begin{pmatrix} 2 \\ -1 \\ 1 \\ 0 \end{pmatrix}, \begin{pmatrix} 1 \\ 0 \\ -2 \\ 1 \end{pmatrix} \right\rangle = 2 + 0 - 2 + 0 = 0$$

We will use the results of Theorem 4.22 to help in calculating these bases and in checking our answers. ☐

Exercises 4.4

1. Let M be the subspace of $\mathbf{R}^2$ spanned by $v_1 = (2, 1)$. Find w in M and u in $M^\perp$ so that $w + u = (1, 2)$. (Compare your answer with the sketch you made for Exercise 4.3.2.)

2. Let M be the subspace of $\mathbf{R}^3$ spanned by $(2, 1, 0)$ and $(1, 1, 1)$. Find w in M and u in $M^\perp$ so that $w + u = (1, 2, -1)$.

3. Let M be the subspace of $\mathbf{R}^4$ spanned by $(2, 1, 0, -1)$, $(1, 1, 1, 0)$, and $(-1, 0, 1, 1)$. Find w in M and u in $M^\perp$ so that $w + u = (3, 1, 0, 0)$.

4. Let M be span$\{(0, 1, -1, 1); (1, 0, 1, -1); (1, 1, 0, 0); (-3, 2, -5, 5)\}$.

  (a) Find a basis for $M^\perp$.

  (b) Find w in M and u in $M^\perp$ so that $w + u = (-1, 0, 1, 2)$.

5. Let W be the subspace of $\mathbf{R}^5$ spanned by $(1, 1, 0, -1, 1)$, $(0, 2, -1, 1, 1)$, and $(-1, 1, 1, -2, 1)$. Find w in W and u in $W^\perp$ so that $w + u = (1, 1, -1, 1, 1)$.

6. Explain the answers you obtained in Exercise 4.3.12., parts (c) and (f).

7.

$$\text{Let } D = \begin{pmatrix} 1 & 0 & -1 \\ 0 & 1 & 2 \end{pmatrix}$$

The vector $v = (1, -2, 1)$ forms a basis for $\mathcal{N}(D)$, the nullspace of D.

  (a) Find a basis for the range of D.

  (b) Find a basis for the range of D'.

  (c) Find a basis for the orthogonal complement of the range of D'.

  (d) Find a basis for the nullspace of D'.

8.

$$\text{Let } A = \begin{pmatrix} 1 & 0 & 3 & -1 \\ 0 & 1 & 1 & 2 \\ 2 & -1 & 5 & -4 \end{pmatrix}$$

The vectors $v_1 = (-3, -1, 1, 0)$ and $v_2 = (1, -2, 0, 1)$ are a basis for $\mathcal{N}(A)$, the nullspace of A.

  (a) Find a basis for the range of A.

  (b) Find a basis for the range of A'.

  (c) Find a basis for the orthogonal complement of the range of A'.

  (d) Find a basis for the nullspace of A'.

9. The matrix B is an 8×11 matrix and the dimension of $\mathcal{N}(B)$, the nullspace of B, is 5.

  (a) What is the dimension of $\mathcal{R}(B)$, the range of B?

  (b) What is the dimension of $\mathcal{R}(B')$, the range of B'?

  (c) What is the dimension of $\mathcal{N}(B')$, the nullspace of B'?

  (d) What is the dimension of $\mathcal{R}(B')^\perp$, the orthogonal complement of $\mathcal{R}(B')$?

10.(a) Find a basis for the nullspace of $C = \begin{pmatrix} 1 & -1 & 1 & 0 \\ 2 & 0 & 1 & 1 \\ 1 & -3 & 2 & -1 \end{pmatrix}$.

  (b) Find vectors v_0 in the nullspace of C and v_r in the row space of C so that $(1, 1, 1, 1) = v_0 + v_r$.

11.

$$\text{Let } A = \begin{pmatrix} 1 & 1 & -1 & 2 \\ -1 & 0 & 2 & -3 \\ 1 & -1 & -3 & 4 \end{pmatrix}$$

The vectors $v_1 = (2, -1, 1, 0)$ and $v_2 = (-3, 1, 0, 1)$ are a basis for the nullspace of A.

  (a) Find a basis for the range of A.

(b) Find a basis for the range of A'.

(c) Find a basis for the orthogonal complement of the range of A'.

(d) Find a basis for the nullspace of A'.

12. Prove Theorem 4.17: Let M be the subspace of $\mathbf{R}^n$ spanned by $w_1, w_2, \cdots, w_k$ and let

$$N = \{v : \langle w_j, v \rangle = 0 \text{ for } j = 1, 2, \cdots, k\}$$

Show that $N = M^\perp$ by showing $M^\perp \subset N$ and $N \subset M^\perp$.

13. Prove the classical version of Fredholm's Alternative:

Let A be an $m \times n$ matrix and let b be any vector in $\mathbf{R}^m$. Show that one and only one of the following systems has a solution:

$$(i) \ Ax = b \qquad \text{or} \qquad (ii) \ A'y = 0 \text{ and } \langle b, y \rangle \neq 0.$$

(Hint: Show that if (i) has a solution then (ii) does not, and that if (i) does not have a solution then (ii) does have a solution.)

14. Prove that if E is an $n \times n$ matrix satisfying $E = E'$, then the nullspace of E is the same as the nullspace of E^2.

4.5 The Matlab Commands "orth" and "null"

The thesis of Section 4.3 was that finding bases consisting of orthonormal vectors is an important issue. The Gram–Schmidt approach of that section is one suited to hand calculation, but MATLAB and other numerical programs have ways of getting orthonormal bases for subspaces as well. Since we frequently encounter sets of vectors as the columns of particular matrices, our first step is to give a matrix interpretation of orthonormality.

THEOREM 4.24

The columns of the $m \times k$ matrix Q are an orthonormal set if and only if $Q'Q = I$.

PROOF The proof of this result relies on the fact that, as we have seen in other instances, multiplication of blocked matrices as if they were matrices of matrices makes sense (see Appendix B). If we regard Q as a being blocked into columns:

$$Q = (q_1 \ q_2 \ \cdots \ q_k)$$

then Q' is blocked into rows:

$$Q' = \begin{pmatrix} q'_1 \\ q'_2 \\ \vdots \\ q'_k \end{pmatrix}$$

Thus, multiplying $Q'Q$ as if they were a column vector and a row vector gives

$$Q'Q = \begin{pmatrix} q'_1 \\ q'_2 \\ \vdots \\ q'_k \end{pmatrix} (q_1 \; q_2 \; \cdots \; q_k) = \begin{pmatrix} q'_1 q_1 & q'_1 q_2 & \cdots & q'_1 q_k \\ q'_2 q_1 & q'_2 q_2 & \cdots & q'_2 q_k \\ \vdots & & \ddots & \vdots \\ q'_k q_1 & q'_k q_2 & \cdots & q'_k q_k \end{pmatrix}$$

We can identify the entries of $Q'Q$ as inner products

$$(Q'Q)_{i,j} = q'_i q_j = \langle q_i, q_j \rangle$$

so

$$Q'Q = \begin{pmatrix} \langle q_1, q_1 \rangle & \langle q_1, q_2 \rangle & \cdots & \langle q_1, q_k \rangle \\ \langle q_2, q_1 \rangle & \langle q_2, q_2 \rangle & \cdots & \langle q_2, q_k \rangle \\ \vdots & & \ddots & \vdots \\ \langle q_k, q_1 \rangle & \langle q_k, q_2 \rangle & \cdots & \langle q_k, q_k \rangle \end{pmatrix}$$

This shows the matrix $Q'Q$ is the $k \times k$ identity matrix if these numbers are 0 for $i \neq j$ and 1 for $i = j$, that is, exactly when $\{q_j\}$ is an orthonormal set.   ▌

Of course, it is impossible for a matrix Q whose columns are orthonormal to have more columns than rows because the columns are linearly dependent in a matrix with more columns than rows and this would contradict their orthonormality. If the columns of the $m \times k$ matrix Q are orthonormal and $k = m$, then $Q'Q = I$, and for square matrices, this means $QQ' = I$ as well by Theorem 3.63. On the other hand, if k is less than m, then QQ' is an $m \times m$ matrix of rank k, and it cannot be the identity. As an exercise (Exercise 5.4.9.), you may prove that QQ' is the orthogonal projection onto the range of Q.

The software package MATLAB has commands to find orthonormal bases for ranges and nullspaces of matrices. Indeed, the descriptions of the commands `orth` and `null` given by `help`, say that they implement precisely what we were seeking. Notice that the explanations of the commands reiterate some of the definitions and theorems about range and nullspace, as well as Theorem 4.24.

```
> help orth

  ORTH Orthogonalization.
    Q = orth(A) is an orthonormal basis for the range of A.
    Q'*Q = I, the columns of Q span the same space as the
      columns of A and the number of columns of Q is the
      rank of A.

    See also SVD, ORTH, RANK.

> help null

  NULL Null space.
    Z = null(A) is an orthonormal basis for the null space of A.
    Z'*Z = I, A*Z has negligible elements, and the number
      of columns of Z is the nullity of A.

    See also SVD, ORTH, RANK.
```

EXAMPLE 4.25

Find bases for the range and the nullspace of the matrix

$$B = \begin{pmatrix} -1 & 2 & 0 & 3 & 4 \\ 1 & -1 & 1 & 1 & 1 \\ 2 & 1 & -1 & 0 & 1 \\ 1 & 1 & 0 & 2 & 3 \end{pmatrix}$$

SOLUTION Although the problem does not specifically ask for an orthonormal basis for the range and the nullspace, we can use "orth" and "null" in MATLAB to find the bases: after all, an orthonormal basis is a basis.

```
> B=[-1 2 0 3 4; 1 -1 1 1 1; 2 1 -1 0 1; 1 1 0 2 3]

B =
    -1     2     0     3     4
     1    -1     1     1     1
     2     1    -1     0     1
     1     1     0     2     3

> rank(B)
```

```
ans =
     3
```

so the range of B is a 3–dimensional subspace of $\mathbf{R}^4$ and, by the rank–nullity theorem, the nullspace is a 2–dimensional subspace of $\mathbf{R}^5$.

```
> Ra = orth(B)

Ra =
    0.7938     0.4408    -0.1808
    0.1536    -0.2392     0.8811
    0.1758    -0.8099    -0.4127
    0.5616    -0.3041     0.1438
```

so the vectors that form the columns of this matrix, $v_1 = (.79, .15, .18, .56)$, $v_2 = (.44, -.24, -.81, -.30)$, and $v_3 = (-.18, .88, -.41, .14)$, are a basis for the range of B. subspace of $\mathbf{R}^5$. Similarly, since

```
> Nu = null(B)

Nu =
   -0.2024     0.0745
   -0.6152    -0.1105
   -0.6233    -0.4445
   -0.1863     0.7425
    0.3967    -0.4830
```

the vectors that form the columns of this matrix, $w_1 = (-.20, -.62, -.62, -.19, .40)$ and $w_2 = (.07, -.11, -.44, .74, -.48)$ are a basis for the nullspace of B.

If we want to check these answers,

```
> Ra'*Ra

ans =
    1.0000    -0.0000    -0.0000
   -0.0000     1.0000     0.0000
   -0.0000     0.0000     1.0000
```

so the columns of Ra are an orthonormal set (Theorem 4.24) and are therefore linearly independent (Proposition 4.8). Moreover, the rank of B is 3 and

```
> rank([B Ra])

ans =
      3
```

so the subspace spanned by the columns of B is the same as the subspace spanned by the columns of B together with v_1, v_2, and v_3. This means that the vectors v_1, v_2, and v_3 are in the range of B. Since they are linearly independent and the range of B is three dimensional, they must be a basis for the range of B.

Similarly,

```
> Nu'*Nu

ans =

     1.0000     0.0000
     0.0000     1.0000

> B*Nu

ans =

   1.0e-14  *

   -0.2220    -0.0888
   -0.0222    -0.0833
   -0.0555     0.0555
   -0.1443    -0.0444
```

so the columns of Nu are an orthonormal set, hence independent, and BNu is (nearly) zero, so the columns of Nu are (nearly) in the nullspace of B. Finally, since the nullspace of B is a 2–dimensional subspace, w_1 and w_2 are a basis for the nullspace.   ▯

These computations allow us to easily find all solutions to a system with infinitely many solutions.

EXAMPLE 4.26

Find all solutions of the system

$$\begin{cases} -a & + & 2b & & & + & 3d & + & 4e & = & -1 \\ a & - & b & + & c & + & d & + & e & = & 5 \\ 2a & + & b & - & c & & & + & e & = & -2 \\ a & + & b & & & + & 2d & + & 3e & = & 1 \end{cases}$$

SOLUTION We first need to find a solution to this system.

```
> h=[-1;5;-2;1]

h =

    -1
     5
    -2
     1

> xp=B\h

Warning: Rank deficient, rank = 3  tol =     5.7689e-15

xp =
   -0.0000
   -3.5000
        0
        0
    1.5000
```

Note that MATLAB is warning us that the rank of B is not as large as expected, namely, that the rank is less than the number of rows and columns. But to check that this is really a solution of the system, we compare Bxp with h.

```
> B*xp-h

ans =
   1.0e-14  *

         0
    0.1776
   -0.2220
   -0.0444
```

and we see that, indeed, xp is a solution of the system.

Since we found in Example 4.25 $w_1 = (-.20, -.62, -.62, -.19, .40)$ and $w_2 = (.07, -.11, -.44, .74, -.48)$ are a basis for the nullspace of B, every solution of $BX = h$ is of the form $X_p + \alpha w_1 + \beta w_2$ where α and β are numbers.

As a partial check, if we take, for example, $\alpha = 3.81$ and $\beta = -2.44$, then we get

$$X_p + 3.81 * w1 - 2.44 * w2 = (-.95, -5.57, -1.29, -2.52, 4.19)$$

is supposed to be a solution of the system

```
> B*(xp+3.81*w1-2.44*w2)

ans =
    -1.0000
     5.0000
    -2.0000
     1.0000
```

and it is.   ☐

We can use the ideas of this section to find orthonormal bases for subspaces and their orthogonal complements as well.

EXAMPLE 4.27

Let M be the subspace of $\mathbf{R}^4$ spanned by $w_1 = (1, -1, 1, 3)$, $w_2 = (2, -1, 0, 1)$, and $w_3 = (-1, 0, 1, 2)$. Find a basis for $M^\perp$. Find vectors w in M and u in $M^\perp$ so that $w + u = (1, 0, 1, 0)$.

SOLUTION Note this is the same problem as Example 4.20. We begin by finding an orthonormal basis for M by using the fact that M is the range of the matrix whose columns are w_1, w_2, and w_3.

```
> w1=[1; -1; 1; 3]

w1 =
     1
    -1
     1
     3
```

```
> w2=[2; -1; 0; 1]

w2 =
      2
     -1
      0
      1

> w3=[-1; 0; 1; 2]

w3 =
     -1
      0
      1
      2

> qw = orth([w1 w2 w3])

qw =
     0.2887      0.8660
    -0.2887     -0.2887
     0.2887     -0.2887
     0.8660     -0.2887
```

The columns of qw are supposed to be orthornormal and be a basis for the
subspace spanned by w_1, w_2, and w_3. We can check this easily.

```
> qw'*qw

ans =
     1.0000      0.0000
     0.0000      1.0000

> rank([w1 w2 w3 qw])

ans =
      2
```

Thus, the columns of qw are orthornormal and they lie in the subspace spanned
by w_1, w_2, and w_3. Since this subspace is 2–dimensional, the two columns of qw

form a basis for it.

We were asked to find a basis for $M^\perp$. The easiest way to do this is to use the Duality Theorem (Theorem 4.22). This theorem says that $M^\perp$ is the nullspace of the adjoint of the matrix whose columns are w_1, w_2, and w_3.

```
> qu = null([w1 w2 w3]')

qu =
    0.4006    -0.0785
    0.7227    -0.5577
    0.5577     0.7227
   -0.0785    -0.4006
```

The columns of qu are supposed to be orthonormal and they are supposed to be a basis for $M^\perp$. We can check this as well.

```
> qu'*qu

ans =
    1.0000    0.0000
    0.0000    1.0000

> qu'*[w1 w2 w3]

ans =
   1.0e-15 *

    0.5274    0.1388    0.3886
    0.1665   -0.0555    0.3331
```

The first calculation shows that the columns of qu are orthonormal. If we write u_1 and u_2 for the two columns of qu, then the second calculation has computed

$$
\begin{pmatrix} u_1' \\ u_2' \end{pmatrix}
\begin{pmatrix} w_1 & w_2 & w_3 \end{pmatrix}
=
\begin{pmatrix}
\langle u_1, w_1 \rangle & \langle u_1, w_2 \rangle & \langle u_1, w_3 \rangle \\
\langle u_2, w_1 \rangle & \langle u_2, w_2 \rangle & \langle u_2, w_3 \rangle
\end{pmatrix}
$$

The fact that this is (nearly) zero shows that u_1 and u_2 are each orthogonal to the w's as we want to be the case. Since we know $M^\perp$ is 2–dimensional, these vectors form an orthonormal basis for $M^\perp$ as we intended.

Now to break up $v = (1, 0, 1, 0)$, we proceed much as before. The columns of qw and the columns of qu form a basis for $\mathbf{R}^4$ and we can find the coordinates, x, of the vector v with respect to this basis. To get w and u, we write the linear combinations of the columns of qw and qu separately.

```
> v=[1;0;1;0]

v =
        1
        0
        1
        0

> x=[qw qu]\v

x =
      0.5774
      0.5774
      0.9583
      0.6442

> w= x(1)*qw(:,1)  + x(2)*qw(:,2)

w =
      0.6667
     -0.3333
     -0.0000
      0.3333

> u= x(3)*qu(:,1)  + x(4)*qu(:,2)

u =
      0.3333
      0.3333
      1.0000
     -0.3333

> w+u

ans =
      1.0000
           0
      1.0000
           0
```

The last calculation is a check on our work. Even though we used different bases for M and $M^\perp$, this is the same answer as we obtained by the calculations in Example 4.20 because, as we saw at that time, there is only one way to split v into a sum of vectors, one from M and one from $M^\perp$. ⬚

Exercises 4.5

Use "orth" and "null" to find orthonormal bases for the ranges and the nullspaces of the following matrices. Compare your answers with your answers for Exercises 4.4.8., 4.6.4., and 4.6.5.

◇ 1.

$$A = \begin{pmatrix} 1 & 0 & 3 & -1 \\ 0 & 1 & 1 & 2 \\ 2 & -1 & 5 & -4 \end{pmatrix}$$

◇ 2.

$$D = \begin{pmatrix} 1 & 2 & -1 & 3 & 4 \\ 3 & 1 & -1 & 2 & -2 \\ 1 & 3 & -1 & 2 & 0 \\ 2 & -1 & 1 & 0 & 1 \\ 2 & 0 & 1 & -1 & -3 \end{pmatrix}$$

◇ 3.

$$E = \begin{pmatrix} 2 & 1 & 1 & 1 & -1 & 0 & 2 \\ 3 & 1 & 2 & 0 & 1 & 2 & 1 \\ 1 & 2 & -1 & -1 & -1 & 1 & 1 \\ 1 & -1 & 2 & 2 & 0 & -1 & 1 \end{pmatrix}$$

◇ 4.

(a) Find a basis for the nullspace of $B = \begin{pmatrix} 1 & -1 & 0 & 2 & 1 \\ 1 & -2 & -1 & 3 & 0 \\ 2 & -1 & 2 & 4 & 4 \\ -1 & 1 & 1 & -1 & 0 \end{pmatrix}$

(b) Find vectors v_0 in the nullspace of B and v_r in the row space of B so that $(1, 1; 0, 1, 1) = v_0 + v_r$.

◇ 5.

(a) Find a basis for the nullspace of $C = \begin{pmatrix} 1 & -1 & 2 & 0 & 1 & 1 \\ -1 & 2 & -3 & 2 & -1 & 0 \\ 1 & -2 & 4 & -1 & 2 & 1 \\ -2 & 3 & -2 & 5 & 1 & 2 \end{pmatrix}$

(b) Find all solutions of the system

$$\begin{cases} a & - & b & + & 2c & & & + & e & + & f & = & 0 \\ -a & + & 2b & - & 3c & + & 2d & - & e & & & = & 1 \\ a & - & 2b & + & 4c & - & d & + & 2e & + & f & = & -1 \\ -2a & + & 3b & - & 2c & + & 5d & + & e & + & 2f & = & 1 \end{cases}$$

$\Diamond$ 6. Let M be the subspace of $\mathbf{R}^5$ spanned by $w_1 = (1,1,0,-1,1)$, $w_2 = (1,0,2,0,-1)$, and $w_3 = (0,1,1,1,-1)$. Find vectors w in M and u in $M^\perp$ so that $(1,0,0,1,1) = w + u$.

$\Diamond$ 7. Let M be the subspace of $\mathbf{R}^5$ spanned by $(2,1,0,-1,1),(1,1,1,0,2)$, and $(-1,0,1,1,1)$. Find w in M and u in $M^\perp$ so that $w + u = (3,1,0,0,-1)$.

$\Diamond$ 8. Let M be the subspace of $\mathbf{R}^5$ spanned by $w_1 = (1,2,1,1,1)$, $w_2 = (1,1,0,2,-1)$, and $w_3 = (1,0,-1,3,-3)$. Find vectors w in M and u in $M^\perp$ so that $(0,1,0,0,0) = w+u$.

$\Diamond$ 9. The vectors $v_1 = (1,0,-1)$ and $v_2 = (2,1,-1)$ span the subspace $\mathcal{U}$ in $\mathbf{R}^3$. Use "orth" to find an orthonormal basis for $\mathcal{U}$. Compare your answer to your answer to problem 4.3.6.

$\Diamond$ 10. The vectors $v_1 = (1,0,-1,1)$; $v_2 = (2,1,1,-1)$; and $v_3 = (1,-1,-1,0)$ span the subspace $\mathcal{W}$ in $\mathbf{R}^4$. Use "orth" to find an orthonormal basis for $\mathcal{W}$. Compare your answer to your answer to problem 4.3.7.

$\Diamond$ 11. Let $\mathcal{W}$ be the hyperplane (i.e. 3-dimensional subspace) in $\mathbf{R}^4$ with equation $2a + b - c + 2d = 0$. Find an orthonormal basis for $\mathcal{W}$. Compare your answer to your answer to problem 4.3.10.

$\Diamond$ 12.

$$\text{Let } F = \begin{pmatrix} 1 & 1 & 1 & 1 \\ 0 & 1 & 2 & 3 \\ 2 & 4 & 6 & 8 \\ 3 & 6 & 9 & 12 \end{pmatrix}$$

A technician measures a signal to be $s_m = (1,-2,-1,2)$. In principle, the signal is supposed to be in the range of the filter F. In this case, however, the measured signal s_m is not in the range of F. The engineer in charge decides to use s_a as an estimate of the actual signal where $s_m = s_a + s_n$ for s_a in the range of F and s_n orthogonal to the range of F. Find s_a and s_n, the estimated signal and noise components of s_m.

4.6 The QR–Factorization

Just as the factorization of the augmented matrix into an invertible matrix L and a reduced row echelon matrix E is a way to view Gaussian elimination, the QR–

factorization is a way to view the Gram–Schmidt algorithm. The QR–factorization will be useful in various applications where orthogonality is necessary or naturally present. For example, in the next chapter, we will use it to express in an easier way the normal equations from least squares solution of systems. We will also find better ways to compute the QR–factorization than using Gram–Schmidt and then we can use the factorization as a replacement for the Gram–Schmidt algorithm.

DEFINITION *The factorization $A = QR$ is called a* QR–factorization *of A if the columns of Q are orthonormal and R is upper triangular.*

If A is an $m \times n$ matrix, the factorization $A = QR$ means Q is $m \times k$ and R is $k \times n$ for some k. Since the columns of Q are orthonormal, they are linearly independent and the rank of Q is k, the number of columns. However, the rank of Q cannot be more than the number of rows of Q, so $k \leq m$. Moreover, because $A = QR$, Exercise 3.7.15. implies the rank of A is no more than the rank of Q. That is,

$$\text{rank}(A) \leq k \leq m$$

All integers k satisfying the inequality are possible and, even if the rank of A is m, the QR–factorization is not unique. MATLAB computes a QR–factorization of the matrix A with the command [Q,R]=qr(A) and the resulting Q is square. We will eventually show that all matrices have a QR–factorization but in this section we will assume that the rank of A is r and that the first r columns of A are linearly independent.

The utility of R being upper triangular lies principally in that systems of equations involving it are easily solved. Theorem 4.24 of the previous section explained the significance of the condition that Q have orthonormal columns.

A naive approach to computing the QR–factorization comes directly from the Gram–Schmidt algorithm. Recall that if $v_1, \cdots, v_k$ are linearly independent vectors, none of the vectors constructed in the Gram–Schmidt algorithm are zero and the orthogonal set can be normalized to give an orthonormal set

$$q_j = \frac{1}{\|w_j\|} w_j$$

The defining equations (4.3.5) for the Gram–Schmidt algorithm express each w_{j+1} in terms of v_{j+1} and the previously constructed w's. Replacing w_i by $\|w_i\| q_i$ in this equation and noting that $\langle w_i, w_i \rangle = \|w_i\|^2$, we get

$$\|w_{j+1}\| q_{j+1} = v_{j+1} - \langle q_1, v_{j+1} \rangle q_1 - \cdots - \langle q_j, v_{j+1} \rangle q_j \qquad (4.6.6)$$

or expressing v_{j+1} in terms of the orthonormal basis,

$$v_{j+1} = \langle q_1, v_{j+1} \rangle q_1 + \cdots + \langle q_j, v_{j+1} \rangle q_j + \|w_{j+1}\| q_{j+1} \qquad (4.6.7)$$

This collection of equations expresses a QR–factorization: the columns of Q are q_1, q_2, and so on and the entries of R are the coefficients. To see this, we block A and Q by columns, while R is unblocked, and $A = QR$ is expressed as

$$\begin{pmatrix} v_1 & v_2 & \cdots & v_k \end{pmatrix} = \begin{pmatrix} q_1 & q_2 & \cdots & q_k \end{pmatrix} \begin{pmatrix} r_{11} & r_{12} & \cdots & r_{1k} \\ 0 & r_{22} & \cdots & r_{2k} \\ \vdots & \vdots & \ddots & \vdots \\ 0 & 0 & \cdots & r_{kk} \end{pmatrix}$$

and the formula for multiplying matrices gives

$$v_1 = r_{11} q_1$$

$$v_2 = r_{12} q_1 + r_{22} q_2$$

$$\vdots$$

$$v_k = r_{1k} q_1 + r_{2k} q_2 + \cdots + r_{kk} q_k$$

Comparing this with equation (4.6.7) explains the relationship and gives the coefficients r_{ij} in terms of the data generated by the Gram–Schmidt process. Another interpretation of the coefficients is that they are the inner products of the given vectors with the orthonormal basis: Theorem 4.9 implies

$$r_{ij} = \langle q_i, v_j \rangle \qquad (4.6.8)$$

If the rank of A is r and the first r columns of A are linearly independent, the Gram–Schmidt algorithm will produce an orthonormal basis for the range of A consisting of r vectors. Since each of the columns is a linear combination of the q's and for $j \leq r$, v_j is a linear combination of $q_1, \cdots, q_j$ the above process gives a QR–factorization. The following is a formal statement of this result; the proof results from combining the above observations with equation (4.6.8).

THEOREM 4.28

Suppose A is an $m \times n$ matrix with rank r such that the first r columns of A are linearly independent. Then A has a QR–factorization $A = QR$ such that Q is $m \times r$ and R is $r \times n$.

EXAMPLE 4.29

(Compare with Example 4.13.)   Find a QR–factorization for the matrix

$$\begin{pmatrix} 0 & 1 & 1 \\ 1 & 0 & 1 \\ 1 & 1 & 0 \\ 1 & 1 & 1 \end{pmatrix}$$

SOLUTION Using the calculation of Example 4.13 for an orthonormal basis for the space spanned by the column vectors, we find

$$Q = \begin{pmatrix} 0 & \frac{3}{\sqrt{15}} & \frac{3}{\sqrt{35}} \\ \frac{1}{\sqrt{3}} & -\frac{2}{\sqrt{15}} & \frac{3}{\sqrt{35}} \\ \frac{1}{\sqrt{3}} & \frac{1}{\sqrt{15}} & -\frac{4}{\sqrt{35}} \\ \frac{1}{\sqrt{3}} & \frac{1}{\sqrt{15}} & \frac{1}{\sqrt{35}} \end{pmatrix}$$

and

$$R = \begin{pmatrix} \sqrt{3} & \frac{2}{\sqrt{3}} & \frac{2}{\sqrt{3}} \\ 0 & \frac{5}{\sqrt{15}} & \frac{2}{\sqrt{15}} \\ 0 & 0 & \frac{7}{\sqrt{35}} \end{pmatrix}$$

$\square$

The perspective above is that the Gram–Schmidt algorithm is closely related to QR–factorizations in which the number of columns of Q is the rank of the matrix. We also noted that MATLAB computes a square Q in the QR–factorization. We will see how to get such a factorization in Section 5.4, but now we will see how a QR–factorization in which Q is square can be used to give an orthonormal basis for the nullspace of A.

THEOREM 4.30

Suppose A is an $m \times n$ matrix of rank k and $A' = QR$ is a QR–factorization in which Q is $n \times n$. If the first k columns of A' are linearly independent, then the last $n - k$ columns of Q are a basis for the nullspace of A.

PROOF   The Duality Theorem (Theorem 4.22) says that the nullspace of A is the orthogonal complement of the range of A', so we need to find a basis for the orthogonal complement of the range of A'. Let $q_1, q_2, \cdots, q_n$ be the columns of Q in the factorization $A' = QR$. Since the first k columns of A' are linearly independent, the span of the first k columns of Q is the same as the span of the first k columns of A'. Since the rank of A' is k, the vectors $q_1, q_2, \cdots, q_k$ form a basis for the range of A'. The columns of Q are an orthonormal basis for $\mathbf{R}^n$, so $q_{k+1}, \cdots, q_n$ form a basis for the orthogonal complement of the range of A'. ▮

Note that the first k columns of A' are linearly independent if and only if the last $n - k$ rows of R are zero. Moreover, the rank of A is k and the first k columns of A' are independent if and only if $r_{11}, r_{22}, \cdots, r_{kk}$, the first k diagonal entries of R, are non–zero and the last $n - k$ rows of R are zero. If the matrix A does not have the property needed, notice that the non–zero diagonal entries of R correspond to k linearly independent columns of A', and permuting the columns of A' will not affect the nullspace of A but will produce a QR–factorization that satisfies the statement of the theorem.

EXAMPLE 4.31

Use the QR–factorization to find an orthonormal basis for the range of

$$A = \begin{pmatrix} 1 & -1 & 1 & 1 \\ 3 & 1 & -1 & 3 \\ 1 & 3 & -3 & 1 \end{pmatrix}$$

Use Theorem 4.30 to find an orthonormal basis for the nullspace of A.
SOLUTION

```
> A=[1 -1 1 1; 3 1 -1 3; 1 3 -3 1]

A =
     1    -1     1     1
     3     1    -1     3
     1     3    -3     1

> [Q R]=qr(A)

Q =
    -0.3015     0.4924     0.8165
    -0.9045     0.1231    -0.4082
    -0.3015    -0.8616     0.4082


R =
```

```
-3.3166    -1.5076     1.5076    -3.3166
      0    -2.9542     2.9542    -0.0000
      0          0     0.0000    -0.0000
```

The fact that the last row of R is zero shows that the last column of Q is not involved in writing the columns of A as linear combinations of the columns of Q. In other words, the first two columns of Q are an orthornormal basis for the range of A: $q_1 = (-0.3015, -0.9045, -0.3015)$ and $q_2 = (0.4924, 0.1231, -0.8616)$. We can check this easily.

```
> Q(:,1:2)'*Q(:,1:2)

ans =
    1.0000    0.0000
    0.0000    1.0000

> rank(A)

ans =
    2

> rank([A Q(:,1:2)])

ans =
    2
```

First, we see that the first two columns of Q are orthonormal. Then, the rank of A is 2, so we will need two vectors to have a basis for $\mathcal{R}(A)$. Moreover, since the rank of the matrix with the first two columns of Q adjoined to the matrix A has the same rank as the matrix A does, the first two columns of Q are in the subspace spanned by the columns of A.

To use Theorem 4.30, we need to find the QR–factorization of A'.

```
> [Qa Ra]=qr(A')

Qa =
    -0.5000    -0.5000     0.1000    -0.7000
     0.5000    -0.5000    -0.7000    -0.1000
    -0.5000     0.5000    -0.7000    -0.1000
    -0.5000    -0.5000    -0.1000     0.7000
```

```
Ra =
   -2.0000   -2.0000    2.0000
         0   -4.0000   -4.0000
         0         0   -0.0000
         0         0         0

> v=Qa(:,3:4)

v =
    0.1000   -0.7000
   -0.7000   -0.1000
   -0.7000   -0.1000
   -0.1000    0.7000
```

The theorem says, because A' has rank 2 and the first two columns of A' are linearly independent, that the last $4 - 2 = 2$ columns of Q must be an orthonormal basis for the nullspace of A. We can check this by multiplying and we see that, indeed, the two columns of v are the basis we seek.

```
> A*v

ans =
   1.0e-15 *
       0.0694    0.1110
      -0.3331    0.4441
      -0.3747         0
```

We can contrast this with using `orth` and `null` directly.

```
> u=Q(:,1:2)

u =
   -0.3015    0.4924
   -0.9045    0.1231
   -0.3015   -0.8616

> uu=orth(A)

uu =
   -0.0000   -0.5774
    0.7071   -0.5774
    0.7071    0.5774
```

```
> vv=null(A)

vv =
    0.6864    -0.1701
   -0.1701    -0.6864
   -0.1701    -0.6864
   -0.6864     0.1701

> rank([u uu])

ans =
     2

> rank([v vv])

ans =
     2
```

The columns of u are the basis for $\mathcal{R}(A)$ we found from the QR–factorization and these are quite different from uu that comes from orth. Similarly, the columns of v found by the QR–factorization of A' are very different from the columns of vv that come from using null. On the other hand, because the rank of the matrix containing both u and uu is 2, the subspaces spanned by their columns are the same. Similarly, because the rank of the matrix containing both v and vv is 2, the subspaces spanned by their columns are the same. Thus, both solutions of the problems are correct, they just give different bases.   ⧫

In fact, until about 1994, the algorithms for "null" and "orth" in MATLAB did more or less exactly what was described in the theorems above with the exception that the columns of A were, in effect, rearranged to automatically handle the case in which A has rank r but the first r columns of A or A' are linearly dependent. Since about 1994, the algorithms have been different and are now based on the singular value decomposition. According to Cleve Moler, Chief Scientist at the Mathworks which publishes MATLAB, the change was made to improve reliability and provide consistent rank determination, in spite of the fact that the new subroutines run more slowly than those using QR–factorization.

Exercises 4.6

1.

$$A = \begin{pmatrix} 2 & 1 & 0 & 1 \\ 1 & 1 & -1 & 0 \\ 2 & 0 & 2 & 2 \end{pmatrix} = \begin{pmatrix} \frac{2}{3} & -\frac{1}{3} & \frac{2}{3} \\ \frac{1}{3} & -\frac{2}{3} & -\frac{2}{3} \\ \frac{2}{3} & \frac{2}{3} & -\frac{1}{3} \end{pmatrix} \begin{pmatrix} 3 & 1 & 1 & 2 \\ 0 & -1 & 2 & 1 \\ 0 & 0 & 0 & 0 \end{pmatrix}$$

is a QR–factorization of A. Find an orthonormal basis for the range of A.

2. Find a QR–factorization of the matrix $A = \begin{pmatrix} 1 & 0 & 0 & 1 \\ 1 & 1 & 0 & 0 \\ 0 & 1 & 1 & 0 \\ 0 & 0 & 1 & 1 \end{pmatrix}$

    You may find your work in problem 4.3.11. to be helpful.

◇ 3.

(a) Find a QR–factorization of $B = \begin{pmatrix} 1 & 2 & -1 & 3 & 4 \\ 3 & 1 & -1 & 2 & -2 \\ 1 & 3 & -1 & 2 & 0 \\ 2 & -1 & 1 & 0 & 1 \\ 2 & 0 & 1 & -1 & -3 \end{pmatrix}$

(b) Find an orthonormal basis for the range of B.

(c) Is $w = (5, 1, 1, 2, -2)$ in the range of B? (Justify your answer.)

    Use QR–factorization to find orthonormal bases for the ranges and the nullspaces of the following matrices:

◇ 4.

$$D = \begin{pmatrix} 1 & 2 & -1 & 3 & 4 \\ 3 & 1 & -1 & 2 & -2 \\ 1 & 3 & -1 & 2 & 0 \\ 2 & -1 & 1 & 0 & 1 \\ 2 & 0 & 1 & -1 & -3 \end{pmatrix}$$

◇ 5.

$$E = \begin{pmatrix} 2 & 1 & 1 & 1 & -1 & 0 & 2 \\ 3 & 1 & 2 & 0 & 1 & 2 & 1 \\ 1 & 2 & -1 & -1 & -1 & 1 & 1 \\ 1 & -1 & 2 & 2 & 0 & -1 & 1 \end{pmatrix}$$

◇ 6.   The vectors $v_1 = (1, 1, -1, 1)$; $v_2 = (1, 0, 1, 2)$; $v_3 = (1, -2, -1, 0)$; and $v_4 = (0, 2, 1, -1)$ are a basis for $\mathbf{R}^4$. Use a QR–factorization of the matrix whose columns are v_1, v_2, v_3, and v_4 to find an orthonormal basis for $\mathbf{R}^4$. Compare your answer with your answer to problem 4.3.5.

◇ 7.   The vectors $v_1 = (1, 0, -1)$ and $v_2 = (2, 1, -1)$ span the subspace $\mathcal{U}$ in $\mathbf{R}^3$. Use a QR–factorization to find an orthonormal basis for $\mathcal{U}$. Compare your answer with your answer to problem 4.3.6. and problem 4.5.9.

◇ 8.   The vectors $v_1 = (1, 0, -1, 1)$; $v_2 = (2, 1, 1, -1)$; and $v_3 = (1, -1, -1, 0)$ span the subspace $\mathcal{W}$ in $\mathbf{R}^4$. Use a QR–factorization to find an orthonormal basis for $\mathcal{W}$. Compare your answer to your answer to problem 4.3.7. and problem 4.5.10.

$\Diamond$ 9. Let $\mathcal{W}$ be the hyperplane (i.e. 3-dimensional subspace) in $\mathbf{R}^4$ with equation $2a + b - c + 2d = 0$. Use a QR–factorization of the 4×1 matrix

$$\begin{pmatrix} 2 \\ 1 \\ -1 \\ 2 \end{pmatrix} = QR$$

where Q is 4×4 to find an orthonormal basis for $\mathcal{W}$. Compare your answer to your answer to problem 4.3.10. and problem 4.5.11.

10. An $n \times n$ matrix U is called *unitary* if $U' = U^{-1}$.
  (a) Show U is unitary if and only if its columns form an orthonormal basis for $\mathbf{R}^n$.
  (b) Show that if U is unitary, then U^{-1} is also unitary.
  (c) Show that if U is a real unitary matrix, the map $x \mapsto Ux$ is a rigid motion of $\mathbf{R}^n$ by showing that if U is unitary and v and w are vectors in $\mathbf{R}^n$, then
   i. $\langle Uv, Uw \rangle = \langle v, w \rangle$,
   ii. $\|Uv\| = \|v\|$,
   iii. the angle between Uv and Uw is the same as the angle between v and w.

4.7 The Geometry of Subspaces

Geometric intuition gained from living in a three dimensional Euclidean world is a big help in understanding facts about the vector space $\mathbf{R}^3$. We can extend this thinking by analogy to other settings. In this section, we study some theorems about subspaces that have a strong geometric component.

We begin with two definitions about subspaces that are analogous to intersection and unions of sets.

DEFINITION *If M and N are subspaces of the vector space V, the* intersection *of M and N, written $M \cap N$, is the set of vectors*

$$M \cap N = \{v : v \in M \text{ and } v \in N\}$$

The sum *of M and N, written $M + N$, is the set of vectors*

$$M + N = \{v : v = x + y \text{ for some } x \in M \text{ and some } y \in N\}$$

PROPOSITION 4.32

If M and N are subspaces of the vector space V, then $M \cap N$ and $M + N$ are also subspaces of V.

The proof of this proposition is an exercise in the meaning the word "subspace". However, it should be noted that this property is a crucial one in justifying the definition. For example, it makes sense to define the union of two subspaces as a set, but it is not usually a subspace, so the union is an uninteresting set from the point of view of Linear Algebra. Just as the union of two sets is the smallest set that contains them both, the sum of two subspaces is the smallest *subspace* that contains them both. This concept replaces the idea of unions in many arguments.

The size of $M \cap N$ is a measure of how different M and N are. Since 0 is in every subspace, $M \cap N$ can never be empty, but if $M \cap N = (0)$, the subspaces are as different as they can be. This relationship is somewhat like linear independence of vectors, as you will see in Exercise 4.7.3.

The first result relates the sizes of the sum and intersection to the sizes of the subspaces. It is analogous to a counting formula you may know for unions and intersections of finite sets.

THEOREM 4.33

Let M and N be subspaces of the vector space V. Then

$$\dim(M) + \dim(N) = \dim(M \cap N) + \dim(M + N)$$

Although we prefer to work in a coordinate free setting, since this theorem is about dimensions, we will need to use bases. The proof constructs bases in a special way, then the dimension results follow by counting how many vectors there are in each basis.

PROOF   Let $u_1, u_2, \cdots u_k$ be a basis for $M \cap N$; in particular, the dimension of $M \cap N$ is k. Since $M \cap N$ is contained in M and the vectors $u_1, u_2, \cdots, u_k$ are linearly independent in M, there are vectors $v_1, v_2, \cdots, v_m$ so that $u_1, u_2, \cdots, u_k, v_1, v_2, \cdots, v_m$ form a basis for M; in particular, the dimension of M is $k + m$. In the same way, there are $w_1, w_2, \cdots, w_n$ in N so that $u_1, u_2, \cdots, u_k, w_1, w_2, \cdots, w_n$ form a basis for N so the dimension of N is $k + n$.

We claim that $u_1, u_2, \cdots, u_k, v_1, v_2, \cdots, v_m, w_1, w_2, \cdots, w_n$ is a basis for $M + N$. Indeed, if z is a vector in $M + N$, then $z = x + y$ for some vectors in M and N respectively. Since $u_1, u_2, \cdots, u_k, v_1, v_2, \cdots, v_m$ is a basis for M, there are scalars so that

$$x = \alpha_1 u_1 + \cdots + \alpha_k u_k + \beta_1 v_1 + \cdots + \beta_m v_m$$

Similarly, there are scalars so that

$$y = \gamma_1 u_1 + \cdots + \gamma_k u_k + \delta_1 w_1 + \cdots + \delta_n w_n$$

This means

$$z = (\alpha_1 + \gamma_1)u_1 + \cdots + (\alpha_k + \gamma_k)u_k + \beta_1 v_1 + \cdots + \beta_m v_m + \delta_1 w_1 + \cdots + \delta_n w_n$$

so the u's, v's, and w's span $M + N$.

On the other hand, suppose there are scalars $a_1, \cdots, a_k, b_1, \cdots, b_m, c_1, \cdots, c_n$ so that

$$a_1 u_1 + \cdots + a_k u_k + b_1 v_1 + \cdots + b_m v_m + c_1 w_1 + \cdots + c_n w_n = 0$$

Rewriting, this means

$$a_1 u_1 + \cdots + a_k u_k + b_1 v_1 + \cdots + b_m v_m = -c_1 w_1 - \cdots - c_n w_n \qquad (4.7.9)$$

This equation represents two ways of writing the same vector. The expression on the left makes it clear that this vector is in M. On the other hand, the expression on the right makes it clear the vector is in N. Since this vector is in *both* M and N, it is in $M \cap N$. But the u's are a basis for $M \cap N$, so there must be an expression for this vector as a linear combination of the u's, say,

$$d_1 u_1 + \cdots + d_k u_k = -c_1 w_1 - \cdots - c_n w_n$$

Rewriting this, we see this means

$$d_1 u_1 + \cdots + d_k u_k + c_1 w_1 + \cdots + c_n w_n = 0 \qquad (4.7.10)$$

By construction, the u's and w's are a basis for N, so they are linearly independent. This implies that all the coefficients in equation (4.7.10) are 0. In particular, the c's are all 0. Now, equation (4.7.9) becomes

$$a_1 u_1 + \cdots + a_k u_k + b_1 v_1 + \cdots + b_m v_m = 0$$

which, since the u's and the v's are a basis for M, means that all the a's and b's are 0. To sum up, if 0 is expressed as a linear combination of the u's, v's, and w's, then the coefficients are all 0, which means the u's, v's, and w's are linearly independent.

Since the u's, v's, and w's span $M + N$ and are linearly independent, they are a basis for $M + N$ and the dimension of $M + N$ is $k + m + n$.

This proves the relationship we wanted

$$\dim(M) + \dim(N) = (k+m) + (k+n) = k + (k+m+n) = \dim(M \cap N) + \dim(M+N)$$

∎

A geometric application of this theorem in three dimensional space states a principle that is well known from solid geometry.

COROLLARY 4.34

Two planes that pass through the origin in $\mathbf{R}^3$ either are the same plane or intersect in a line.

PROOF   Planes M and N that pass through the origin in $\mathbf{R}^3$ are subspaces of dimension 2. Thus $\dim M + \dim N = 4$. Now $M \subset M + N \subset \mathbf{R}^3$ so $2 \leq \dim(M + N) \leq 3$. If $\dim(M + N) = 2$, then $\dim(M \cap N) = 2$ and $M = M \cap N = N$, so they are the same subspace. If $\dim(M + N) = 3$, then $\dim(M \cap N) = 1$, that is, $M \cap N$ is a line through the origin.   ∎

The proof of Theorem 4.33 does more than just prove the theorem, it gives a recipe for finding bases of the subspaces in an efficient way.

EXAMPLE 4.35

Let
$$x_1 = (1, 1, -1, 1); \; x_2 = (1, -1, 1, 0); \; x_3 = (0, 2, -2, 1)$$

let
$$y_1 = (1, 0, 1, 0); \; y_2 = (0, 0, -2, 1)$$

and let $M = \text{span}\{x_1, x_2, x_3\}$ and $N = \text{span}\{y_1, y_2\}$. Find bases for $M \cap N$ and $M + N$.

The proof of the theorem suggests that we begin with $M \cap N$. It is not clear that there *are* any non-zero vectors in $M \cap N$! Throughout this discussion, it is important to remember that every non-zero subspace contains an *infinite* number of vectors, and that the lists above are not an exhaustive list of vectors in the subspaces. Indeed the idea of basis is to describe subspaces in an efficient manner! We will find our work easier if we use bases everywhere. Clearly, y_1 and y_2 are linearly independent, so they form a basis for N. It is less clear if x_1, x_2, and x_3 are linearly independent, and in fact, they are not: $x_3 = x_1 - x_2$. It follows that x_1 and x_2 are a basis for M.

Now any vector in $M \cap N$ is in both M and in N. Since it is in M, it is a linear combination of x_1 and x_2, and since it is in N, it is a linear combination of y_1 and y_2. Thus, the vectors in $M \cap N$ are precisely the vectors that can be written both as a linear combination of x_1 and x_2 and of y_1 and y_2. To find these vectors, we solve the vector equation

$$ax_1 + bx_2 = cy_1 + dy_2$$

This vector equation can be rewritten as the system

$$\begin{cases} a & + & b & - & c & & & = & 0 \\ a & - & b & & & & & = & 0 \\ -a & + & b & - & c & + & 2d & = & 0 \\ a & & & & & - & d & = & 0 \end{cases}$$

This system is easily solved and has a one dimensional solution space: for example, d may be chosen arbitrarily, then $a = b = d$ and $c = 2d$. The solutions of this system correspond to vectors in $M \cap N$, so this subspace must be one dimensional. Since we want a basis for $M \cap N$, we only need one solution, say $a = b = d = 1$ and $c = 2$. The interpretation of this solution is that

$$x_1 + x_2 = 2y_1 + y_2$$

and this assertion is easily checked directly; both sides give $(2, 0, 0, 1)$. Thus, $u_1 = (2, 0, 0, 1)$ is a basis for $M \cap N$.

In the proof of the theorem, the next step was to extend this basis to a basis for M. Since M is 2–dimensional, we need to find one vector so that the pair is linearly independent. It is easy to see $v_1 = x_1$ will do, so that $\{u_1, x_1\}$ is a basis for M. Similarly, $\{u_1, y_1\}$ is a basis for N. The proof of the theorem shows that $\{u_1, x_1, y_1\}$ is a basis for $M + N$.

Note that there are many other choices possible: for example, $\{u_1, x_3\}$ for M and $\{u_1, y_2\}$ for N which leads to $\{u_1, x_3, y_2\}$ for $M + N$. Indeed, there are many choices that do not follow the structure of the proof of the theorem: for example

$$\{(3, 1, -1, 2); \ (4, -2, 2, 1); \ (3, 0, 1, 1)\}$$

is another basis for $M + N$. (Can you check this???) But this is hardly surprising as there are infinitely many different bases for any non–zero subspace. $\square$

Exercises 4.7

1. Let M be span$\{(0, 1, -1, 1); (1, 0, 1, -1); (1, 1, 0, 0); (-3, 2, -5, 5)\}$
   and let N be span$\{(-3, 2, -3, 4); (-1, 1, 0, 1); (0, -1, -3, 1)\}$ in $\mathbf{R}^4$.
   Find bases for $M + N$ and $M \cap N$.

2. Let $M =$ span$\{(-1, 1, 1, 0); (1, -1, 0, 1); (0, 0, 1, 1)\}$
   and let $N =$ span$\{(1, 1, 0, 1); (0, 1, 1, 1); (1, 3, 2, 3)\}$.
   Find bases for $M + N$ and $M \cap N$.

3. Suppose $v_1, v_2, \cdots, v_m$ is a basis for M and $w_1, w_2, \cdots, w_n$ is a basis for N. Show that
   if the set $\{v_1, v_2, \cdots, v_m, w_1, w_2, \cdots, w_n\}$ is linearly independent, then $M \cap N = (0)$.

5

Least Squares and Projections

5.1   A Traffic Control Example

Experience in driving a car in a city quickly leads one to believe there is a relation between how fast people drive, on average, and how heavy the traffic is. Clearly, an understanding of this relationship is important in the design of highway systems[1] if the design engineers are to predict the volume of traffic a particular road can carry. In an early study of the relationship of traffic density and speed, Greenshields[2] proposed that the average speed s and the average density d satisfy an equation of the form

$$s = f - \beta d$$

where f is the mean free speed, the average speed of vehicles when there is very light traffic, and β is the quotient of f and the 'jam density', the traffic density when traffic stops flowing. Speed s will be measured in miles per hour and represents an average of the speeds of the various vehicles on a particular length of road at a particular time. The traffic density d will be measured in vehicles per mile and represents an density for the length of road averaged over a short enough time interval that the average speed is constant.

In principle, the volume of traffic that will flow past a particular point in an hour is sd. If the relationship of speed and density is known, the volume can be computed as a function of the traffic density. For example, the Greenshields model would give the volume as $(f - \beta d)d$. Under reasonable assumptions about speed as a function of traffic density, the volume can be seen to have a maximum value

---

[1]M. Wohl and B. V. Martin, *Traffic System Analysis for Engineers and Planners*, McGraw-Hill, New York, 1967.

[2]B. D. Greenshields, "A Study of Highway Capacity," *Highway Res. Board Proc.*, vol. 14, 1934.

that will represent the maximum capacity of the road under ideal conditions. In the Greenshields model, the volume function has a parabolic graph and has a maximum value of $f^2/(4\beta)$ at a traffic density of $d = f/(2\beta)$. Clearly then, if this model is considered correct, the traffic engineers in charge of a project must estimate the parameters f and β that would apply to a situation. Moreover, to compare the correctness of Greenshields' model with other proposed models, one must estimate the parameters f and β.

One plausible way to find these parameters in a particular case would be to observe traffic flow and use the data to find f and β. For example, the following data were given in a field study by Greenberg[3]

**Speed and density observations
in the North Tube of the Lincoln Tunnel**

Speed (mph)	Density (veh/mi)
32	34
28	44
25	53
23	60
20	74
19	82
17	88
15	94
15	94
14	96
13	102
12	112
11	108
10	129
9	132
8	139
7	160
6	165

To get f and β from this data, we want to find f and β so that $s = f - \beta d$ for each of the choices of s and d given in the table. For example, from the first few

---

[3]H. Greenberg, "A Mathematical Analysis of Traffic Flow", Tunnel Traffic Capacity Study, the Port of New York Authority, New York, 1958. Actually, Greenberg used the data to support an exponential model for the relationship between speed and density.

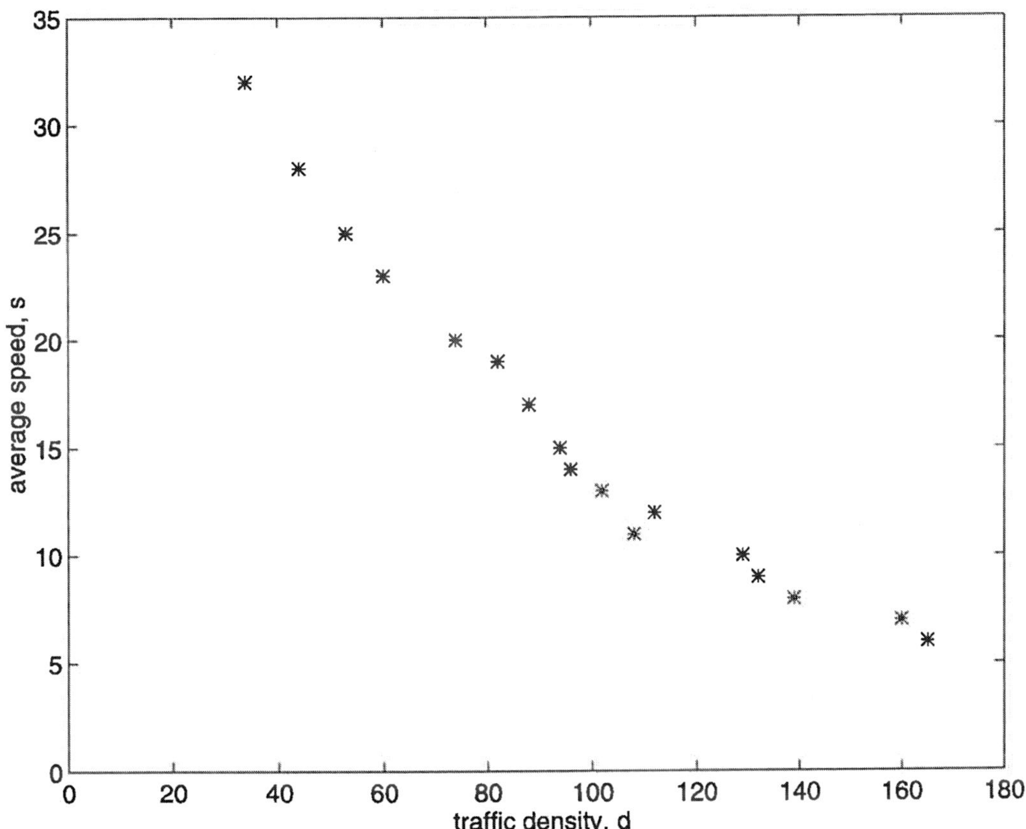

FIGURE 5.1

Lincoln Tunnel speed and density observations

entries of the table, we get

$$\begin{cases} 32 &= f - 34\beta \\ 28 &= f - 44\beta \\ 25 &= f - 53\beta \\ 23 &= f - 60\beta \\ 20 &= f - 74\beta \\ &\vdots \end{cases}$$

We recognize this as a system of linear equations in f and β, but unfortunately, the system of equations is inconsistent! This is hardly surprising since we have 18 equations in 2 unknowns, but what can we do about it? One obvious way to deal with the problem would be to throw away 16 of the equations and find f and β from the other two. This seems silly because, intuitively, we should be able to make a better estimate of f and β the more data we use. Indeed, the key here is to realize that we do not believe f and β should exist that would fit the model perfectly, rather, we believe that it is likely both that errors arise in data collection and that the model $s = f - \beta d$ does not fit the situation exactly. In spite of this, we want to find f and β that will produce a model that fits the data best.

Indeed, a plot (Figure 5.1) shows that the data points from Greenberg's study do not exactly lie on any straight line. Moreover, the folly of choosing two points and throwing the others away is clear because we would get vastly different lines by different choices of two points. We must find a systematic way to choose the 'best' line and find the f and β that describe it.

So far, in our study of linear equations, we have given careful consideration only to consistent systems, however, inconsistent systems arise in practice in a number of ways. As in the example above, the system may be 'overdetermined', that is, it may be a system with more equations than unknowns because there is more data available than needed to determine the unknown parameters. In other cases, it may be a consistent system in which small errors have been made in the measurements of some of the coefficients. In this chapter, we will find a way to deal with such systems.

5.2   Inconsistent Systems

In the previous section, we saw an example that shows why we might want to find a "solution" for a system of equations that is inconsistent. In this section, we examine the situation more completely.

If A is an $m \times n$ matrix, then the range of A is the subspace of $\mathbf{R}^m$ consisting of the righthand sides b for which $AX = b$ has a solution. In other words, $AX = b$ is consistent if and only if b is in the range of A. But if b is not in the range of A, what should we mean by the 'best approximate solution' of $AX = b$? One reasonable answer is that we should choose X so that AX is as close as possible to b, that is, we should find X so that $\|AX - b\|$ is minimized. Since AX is always in the range of A, we should choose a solution of $AX = b_0$, where b_0 is the vector in the range of A that is closest to b. In the following definition, we choose a 'best approximate solution' for a system of equations. The reason for the terminology introduced will become more clear in the following sections. We should note that if the system $AX = b$ has a solution so that b is already in the range of A, then the 'best approximate solution' according to the definition is actually a solution.

DEFINITION *If A is an $m \times n$ matrix and b is a vector in $\mathbf{R}^m$, let b_0 be the vector in the range of A that is closest to b. The* least squares solution *of the system $AX = b$ is the solution X_0 of the system $AX = b_0$ with smallest norm.*

It is clear from the above discussion that there is a least squares solution for any system. Although we will not prove it, since the solution space of $AX = b_0$ is a convex set, there is exactly one solution X_0 of smallest norm, so there is always exactly one least squares solution of a system.

Before we can proceed with finding the least squares solution to systems, we must address the issue of finding the point b_0 of the range of A that is closest to the point b. More generally, we will find the closest point of a subspace to a given point of the space.

We begin with a descriptive solution to the problem. This solution is the one that solves the analogous problem in elementary geometry: if b is a point in the plane and l is a line in the plane, to find the closest point on the line to b, we construct the line through b that is perpendicular to l; the point where the perpendicular meets l is the closest point to b.

THEOREM 5.1

If W is a subspace of $\mathbf{R}^m$ and b is in $\mathbf{R}^m$, the closest point of W to b is the unique point b_0 of W such that $b - b_0$ is in $W^\perp$.

PROOF The proof is broken into three steps: first, we show that there is a vector satisfying the condition, second, we show that there is only one such vector, and finally, we show that this vector solves the problem.

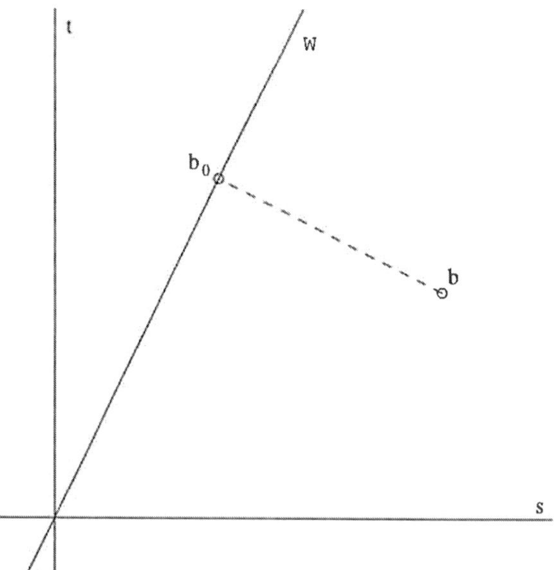

FIGURE 5.2

Closest point, b_0, of line $t = 2s$ to point $b = (7,4)$

1. *There is a point b_0 in $\mathcal{W}$ such that $b - b_0$ is in $\mathcal{W}^\perp$:*
By part *(ii)* of Theorem 4.19, there is b_0 in $\mathcal{W}$ and u in $\mathcal{W}^\perp$ so that $b_0 + u = b$. Thus $u = b - b_0$ is in $\mathcal{W}^\perp$.

2. *There is only one point b_0 in $\mathcal{W}$ such that $b - b_0$ is in $\mathcal{W}^\perp$:*
Suppose b_0 and b_1 are in $\mathcal{W}$ and both $b - b_0$ and $b - b_1$ are in $\mathcal{W}^\perp$. Let $b - b_0 = u_0$ and $b - b_1 = u_1$. Thus, $b_0 + u_0 = b = b_1 + u_1$ which implies $b_0 - b_1 = u_1 - u_0$. Since $b_0 - b_1$ is in $\mathcal{W}$ and $u_1 - u_0$ is in $\mathcal{W}^\perp$, the fact that $\mathcal{W} \cap \mathcal{W}^\perp = (0)$ (part *(i)* of Theorem 4.19) implies that $b_0 - b_1 = 0$, that is, that $b_0 = b_1$.

3. *The point b_0 is the closest point of $\mathcal{W}$ to b:*
Suppose v, with $v \neq b_0$, is a vector in $\mathcal{W}$. Let $b = b_0 + u$ as above, so that $u = b - b_0$ is in $\mathcal{W}^\perp$. Since $b_0 - v$ is in $\mathcal{W}$, the Pythagorean theorem implies that $\|b - v\|^2 = \|b_0 + u - v\|^2 = \|b_0 - v\|^2 + \|u\|^2 > \|u\|^2 = \|b - b_0\|^2$, where the strict inequality holds because $b_0 - v \neq 0$. That is, the distance from b to b_0 is less than the distance from b to any other point v of $\mathcal{W}$, as was to be proved.   ∎

This result holds without alteration for subspaces of $\mathbf{C}^m$ as well.

EXAMPLE 5.2

The point $b_0 = (3, 6)$ is the closest point of the line $t = 2s$, that is, the subspace $W = \text{span}\{(1, 2)\}$, to the point $b = (7, 4)$ because $b - b_0 = (4, -2)$ which is in $W^\perp = \text{span}\{(-2, 1)\}$. (See Figure 5.2.) Moreover, the distance from $(7, 4)$ to the line is the distance from $(7, 4)$ to the closest point on the line, so it is

$$\left\| \begin{pmatrix} 7 \\ 4 \end{pmatrix} - \begin{pmatrix} 3 \\ 6 \end{pmatrix} \right\| = \sqrt{(7-3)^2 + (4-6)^2} = \sqrt{20}$$

$\Box$

Notice that the discussion on page 222 shows that if v is a vector in $\mathbf{R}^n$ and w_1, $w_2, \cdots, w_j$ are orthogonal, non-zero vectors that span the subspace W, then the vector given in Theorem 4.3.4

$$\frac{\langle w_1, v \rangle}{\langle w_1, w_1 \rangle} w_1 + \frac{\langle w_2, v \rangle}{\langle w_2, w_2 \rangle} w_2 + \cdots + \frac{\langle w_j, v \rangle}{\langle w_j, w_j \rangle} w_j$$

is the closest point in W to v.

We will suppose for the remainder of this section that A is an $m \times n$ matrix that has rank n. (That is, we assume the system is overdetermined but has full rank.) As a consequence of this simplifying assumption, the nullspace of A is (0) so the solution of the system $AX = b_0$ is unique. This hypothesis has other consequences that we need as in the result below.

LEMMA 5.3

If A is a matrix then the nullspace of $A'A$ is the same as the nullspace of A.

PROOF If x is in $\mathcal{N}(A)$, then $A'Ax = A'0 = 0$, so x is in $\mathcal{N}(A'A)$. Thus, $\mathcal{N}(A'A) \supset \mathcal{N}(A)$. Conversely, if $A'Ax = 0$ then Ax is in $\mathcal{N}(A')$. But, by the Duality Theorem, $\mathcal{N}(A') = \mathcal{R}(A)^\perp$ so $A'Ax = 0$ means Ax is in $\mathcal{R}(A)^\perp \cap \mathcal{R}(A)$. Now the only vector in both a subspace and its orthogonal complement is 0, so $Ax = 0$. Thus x in $\mathcal{N}(A'A)$ implies x is in $\mathcal{N}(A)$ which means $\mathcal{N}(A'A) \subset \mathcal{N}(A)$. Combining the two containments gives the conclusion $\mathcal{N}(A'A) = \mathcal{N}(A)$. $\blacksquare$

COROLLARY 5.4

If A is a matrix, then $\text{rank}(A'A) = \text{rank}(A)$. In particular, if A is an $m \times n$ matrix with rank n, then $A'A$ is invertible.

PROOF Notice that since A is an $m \times n$ matrix, $A'A$ is a $n \times n$ matrix. The rank–nullity theorem (Theorem 3.50) says that

$$\text{rank}(A'A) = n - \dim\left(\mathcal{N}(A'A)\right) = n - \dim\left(\mathcal{N}(A)\right) = \text{rank}(A)$$

If A has rank n, then $A'A$ is a $n \times n$ matrix with rank n, so it is invertible. ∎

We are now ready to state and prove the result that gives the least squares solution of a system of linear equations.

THEOREM 5.5

If A is an $m \times n$ matrix with rank n, then the least squares solution of the system $AX = b$ is the unique solution of the system

$$(A'A)X_0 = A'b$$

The system $(A'A)X_0 = A'b$ in the conclusion of the theorem is called the system of *normal equations* for the system $AX = b$.

PROOF Let A be an $m \times n$ matrix with linearly independent columns, that is, the rank of A is n.

Let b_0 be the point of the range of A that is closest to b. According to Theorem 5.1, b_0 is the unique point of $\mathcal{R}(A)$ that satisfies $b - b_0$ is orthogonal to $\mathcal{R}(A)$.

But $\mathcal{R}(A)^\perp = \mathcal{N}(A')$, so this means $b - b_0$ is in the nullspace of A', or $A'(b - b_0) = 0$.

Rewriting this, we get $A'b_0 = A'b$. The condition the b_0 is in the range of A means that there is X_0 in $\mathbf{R}^n$ so that $b_0 = AX_0$. Substituting this into the equation above gives $A'AX_0 = A'b$ which is the equation of the conclusion.

Since A is an $m \times n$ matrix with rank n, Corollary 5.4 implies $A'A$ is invertible. In particular, there is a unique X_0 so that $(A'A)X_0 = A'b$. ∎

EXAMPLE 5.6

Find the least squares solution of the inconsistent system

$$\begin{cases} x + 2y &= 3 \\ 2x + 3y &= -1 \\ 2x + 2y &= 1 \end{cases}$$

SOLUTION We have

$$A = \begin{pmatrix} 1 & 2 \\ 2 & 3 \\ 2 & 2 \end{pmatrix}$$

which has rank 2, so

$$(A'A) = \begin{pmatrix} 1 & 2 & 2 \\ 2 & 3 & 2 \end{pmatrix} \begin{pmatrix} 1 & 2 \\ 2 & 3 \\ 2 & 2 \end{pmatrix} = \begin{pmatrix} 9 & 12 \\ 12 & 17 \end{pmatrix}$$

and

$$A'b = \begin{pmatrix} 1 & 2 & 2 \\ 2 & 3 & 2 \end{pmatrix} \begin{pmatrix} 3 \\ -1 \\ 1 \end{pmatrix} = \begin{pmatrix} 3 \\ 5 \end{pmatrix}$$

We must solve the system

$$\begin{pmatrix} 9 & 12 \\ 12 & 17 \end{pmatrix} \begin{pmatrix} x_0 \\ y_0 \end{pmatrix} = \begin{pmatrix} 3 \\ 5 \end{pmatrix}$$

This easily implies that $X_0 = (-1, 1)$ is the least squares solution of the given system. $\square$

Overdetermined systems are so important in science and engineering applications that the MATLAB command A\b gives the least squares solution of $AX = b$ whenever A is not square. (When A is square, MATLAB assumes the system has a unique solution and computes the solution using Gaussian elimination, with a warning if A is "close to singular.") From the perspective of Chapter 2 when we were just learning about systems, it seemed somewhat strange that the "equation solver" in MATLAB did not always give solutions (forcing us to check whether a "solution" really was a solution), but the utility and importance of the least squares solution of systems justifies this point of view.

Exercises 5.2

1.

$$\begin{cases} x + 2y &= 3 \\ 2x + 3y &= -1 \\ 2x + 2y &= 1 \end{cases}$$

(a) Explain, in terms of ranges of matrices, why this system is inconsistent.

(b) Find the least squares solution to the system.

(c) Explain the relationship of your answer to (b) above and ranges of matrices.

2. The vectors $(1, -1, 0)$ and $(1, 1, 1)$ form an orthogonal basis for the subspace $\mathcal{W}$ of $\mathbf{R}^3$. Use the formula of Theorem 4.9 as in the discussion on page 222 to find the point of $\mathcal{W}$ that is closest to the point $(2, 1, 3)$.

3. The vectors $(1, -1, 1, -1)$, $(1, 1, -1, -1)$, and $(1, -1, -1, 1)$ form an orthogonal basis for the subspace $\mathcal{W}$ of $\mathbf{R}^4$. Use the formula of Theorem 4.9 as in the discussion on page 222 to find the point of $\mathcal{W}$ that is closest to the point $(0, 1, 2, 3)$.

Use the normal equations to find the "best" (i.e. least squares) solution to the following overdetermined systems. Decide whether each system is consistent or inconsistent. If you are using MATLAB or other software that gives the least squares solution directly, compare your answers to the solution given by your software.

4. $\begin{cases} 2x - 3y = 1 \\ -x + 2y = 3 \\ -x + y = -1 \end{cases}$

◇ 8. $\begin{cases} r - 2s = 3 \\ -r + 3s = 2 \\ r - 2s = -1 \end{cases}$

5. $\begin{cases} a + b = 2 \\ 2a + b = 4 \\ a + 3b = 0 \\ 2a - b = 6 \end{cases}$

◇ 9. $\begin{cases} x + 2y = -1 \\ x + y = 2 \\ 2x - y = 13 \\ 3x + y = 12 \end{cases}$

6. $\begin{cases} a - b + 2c = 2 \\ a + c = 1 \\ b - 2c = -1 \\ a + 2b - c = 0 \end{cases}$

◇ 10. $\begin{cases} r - 2s + 3t = -1 \\ 2r + s - t = 2 \\ s + 2t = 0 \\ -r + 2s + t = 3 \end{cases}$

7. $\begin{cases} 2p - q = 1 \\ p + q - r = 0 \\ -2p + q + r = 0 \\ p - 2q + r = 1 \end{cases}$

◇ 11. $\begin{cases} -2r + 4s - 6t = 2 \\ 2r + s - t = 2 \\ s + 2t = 0 \\ -r + 2s + t = 3 \end{cases}$

(Compare with Exercise 5.2.10.)

5.3 Least Squares Fitting of Data

Frequently, in dealing with experimental data, a worker knows the general nature of the relationship of the quantities measured, but does not know the values of some of the parameters involved. Indeed, the goal of performing the experiment may be to determine these parameters. As in the example of Section 5.1, it might be known that the data points ought to lie on a straight line, and the experimenter wishes to find the slope and intercept of the line. In fact, of course, there may be a rather large number of data points, and there may be *no* line through all the data points. In this case, the worker wants to find a line that 'nearly goes through' the data points, and wants to choose the 'best' line to describe the data points. The process of finding the best line is sometimes called linear regression and the line is called the regression line. In this section, we will describe a common technique for doing this.

Before we can hope to do such a thing, we must decide what we mean by 'nearly goes through' and 'best'. If the data points are $(s_1, t_1), \ldots, (s_m, t_m)$ and we are approximating these data points by the graph $t = g(s)$, the error at $s = s_k$ is $g(s_k) - t_k$ (see Figure 5.3). However, since this error could be positive or negative and we do not want the errors at different points to cancel out, we want to consider a positive function of the errors, for example, the functions $\sum |g(s_k) - t_k|$, $\sum |g(s_k) - t_k|^2$, $\sum |g(s_k) - t_k|^7$, or $\max\{|g(s_k) - t_k| : k = 1, 2, \cdots, m\}$ might be reasonable in various contexts. We will use $\sum |g(s_k) - t_k|^2$; if the data is being approximated as above, we will measure the error by the square root of the total of the squares of the errors, as in the definition below. This turns out to be the most convenient computationally, and is closely connected with the *arithmetic mean* as a measure of central tendency (see Exercise 5.3.10.). Moreover, because it the easiest to handle computationally, it is the measure most commonly used in elementary statistics. The total of the absolute values of the errors is also a reasonable measure, and is connected with the *median* as a measure of central tendency (see Exercise 5.3.11.), but is, perhaps surprisingly, much harder to deal with computationally. One of the reasons the "least squares" criterion is more easily dealt with than others is that it is intimately connected with the geometry of the Euclidean space $\mathbf{R}^m$ and the tools from linear algebra we have developed can be applied.

DEFINITION *Suppose g is a function whose domain includes $s_1, s_2, \ldots, s_m$ and suppose $(s_1, t_1), (s_2, t_2), \ldots, (s_m, t_m)$ are data points. Then the* distance *(in the*

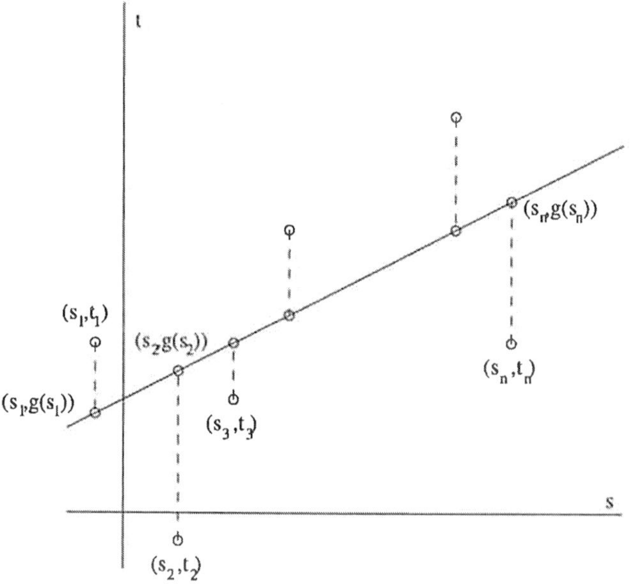

FIGURE 5.3

Errors in fitting a line to data points

sense of least squares) of g from the data is

$$\left(|g(s_1) - t_1|^2 + |g(s_2) - t_2|^2 + \ldots + |g(s_m) - t_m|^2 \right)^{\frac{1}{2}}$$

The general problem of least squares fitting of data is: Given a family of 'nice' functions and m data points, find the function in the family that minimizes the least squares distance from the data. The most common family of 'nice' functions is $\{g : g(s) = \alpha s + \beta$, where α and β are numbers$\}$, and the problem is then one of linear regression. In any case, if the family is a finite dimensional subspace of the vector space of functions, then this can be solved as a linear algebra problem of the sort we have been considering. Many of the most common families of functions fall into this category; what is needed is that the functions depend linearly on the parameters that define the family. For example, the family of second degree polynomials is $\{p : p(s) = \alpha s^2 + \beta s + \gamma\}$; if we double each of the numbers α, β, and γ, then the function is doubled. On the other hand, the family of exponential curves, $\{f : f(s) = ae^{bs}\}$, is not in this category; if we double a and b, we do not double the function. We will not discuss the problem of choosing the best fitting function from a family that does not form a finite dimensional subspace of the space of functions.

To see how we can apply what we know about minimizing the distance from a point to a subspace to the problem of choosing the "best" function to fit some data, we look more closely at the case of linear regression. If it happens that the data points actually lie on a line, we would have $t_k = \alpha s_k + \beta$ for each k and finding the coefficients just involves solving the system

$$\begin{cases} \alpha s_1 + \beta &= t_1 \\ \alpha s_2 + \beta &= t_2 \\ &\vdots \\ \alpha s_m + \beta &= t_m \end{cases}$$

for the unknowns α and β. We can rewrite this system as $AX = b$ where

$$A = \begin{pmatrix} s_1 & 1 \\ s_2 & 1 \\ \vdots & \vdots \\ s_m & 1 \end{pmatrix}, \quad X = \begin{pmatrix} \alpha \\ \beta \end{pmatrix}, \quad \text{and } b = \begin{pmatrix} t_1 \\ t_2 \\ \vdots \\ t_m \end{pmatrix}$$

Now if the points *don't* lie on a line, this system will be inconsistent and the least squares distance of $g = g(s) = \alpha s + \beta$ from the data is

$$\left(|g(s_1) - t_1|^2 + |g(s_2) - t_2|^2 + \ldots + |g(s_m) - t_m|^2 \right)^{\frac{1}{2}} = \|AX - b\|$$

Thus, the problem of finding the coefficients α and β in order to minimize the least squares distance from the data points is just the problem of finding the least squares solution of the inconsistent system $AX = b$. More generally, if the family of functions g depends linearly on the unknown coefficients, this procedure applies.

EXAMPLE 5.7

Find the regression line that best fits the data points $(2, 5)$; $(3, 9)$; $(4, 15)$; and $(5, 21)$.

SOLUTION Since the line has equation $t = g(s) = \alpha s + \beta$ for some numbers α and β, we may restate the problem: Find coefficients α and β that are the best (least squares) solution of the system

$$\begin{cases} 2\alpha + \beta &= 5 \\ 3\alpha + \beta &= 9 \\ 4\alpha + \beta &= 15 \\ 5\alpha + \beta &= 21 \end{cases}$$

That is, as above,

$$A = \begin{pmatrix} 2 & 1 \\ 3 & 1 \\ 4 & 1 \\ 5 & 1 \end{pmatrix} \quad \text{and } b = \begin{pmatrix} 5 \\ 9 \\ 15 \\ 21 \end{pmatrix}$$

By Theorem 5.5, we must solve the equations $A'AX = A'b$. Since

$$A'A = \begin{pmatrix} 54 & 14 \\ 14 & 4 \end{pmatrix} \quad \text{and } A'b = \begin{pmatrix} 202 \\ 50 \end{pmatrix}$$

it follows that $\alpha = 5.4$ and $\beta = -6.4$. This means that the line that best fits the data (using the least squares criterion) is $t = 5.4s - 6.4$. $\square$

There is no necessity of working only with lines; if we seek a best approximant from any subspace of functions, the same procedure applies: we recognize the problem as one of finding the best solution to a (presumably) inconsistent system of equations, and apply Theorem 5.5.

EXAMPLE 5.8

Find the least squares quadratic for the data points $(-1, 4)$; $(0, 4)$; $(1, 6)$; and $(2, 10)$.

SOLUTION We are looking for the function of the form $t = g(s) = \alpha s^2 + \beta s + \gamma$ that best fits the data. Restating the problem, we want to find coefficients α, β, and γ that are the best (least squares) solution of the system

$$\begin{cases} \alpha - \beta + \gamma &=& 4 \\ \gamma &=& 4 \\ \alpha + \beta + \gamma &=& 6 \\ 4\alpha + 2\beta + \gamma &=& 10 \end{cases}$$

That is, as above,

$$A = \begin{pmatrix} 1 & -1 & 1 \\ 0 & 0 & 1 \\ 1 & 1 & 1 \\ 4 & 2 & 1 \end{pmatrix} \quad \text{and } b = \begin{pmatrix} 4 \\ 4 \\ 6 \\ 10 \end{pmatrix}$$

By Theorem 5.5, we must solve the equations $A'AX = A'b$. Since

$$A'A = \begin{pmatrix} 18 & 8 & 6 \\ 8 & 6 & 2 \\ 6 & 2 & 4 \end{pmatrix} \quad \text{and } A'b = \begin{pmatrix} 50 \\ 22 \\ 24 \end{pmatrix}$$

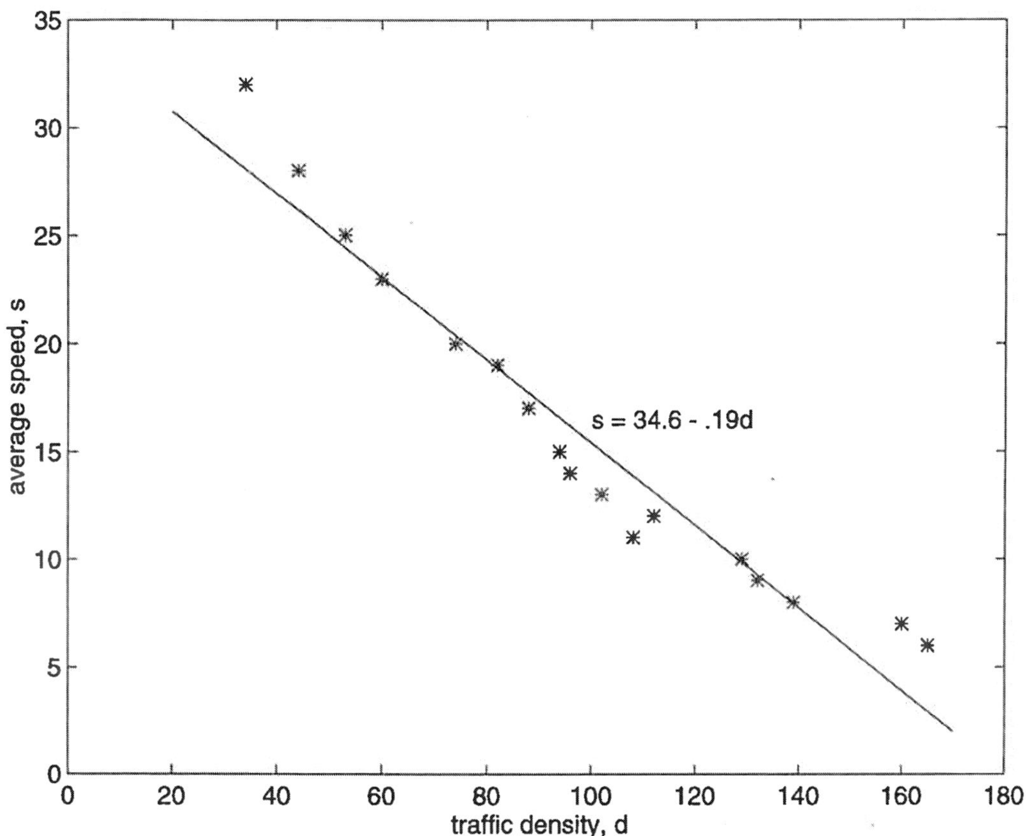

FIGURE 5.4

Lincoln Tunnel speed and density observations

which gives $\alpha = 1$, $\beta = 1$, and $\gamma = 4$. Thus the least squares quadratic is $t = s^2 + s + 4$. $\quad\Box$

EXAMPLE 5.9

Find the least squares estimates for the parameters f and β in Greenshields' model for the Lincoln Tunnel traffic flow data on page 272.

SOLUTION According to Section 5.1, Greenshields' model for traffic flow is $s = f - \beta d$ where s is the average speed and d is the average traffic density. We want to find the mean free speed f and the quotient β from the speed and density data. This system is $AX = s$ where A is the matrix whose columns are the coefficients of f (all 1's) and the coefficients of β (always $-d$). To find the least squares estimates for f and β, we can use MATLAB's \ command:

```
> s=[32; 28; 25; 23; 20; 19; 17; 15; 15; 14; 13; 12; 11; ...
    10; 9; 8; 7; 6]

s =
    32
    28
    25
    23
    20
    19
    17
    15
    15
    14
    13
    12
    11
    10
     9
     8
     7
     6

> d=[34; 44; 53; 60; 74; 82; 88; 94; 94; 96; 102; 112; 108; ...
    129; 132; 139; 160; 165];
> A=[ones(18,1) -d]

A =
    1    -34
    1    -44
    1    -53
    1    -60
    1    -74
    1    -82
    1    -88
    1    -94
    1    -94
    1    -96
    1   -102
```

```
1   -112
1   -108
1   -129
1   -132
1   -139
1   -160
1   -165

> x=A\s

x =
   34.5960
    0.1918
```

So the least squares estimate of the mean free speed f is 34.5960 and the least squares estimate for the parameter β is .1918.

The data points are plotted together with the line $s = 34.5960 - .1918d$ in Figure 5.4.

With these values for the parameters, the Greenshields model predicts a maximum traffic flow through the North Tube of the Lincoln Tunnel of $f^2/(4\beta) \approx 1560$ vehicles per hour at a traffic density of $d = f/(2\beta) \approx 90$ vehicles per mile. Greenberg's data on the Lincoln Tunnel is in reasonable agreement with this maximum traffic flow. Since the parameter β is the quotient of f and the 'jam density', we see that these values for the parameters predict that the tunnel would be jammed with essentially no movement if the traffic density reaches as high as $f/\beta \approx 180$ vehicles per mile. Greenberg's data suggests that this is not actually correct; presumably, this is one reason Greenberg sought to explain the data using an exponential model rather than Greenshields' straight line model. $\square$

Exercises 5.3

1. A thermocouple is supposed to put out a voltage in response to a temperature so that the graph of temperature versus voltage is a straight line. In calibrating a thermocouple for their ChemE lab, Mary and John obtain the data $p_1 = (5, -8)$; $p_2 = (13, -6)$; $p_3 = (22, -3)$; and $p_4 = (39, 0)$ where the first coordinate is the output voltage in microvolts and the second coordinate is the temperature in degrees Celsius. Find the line that best fits this data (i.e., the least squares regression line).

$\Diamond$ 2. An electrical engineer has developed a new low temperature resistor in which the resistance R can be controlled by the strength of the ambient magnetic field S. The results of the calibration experiments at $24°$ Kelvin were:

S	.05	.10	.15	.20	.25	.30	.35	.40	.45	.50
R	.21	.37	.43	.53	.60	.69	.77	.86	.91	1.01

What is the best (least squares criterion) line that fits this data?

3. Dr. Juanita Hernandez is doing research into wear levels in stationary engines. She believes that the wear level depends on the ambient temperature and the dust level in the environment and further that the wear level (w) is a linear function of the excess temperature (t) and the dust level (d), i.e. that $w = \alpha t + \beta d$ where α and β are constants. Estimate the values of the constants from her data:

> Engine site 1:   50 degrees   dust level: 4   wear level: 21
> Engine site 2:   30 degrees   dust level: 5   wear level: 18
> Engine site 3:   40 degrees   dust level: 6   wear level: 25
> Engine site 4:   80 degrees   dust level: 3   wear level: 32

4. A researcher obtains data points $p_1 = (1,3)$; $p_2 = (2,2)$; and $p_3 = (3,2)$. The theoretical model that applies to the situation says that the data points ought to lie on a curve of the form

$$y = a + b\frac{1}{x}.$$

Find the a and b that best (least squares criterion) fit this data.

5. A large meteorite is following an orbit that is nearly parabolic with the sun at the focus of the parabola. Measurements have been made which have very accurately determined the direction of the axis, and it is desired to predict the future course of the meteorite to determine if any significant collisions are likely. In a suitable coordinate system, the path of the meteorite should be $d = \alpha + \beta t + \gamma t^2$ for some choice of α, β, and γ. The four measurements that are to be used in the prediction are $d = -1.6$ when $t = -2$; $d = 2.1$ when $t = -1$; $d = 3.4$ when $t = 0$; and $d = 1.8$ when $t = 1$. Find an optimal choice (i.e. least squares) for the values of α, β, and γ and use them to predict the value of d for $t = 5$.

$\Diamond$ 6. The City of Chicago plans to build a new building for the Department of Tourism and needs to include the effect of the wind in the design of the building. Experience has shown that the effective force E of the wind on a very windy day is related to the height of the building and the area of the west wall by a function of the form $E = ah + bwh$ where h is the height of the building and w is the width of the west wall and a and b are constants that depend on the wind patterns in the area. Find the best (least squares) estimates for a and b based on the data below:

Effective Wind Force Data

Force	Height	Width
10	40	75
12	60	40
15	50	85
20	60	105
25	100	40

$\Diamond$ 7. A simple model for the air resistance of a moving body is $R = av + bv^2$ where R is the force due to air resistance, v is the velocity of the object, and a and b are constants that depend on the shape of the object.

(a) Find the best (least squares) estimates for a and b based on the data below:

Air Resistance Data

Force	Velocity
2	1
7	2
18	5
45	10
80	20
144	30

(b) Use the parameters you have computed to estimate the force due to the air resistance of the object whose velocity is 40.

$\Diamond$ 8. In the discussion of the Greenshields model as applied to the Greenberg data, we wrote the speed as a function of the density: $s = f - \beta d$. We could just as easily have written the density as a function of the speed: $d = f/\beta - (1/\beta)s$. Use the Greenberg data on the Lincoln Tunnel to find the least squares estimates of f/β and $1/\beta$. Compare this with the estimates of f and β obtained in Example 5.9. In using least squares estimates for parameters, you need to be aware of the sensitivity of least squares estimates to the form of the original system.

9. Suppose q_1, q_2, q_3, q_4, and q_5 form an orthonormal basis for $\mathbf{R}^5$ and suppose Q is the 5×3 matrix with columns q_1, q_2, and q_3, that is, $Q = [q_1\ q_2\ q_3]$.

(a) Describe the nullspace of Q'.

(b) Describe the nullspace of Q.

(c) Consider the vector $b = 3q_1 - q_2 + 2q_3 - 2q_4 + q_5$. Find the point w in the range of Q that is closest to b and find the distance from w to b.

(d) Find the vector X in $\mathbf{R}^3$ that is the least squares solution of $QX = b$.

10.(a) Use calculus to find the constant c that is the best (least squares) estimate of the (real) data values $y_1, y_2, \ldots, y_n$; that is, find the value of c that minimizes the function

$$E(c) = [(y_1 - c)^2 + (y_2 - c)^2 + \cdots + (y_n - c)^2]^{\frac{1}{2}}.$$

(b) Repeat part (a) using the projection technique from this course. It might be easier to think about this problem by viewing it as follows: given the data points

$(0, y_1), (0, y_2), \ldots, (0, y_n)$, find the function in the subspace

$$W = \{g : g(x) = c, \text{ for } c \text{ in } \mathbf{R} \}$$

that best (least squares) approximates the data.

11. (Note: this is *not* a least squares problem, or even a linear algebra problem!)
    Find the constant c that is the best (absolute error) estimate of the (real) data values
    $y_1, y_2, \ldots, y_n$; that is, find the value of c that minimizes the function

$$F(c) = |y_1 - c| + |y_2 - c| + \cdots + |y_n - c|$$

5.4   Projections as Linear Transformations

In this section, we look more closely at projections as linear transformations. We
saw in Section 4.2 that the Gram–Schmidt algorithm is based on projecting vectors
onto a subspaces. We will also relate the ideas of this chapter to QR–factorizations
of matrices. In Section 4.2 the Gram–Schmidt algorithm was used to give a QR–
factorization in which the number of columns of Q is the rank of A. We will see how
QR–factorizations of matrices in which the number of columns of Q is the rank of A
leads directly to the solution of least squares problems. We saw in Section 4.4 how a
factorization for which Q is square can be used to give an orthonormal basis for the
nullspace of A. Householder transformations lead to QR–factorizations of a matrix
in which Q is square. We will see how to construct Householder transformations
and the resulting QR–factorization from some specially chosen projection matrices.

In Section 5.2 we had to face the question: Suppose W is a subspace of $\mathbf{R}^m$ and
b is a vector in $\mathbf{R}^m$; which point of W is closest to b? Surprisingly, this geometric
question turned out to be at the heart of the analysis of inconsistent systems. We
gave a descriptive solution to the problem analogous to the solution of the problem
in elementary geometry of finding the closest point on the line to b by constructing
the line through b that is perpendicular to the given line.

> **Theorem 5.1** If W is a subspace of $\mathbf{R}^m$ and b is in $\mathbf{R}^m$, the closest point
> of W to b is the unique point b_0 of W such that $b - b_0$ is in $W^\perp$.

If we view this statement as describing an action that can take place for any
point in the space rather than one particular point, the theorem characterizes the
function that takes a vector to the point of a subspace that is closest to it. According

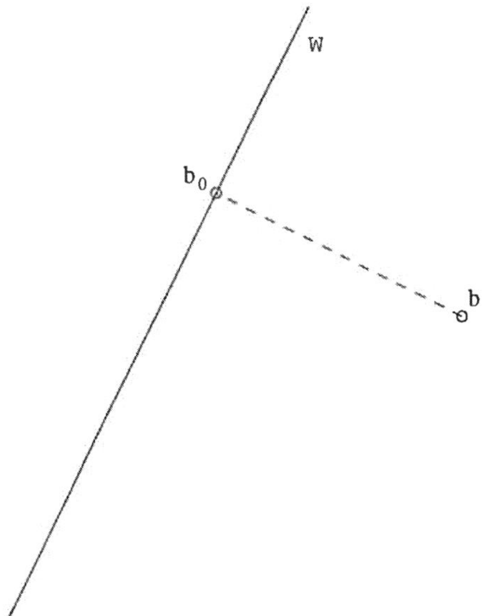

FIGURE 5.5

Closest point, b_0, of a subspace to point b

to the theorem, geometrically, this function projects each point of the space perpendicularly onto the subspace. The action of this function is like the sun when it is directly overhead, projecting shadows onto the ground, and the name "projection" comes from this analogy. The uniqueness assertion in Theorem 5.1 guarantees that, in the following definition, $proj_W$ really is a function.

DEFINITION For W a subspace of $\mathbf{R}^m$, the orthogonal projection of $\mathbf{R}^m$ onto W *is the function on $\mathbf{R}^m$ given by $proj_W(x)$ is the closest point of W to x.*

When we first encounter functions, we are frequently told that "a function is a rule" but we are most comfortable, in calculus for example, with functions that are given by formulas. Comparing the definition of $proj_W$ with Theorem 5.1 gives a rule of sorts that describes the projection. The corollary below makes this explicit. One of the goals of the rest of the section is to get a formula for the projection.

COROLLARY 5.10

If x is in $\mathbf{R}^m$, then

$$proj_W(x) = w \quad \text{if and only if} \quad x = w + u$$

for some w in W and u in $W^{\perp}$.

Note that for all x in $\mathbf{R}^m$, the vector $\mathrm{proj}_{\mathcal{W}}(x)$ lies in $\mathcal{W}$. Thus, the range of $\mathrm{proj}_{\mathcal{W}}$ is a subset of $\mathcal{W}$. Moreover, if w is in $\mathcal{W}$, the closest point of $\mathcal{W}$ to w is clearly itself, so for points of $\mathcal{W}$, we have $\mathrm{proj}_{\mathcal{W}}(w) = w$ and we see that the range of $\mathrm{proj}_{\mathcal{W}}$ is all of $\mathcal{W}$. That is, we are thinking of $\mathrm{proj}_{\mathcal{W}}$ as a function that maps $\mathbf{R}^m$ into $\mathbf{R}^m$, and more specifically, that $\mathrm{proj}_{\mathcal{W}}$ maps $\mathbf{R}^m$ onto $\mathcal{W}$ as a subset of $\mathbf{R}^m$.

The following theorem gives a special property of $\mathrm{proj}_{\mathcal{W}}$ that will allow us to find an easy formula for the function. Recall from page 29 that a function T is a *linear transformation* if, whenever α is a number and x, x_1, and x_2 are vectors, $T(\alpha x) = \alpha T(x)$ and $T(x_1 + x_2) = T(x_1) + T(x_2)$.

THEOREM 5.11

The function $\mathrm{proj}_{\mathcal{W}}$ is a linear transformation, that is, if α is a number and x, x_1, and x_2 are vectors in $\mathbf{R}^m$, then

$$\mathrm{proj}_{\mathcal{W}}(\alpha x) = \alpha\,\mathrm{proj}_{\mathcal{W}}(x) \qquad and \qquad \mathrm{proj}_{\mathcal{W}}(x_1 + x_2) = \mathrm{proj}_{\mathcal{W}}(x_1) + \mathrm{proj}_{\mathcal{W}}(x_2)$$

PROOF   Let $x = w + u$ where w is in $\mathcal{W}$ and u is in $\mathcal{W}^\perp$ and suppose α is a number. It follows that $\alpha x = \alpha w + \alpha u$, and that αw is in $\mathcal{W}$, and αu is in $\mathcal{W}^\perp$. Now, by Corollary 5.10, this means $\mathrm{proj}_{\mathcal{W}} x = w$ and $\mathrm{proj}_{\mathcal{W}}(\alpha x) = \alpha w$. Thus, $\mathrm{proj}_{\mathcal{W}}(\alpha x) = \alpha w = \alpha(\mathrm{proj}_{\mathcal{W}} x)$, as required.

Similarly, suppose $x_1 = w_1 + u_1$ and $x_2 = w_2 + u_2$ where w_1 and w_2 are in $\mathcal{W}$ and u_1 and u_2 are in $\mathcal{W}^\perp$ so that $\mathrm{proj}_{\mathcal{W}} x_1 = w_1$ and $\mathrm{proj}_{\mathcal{W}} x_2 = w_2$. Again, $(x_1 + x_2) = (w_1 + w_2) + (u_1 + u_2)$ where $w_1 + w_2$ is in $\mathcal{W}$ and $u_1 + u_2$ is in $\mathcal{W}^\perp$, so it follows from Corollary 5.10 that $\mathrm{proj}_{\mathcal{W}}(x_1 + x_2) = w_1 + w_2 = \mathrm{proj}_{\mathcal{W}} x_1 + \mathrm{proj}_{\mathcal{W}} x_2$. This means $\mathrm{proj}_{\mathcal{W}}$ is additive as well, so it is a linear transformation.   ∎

The following theorem gives a formula for the orthogonal projection onto a subspace of $\mathbf{R}^m$ in terms of a basis for the subspace. As in the discussion on page 29 and the related examples, this formula says that to compute the projection of a vector onto the subspace, simply multiply the vector by certain matrix.

THEOREM 5.12

Let $\mathcal{W}$ be a subspace of $\mathbf{R}^m$ and let $\{v_1, v_2, \ldots, v_n\}$ be a basis for $\mathcal{W}$. For every vector x, the orthogonal projection of x onto $\mathcal{W}$ is

$$\mathrm{proj}_{\mathcal{W}}(x) = A(A'A)^{-1}A'x$$

where A is the $m \times n$ matrix with columns $v_1, v_2, \ldots, v_n$.

We will prove that this formula is correct after giving an example. Theorem 5.12 gives an explicit solution to the problem of finding the point of a subspace closest to a given point, although it is not always the most efficient one computationally. (Exercise 5.4.9. gives another approach.) The use of Theorem 5.12 is illustrated in the following example.

EXAMPLE 5.13

Let W be the subspace of $\mathbf{R}^3$ spanned by $v_1 = (1, 2, 1)$ and $v_2 = (1, -1, 0)$. Find the point of W that is closest to $x = (1, 0, 0)$. What is the distance from x to W?

SOLUTION The vectors $v_1 = (1, 2, 1)$ and $v_2 = (1, -1, 0)$ are linearly independent so they are a basis for W. The matrix A of Theorem 5.12 is therefore

$$A = \begin{pmatrix} 1 & 1 \\ 2 & -1 \\ 1 & 0 \end{pmatrix}$$

According to the theorem, we need to compute $A(A'A)^{-1}A'x$. We begin by finding $A'A$ and its inverse.

$$A'A = \begin{pmatrix} 1 & 2 & 1 \\ 1 & -1 & 0 \end{pmatrix} \begin{pmatrix} 1 & 1 \\ 2 & -1 \\ 1 & 0 \end{pmatrix} = \begin{pmatrix} 6 & -1 \\ -1 & 2 \end{pmatrix}$$

Thus

$$(A'A)^{-1} = \frac{1}{11} \begin{pmatrix} 2 & 1 \\ 1 & 6 \end{pmatrix}$$

and

$$A(A'A)^{-1}A' = \frac{1}{11} \begin{pmatrix} 1 & 1 \\ 2 & -1 \\ 1 & 0 \end{pmatrix} \begin{pmatrix} 2 & 1 \\ 1 & 6 \end{pmatrix} \begin{pmatrix} 1 & 2 & 1 \\ 1 & -1 & 0 \end{pmatrix} =$$

$$\frac{1}{11} \begin{pmatrix} 1 & 1 \\ 2 & -1 \\ 1 & 0 \end{pmatrix} \begin{pmatrix} 3 & 3 & 2 \\ 7 & -4 & 1 \end{pmatrix} = \frac{1}{11} \begin{pmatrix} 10 & -1 & 3 \\ -1 & 10 & 3 \\ 3 & 3 & 2 \end{pmatrix}$$

So the closest point of W to x is

$$w = \mathrm{proj}_W x = \frac{1}{11} \begin{pmatrix} 10 & -1 & 3 \\ -1 & 10 & 3 \\ 3 & 3 & 2 \end{pmatrix} \begin{pmatrix} 1 \\ 0 \\ 0 \end{pmatrix} = \frac{1}{11} \begin{pmatrix} 10 \\ -1 \\ 3 \end{pmatrix}$$

Notice that this point really is in W since

$$w = \frac{3}{11} v_1 + \frac{7}{11} v_2$$

Moreover, since the vector $x - w = (1/11, 1/11, -3/11)$ is orthogonal to $\mathcal{W}^{\perp}$, it solves our problem (Theorem 5.1):

$$\langle v_1, x - w \rangle = \left\langle \begin{pmatrix} 1 \\ 2 \\ 1 \end{pmatrix}, \begin{pmatrix} \frac{1}{11} \\ \frac{1}{11} \\ -\frac{3}{11} \end{pmatrix} \right\rangle = 0$$

and

$$\langle v_2, x - w \rangle = \left\langle \begin{pmatrix} 1 \\ -1 \\ 0 \end{pmatrix}, \begin{pmatrix} \frac{1}{11} \\ \frac{1}{11} \\ -\frac{3}{11} \end{pmatrix} \right\rangle = 0$$

The distance from x to $\mathcal{W}$ is just the distance from x to w,

$$\|x - w\| = \left\| \begin{pmatrix} \frac{1}{11} \\ \frac{1}{11} \\ -\frac{3}{11} \end{pmatrix} \right\| = \sqrt{\frac{11}{121}} = \frac{1}{\sqrt{11}}$$

$\square$

Theorem 5.12 gives a formula for the projection function that finds the projection of x onto $\mathcal{W}$ by multiplying x by a matrix. The theorem below characterizes the matrices that accomplish this task. Recall from Lemma 4.21 that if A is an $m \times n$ matrix, then $\langle Au, v \rangle = \langle u, A'v \rangle$.

THEOREM 5.14

If P is an $m \times m$ matrix such that $P^2 = P$ and $P' = P$, then the function $x \mapsto Px$ is the orthogonal projection of $\mathbf{R}^m$ onto the range of P. Conversely, if $\mathcal{W}$ is a subspace of $\mathbf{R}^m$ and P is a matrix such that $proj_{\mathcal{W}}(x) = Px$, then $P^2 = P$, $P' = P$, and the range of P is $\mathcal{W}$.

PROOF Suppose P is an $m \times m$ matrix such that $P^2 = P$ and $P' = P$. We need to show that P is $proj_{\mathcal{R}(P)}$, that is, we need to show that if v is in $\mathbf{R}^m$ and $v = w + u$ for w in $\mathcal{R}(P)$, the range of P, and u is in $\mathcal{R}(P)^{\perp}$ then $Pv = w$.

Suppose w is in $\mathcal{R}(P)$. The definition of "w is in $\mathcal{R}(P)$" is that there is x so that $Px = w$. Since $P^2 = P$, we see $w = Px = P^2x = P(Px) = Pw$, so $Pw = w$ for all w in $\mathcal{R}(P)$.

By the Duality Theorem (Theorem 4.22), the nullspace of P is $\mathcal{R}(P')^{\perp}$, but since $P = P'$, the nullspace of P is $\mathcal{R}(P)^{\perp}$, that is, $Pu = 0$ for all u in $\mathcal{R}(P)^{\perp}$.

Now if v is in $\mathbf{R}^m$ and $v = w + u$ where w is in $\mathcal{R}(P)$ and u is in $\mathcal{R}(P)^{\perp}$, then $Pv = Pw + Pu = w + 0 = w$. But w is just the projection of v onto $\mathcal{R}(P)$. Since v was an arbitrary vector in $\mathbf{R}^m$, this gives the conclusion.

Conversely, suppose W is a subspace of $\mathbf{R}^m$ and P is a matrix such that $\text{proj}_W(x) = Px$ for every x. This means that if $x = w + u$ for w in W and u in $W^{\perp}$, then $Px = w$.

Now, if w is in W, then $w = w + 0$ and 0 is in $W^{\perp}$, so $\text{proj}_W(w) = w$. This means that for any vector x, letting $x = w + u$ for w in W and u in $W^{\perp}$, we have

$$P^2 x = P(Px) = Pw = w = Px$$

which means that $P^2 = P$.

Similarly, if u is in $W^{\perp}$, then $u = 0 + u$ and 0 is in W, so $\text{proj}_W(u) = 0$. Letting $x_1 = w_1 + u_1$ and $x_2 = w_2 + u_2$ for w_1 and w_2 in W and u_1 and u_2 in $W^{\perp}$, then

$$\langle Px_1, x_2 \rangle = \langle w_1, w_2 + u_2 \rangle = \langle w_1, w_2 \rangle + \langle w_1, u_2 \rangle$$
$$= \langle w_1, w_2 \rangle + 0 = \langle w_1, w_2 \rangle + \langle u_1, w_2 \rangle$$
$$= \langle w_1 + u_1, w_2 \rangle = \langle x_1, Px_2 \rangle = \langle P'x_1, x_2 \rangle$$

This means

$$\langle Px_1 - P'x_1, x_2 \rangle = \langle Px_1, x_2 \rangle - \langle P'x_1, x_2 \rangle = 0$$

for every x_1 and x_2, so $Px_1 - P'x_1$ is a vector that is perpendicular to every vector x_2. Since the only vector that is perpendicular to every vector is zero, $Px_1 - P'x_1 = 0$ for every x_1. Now, $Px_1 = P'x_1$ for every vector x_1 means $P = P'$.

Since $Px = w$ is in W for every x, then the range of P is a subset of W. On the other hand, since $Pw = w$ for all w in W, every point of W is in the range of P, so $\mathcal{R}(P) = W$. $\blacksquare$

If we imagine proj_W as the sun projecting a shadow, of a tree, say, onto the ground, then we can imagine getting a big piece of black paper, carefully cutting it out in the shape and size of the shadow of the tree, and laying it on the ground on top of the shadow. If we remove the tree, but leave the black paper on the ground, we can then inquire about the shadow of the black paper. Of course, the shadow of the black paper will be completely hidden by the paper; it will coincide with its shadow. The property $P^2 = P$ in the theorem above can be paraphrased as "the shadow of the shadow is the shadow". Although more difficult to see, the property $P' = P$ is associated with the orthogonality of the projection, that is, with the sun's being directly overhead. (Thus, in Indiana, where the latitude is about $40°$, the sun never acts like an orthogonal projection and the associated transformation is not self–adjoint.) The range of P is just the collection of points that are shadows.

Theorem 5.14 justifies the following terminology for matrices.

DEFINITION *An $m \times m$ matrix P such that $P^2 = P$ and $P' = P$ is called an* orthogonal projection matrix *(or sometimes just* projection matrix*).*

In order for the formula in Theorem 5.12 to make sense, we need the matrix $A'A$ to be invertible. Notice that since A is, in general, not square and therefore not invertible, the formula from Chapter 1 on the inverse of a product does not apply. The invertibility of $A'A$ follows from Corollary 5.4 because A is an $m \times n$ matrix with linearly independent columns. The proof of Theorem 5.12 is now built on Theorem 5.14.

PROOF **(of Theorem 5.12)** Let A be the matrix whose columns are $v_1, \cdots, v_n$. Since the v's are a basis for $\mathcal{W}$, the matrix A is an $m \times n$ matrix with linearly independent columns and the rank of A is n. Let P be the $m \times m$ matrix $P = A(A'A)^{-1}A'$, which makes sense by Corollary 5.4. We will verify the hypotheses of Theorem 5.14.

First, we note that $A'A$ is Hermitian (Exercise 1.3.12.) which implies $(A'A)^{-1}$ is Hermitian (Exercise 1.3.10.) which gives

$$P' = \left(A(A'A)^{-1}A'\right)' = (A')'\left((A'A)^{-1}\right)' A' = A(A'A)^{-1}A' = P$$

Moreover

$$P^2 = \left(A(A'A)^{-1}A'\right)^2 = A(A'A)^{-1}A'A(A'A)^{-1}A' = A(A'A)^{-1}A' = P$$

Finally, we need to show that the range of P is the same as the range of A. Clearly, $\mathcal{R}(P) \subset \mathcal{R}(A)$ because if z is in the range of P, there is x so that $z = Px$. Therefore,

$$z = Px = A(A'A)^{-1}A'x = A\left((A'A)^{-1}A'x\right) = Ay$$

for $y = (A'A)^{-1}A'x$ so z is in the range of A. Now suppose z is in the range of A, that is $z = Ax$ for some x. To show that z is in the range of P, we need only show $z = Py$ for some vector y. In fact, we will show that $z = Pz$!

$$Pz = PAx = A(A'A)^{-1}A'Ax = Ax = z$$

Thus, $P = A(A'A)^{-1}A'$ satisfies $P = P^2$, $P = P'$, and the range of P is the range of A: by Theorem 5.14, P is the orthogonal projection matrix of $\mathbf{R}^n$ onto the range of A. ∎

If A is an $m \times n$ matrix with rank n and $A = QR$ is a QR–factorization of A in which Q has n columns, then QQ' orthogonal projection of $\mathbf{R}^m$ onto the range of A. The proof of this fact is Exercise 5.4.9.

It turns out that there is a version of the normal equations using the QR–factorization that is useful if you know the QR–factorization. Instead of expressing the normal equations in terms of A and A', it expresses them in terms of Q and R. The point of the theorem is that, because R is lower triangular, this form of the normal equations is very easy to solve. For a system $AX = b$ in which A is not square, the solution found by MATLAB using the command X=A\b is the least squares solution of the system and the algorithm used is based on the QR–factorization as in the theorem below.

THEOREM 5.15

Suppose A is an $m \times n$ matrix with rank n. If $A = QR$ is a QR–factorization of A in which Q is $m \times n$, then the least squares solution of $AX = b$ is the unique solution of

$$RX_0 = Q'b$$

PROOF The matrix A is $m \times n$ and Q is $m \times n$, so R must be $n \times n$. Since A has rank n and $A = QR$ implies the rank of A is less than or equal to the rank of R, it follows that R is an $n \times n$ matrix with rank n. This means it is invertible so R' is invertible also. Now the least squares solution of $AX = b$ is the solution of the normal equations $A'AX_0 = A'b$. If X_0 is a solution of the normal equations, X_0 satisfies

$$(QR)'QRX_0 = (QR)'b$$

so that

$$R'Q'QRX_0 = R'Q'b$$

Recalling that $Q'Q = I$ gives

$$R'RX_0 = R'Q'b$$

Multiplying by the inverse of R', we get

$$(R')^{-1}R'RX_0 = (R')^{-1}R'Q'b$$

or

$$RX_0 = Q'b$$

as we were to show. Since the rank of R is n, there is only one solution of this equation. ∎

To see how easy the least squares solution is to calculate if we are given the QR–factorization, we consider the following example. From this we can see that the difficult step in the solution is finding the QR–factorization; an efficient method for finding a QR–factorization is discussed later in this section.

EXAMPLE 5.16

Find the least squares solution of the system

$$\begin{cases} 2x - y & = & 4 \\ -2x + 2y & = & 4 \\ x & = & 3 \end{cases}$$

SOLUTION This system is $AX = b$ where

$$A = \begin{pmatrix} 2 & -1 \\ -2 & 2 \\ 1 & 0 \end{pmatrix} \quad \text{and } b = \begin{pmatrix} 4 \\ 4 \\ 3 \end{pmatrix}$$

The matrix A has QR–factorization

$$\begin{pmatrix} 2 & -1 \\ -2 & 2 \\ 1 & 0 \end{pmatrix} = \begin{pmatrix} \frac{2}{3} & \frac{1}{3} \\ -\frac{2}{3} & \frac{2}{3} \\ \frac{1}{3} & \frac{2}{3} \end{pmatrix} \begin{pmatrix} 3 & -2 \\ 0 & 1 \end{pmatrix}$$

To find the least squares solution of this system, we need to solve the normal equations $RX = Q'b$:

$$\begin{pmatrix} 3 & -2 \\ 0 & 1 \end{pmatrix} \begin{pmatrix} x \\ y \end{pmatrix} = \begin{pmatrix} \frac{2}{3} & -\frac{2}{3} & \frac{1}{3} \\ \frac{1}{3} & \frac{2}{3} & \frac{2}{3} \end{pmatrix} \begin{pmatrix} 4 \\ 4 \\ 3 \end{pmatrix}$$

which is

$$\begin{pmatrix} 3 & -2 \\ 0 & 1 \end{pmatrix} \begin{pmatrix} x \\ y \end{pmatrix} = \begin{pmatrix} 1 \\ 6 \end{pmatrix}$$

so $y = 6$ and $x = 13/3$ is the least squares solution of the system. ⬜

Householder Transformations

The construction of the QR–factorization of the $m \times n$ matrix A from the Gram–Schmidt algorithm is problematic from two points of view. First, it is not very stable numerically since it frequently requires division by differences that may be small. Secondly, it leads to a factorization in which Q is an $m \times k$ matrix where k is the rank of A, rather than an $m \times m$ matrix. We will use *Householder transformations* to address both difficulties. The algorithm used by MATLAB to compute the QR–factorization with the command qr is a variation of that given here.

We say that U is *unitary* if U is an $m \times m$ matrix such that $U' = U^{-1}$. It is the case (see Exercise 4.6.10. and Theorem 7.1) that U is unitary if and only if $\|Uv\| = \|v\|$ for every v if and only if $\langle Ux, Uy \rangle = \langle x, y \rangle$ for all x and y. We have already seen that $U'U = I$ if and only if the columns of U are an orthonormal set (Theorem 4.24) and that U is invertible if and only if its columns are linearly independent (Theorem 3.63). Thus, U is unitary if and only if its columns are an orthonormal basis for $\mathbf{C}^m$. Sometimes a real unitary matrix, that is, a matrix whose columns are an orthonormal basis for $\mathbf{R}^m$, is called an *orthogonal matrix*.

It follows from the definition that a product of unitary matrices is unitary (Exercise 5.4.10.). Exercise 5.4.11. shows that if P is an orthogonal projection, then $I - 2P$ is a unitary matrix. Geometrically, this unitary matrix is reflection in the nullspace of P. Householder transformations are special cases of this reflection.

DEFINITION If v is a vector in $\mathbf{R}^n$, the unitary matrix

$$H = I - 2\frac{1}{\|v\|^2}vv'$$

is called a Householder transformation.

In fact, the Householder transformation given by this formula is reflection in the subspace $\mathcal{W}$ of vectors perpendicular to v (see Figure 5.6).

We will use Householder transformations to move vectors around in specific ways. If x is any vector and we want to use a unitary matrix to move x to a multiple of $z = (1, 0, \cdots, 0)$, then since unitaries preserve length, the multiple must be $\pm\|x\|z$ (or $e^{i\theta}\|x\|z$ for some real number θ in the complex case).

For x and z in $\mathbf{C}^n$ with $\|z\| = 1$, let $v = x - \sigma z$ where $|\sigma| = \|x\|$ and $\sigma\langle x, z \rangle$ is real, that is,

$$\sigma = \|x\|\frac{\langle z, x \rangle}{|\langle z, x \rangle|} \tag{5.4.1}$$

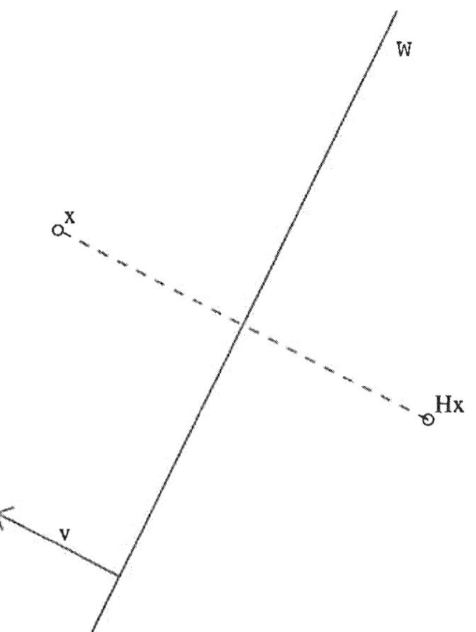

FIGURE 5.6

Householder transformation associated with v

then for H as above, we have

$$Hx = x - \frac{2}{(x - \sigma z)'(x - \sigma z)}(x - \sigma z)(x - \sigma z)'x$$

$$= x - \frac{2}{(x'x - \sigma x'z - \overline{\sigma}z'x + \|x\|^2 z'z)}(xx'x - \sigma zx'x - \overline{\sigma}xz'x + \|x\|^2 zz'x)$$

$$= x - \frac{2\sigma^2 - \overline{\sigma}\langle z, x\rangle - \sigma\langle x, z\rangle}{2(\sigma^2 - \sigma\langle x, z\rangle)}(x - \sigma z) = x - (x - \sigma z) = \sigma z$$

This calculation proves the following theorem.

THEOREM 5.17

If x and z in $\mathbf{C}^m$ with $\|z\| = 1$, then

$$H = I - 2\frac{1}{\|v\|^2}vv'$$

is a Hermitian unitary such that $Hx = \sigma z$ where $v = x - \sigma z$ and $\sigma = e^{i\theta}\|x\|$ for $\theta = \arg(\langle z, x\rangle)$.

EXAMPLE 5.18

Find H unitary so that $H(1,2,2)$ is a multiple of $(1,0,0)$.

SOLUTION Since $\|(1,2,2)\| = 3$, we have $\sigma = 3$ and $v = x - \sigma z = (1,2,2) - 3(1,0,0) = (-2,2,2)$. Thus

$$H = I - \frac{2}{\|v\|^2}vv' = \begin{pmatrix} 1 & 0 & 0 \\ 0 & 1 & 0 \\ 0 & 0 & 1 \end{pmatrix} - \frac{2}{12}\begin{pmatrix} 4 & -4 & -4 \\ -4 & 4 & 4 \\ -4 & 4 & 4 \end{pmatrix} = \begin{pmatrix} \frac{1}{3} & \frac{2}{3} & \frac{2}{3} \\ \frac{2}{3} & \frac{1}{3} & -\frac{2}{3} \\ \frac{2}{3} & -\frac{2}{3} & \frac{1}{3} \end{pmatrix}$$

It is easily checked that this is a Hermitian unitary such that $H(1,2,2) = (3,0,0) = 3(1,0,0)$ as desired. $\quad\square$

We want to use a recursive algorithm to compute the upper triangular matrix R. To do this we first observe that if H_0 is a Hermitian unitary and H is the block matrix

$$\begin{pmatrix} I & 0 \\ 0 & H_0 \end{pmatrix}$$

then H is also a Hermitian unitary. Moreover, if H has this block form and B is blocked in the same way, then

$$\begin{pmatrix} I & 0 \\ 0 & H_0 \end{pmatrix}\begin{pmatrix} B_{11} & B_{12} \\ B_{21} & B_{22} \end{pmatrix} = \begin{pmatrix} B_{11} & B_{12} \\ H_0 B_{21} & H_0 B_{22} \end{pmatrix}$$

Thus, given an $m \times n$ matrix A, to find R, we first find H_1 so that H_1 times the first column of A is a multiple of $(1,0,\cdots,0)$, and let $R_1 = H_1 A$. Now, using the lower right $(m-1) \times (n-1)$ submatrix of R_1, find an $(m-1) \times (m-1)$ Hermitian unitary that moves the first column of the submatrix to a multiple of $(1,0,\cdots,0)$ and let H_2 be the block matrix as above. Letting $R_2 = H_2(H_1 A)$, we continue until we have an upper triangular matrix R factored as $R = H_k H_{k-1} \cdots H_2 H_1 A$ where k is the smaller of m and n. Solving for A, we get $A = H_1^{-1} H_2^{-1} \cdots H_{k-1}^{-1} H_k^{-1} R$. Now, each H_j is Hermitian and unitary, so $H_j^{-1} = H_j' = H_j$, and

$$A = (H_1 H_2 \cdots H_{k-1} H_k)R = QR$$

EXAMPLE 5.19

Find a QR–factorization of

$$A = \begin{pmatrix} 1 & 1 \\ 2 & 1 \\ 2 & 1 \end{pmatrix}$$

SOLUTION Let H_1 be

$$\begin{pmatrix} \frac{1}{3} & \frac{2}{3} & \frac{2}{3} \\ \frac{2}{3} & \frac{1}{3} & -\frac{2}{3} \\ \frac{2}{3} & -\frac{2}{3} & \frac{1}{3} \end{pmatrix}$$

as we computed above. So

$$R_1 = H_1 A = \begin{pmatrix} \frac{1}{3} & \frac{2}{3} & \frac{2}{3} \\ \frac{2}{3} & \frac{1}{3} & -\frac{2}{3} \\ \frac{2}{3} & -\frac{2}{3} & \frac{1}{3} \end{pmatrix} \begin{pmatrix} 1 & 1 \\ 2 & 1 \\ 2 & 1 \end{pmatrix} = \begin{pmatrix} 3 & \frac{5}{3} \\ 0 & \frac{1}{3} \\ 0 & \frac{1}{3} \end{pmatrix}$$

Now find H_0 so that $H_0(1/3, 1/3)$ is a multiple of $(1,0)$. That is, $x = (1/3, 1/3)$, $\sigma = \|x\| = 2/\sqrt{3}$, and

$$v = x - \sigma z = \begin{pmatrix} \frac{1}{3} \\ \frac{1}{3} \end{pmatrix} - \frac{\sqrt{2}}{3} \begin{pmatrix} 1 \\ 0 \end{pmatrix} = \begin{pmatrix} \frac{1-\sqrt{2}}{3} \\ \frac{1}{3} \end{pmatrix}$$

which gives

$$I - \frac{2}{\|v\|^2} vv' = \begin{pmatrix} \frac{1}{\sqrt{2}} & \frac{1}{\sqrt{2}} \\ \frac{1}{\sqrt{2}} & -\frac{1}{\sqrt{2}} \end{pmatrix}$$

and

$$H_2 = \begin{pmatrix} 1 & 0 & 0 \\ 0 & \frac{1}{\sqrt{2}} & \frac{1}{\sqrt{2}} \\ 0 & \frac{1}{\sqrt{2}} & -\frac{1}{\sqrt{2}} \end{pmatrix}$$

It follows that

$$R_2 = H_2(H_1 A) = \begin{pmatrix} 1 & 0 & 0 \\ 0 & \frac{1}{\sqrt{2}} & \frac{1}{\sqrt{2}} \\ 0 & \frac{1}{\sqrt{2}} & -\frac{1}{\sqrt{2}} \end{pmatrix} \begin{pmatrix} 3 & \frac{5}{3} \\ 0 & \frac{1}{3} \\ 0 & \frac{1}{3} \end{pmatrix} = \begin{pmatrix} 3 & \frac{5}{3} \\ 0 & \frac{\sqrt{2}}{3} \\ 0 & 0 \end{pmatrix}$$

Since H_1 and H_2 are Hermitian unitaries, a QR–factorization is given by

$$A = H_1^{-1} H_2^{-2} \begin{pmatrix} 3 & \frac{5}{3} \\ 0 & \frac{\sqrt{2}}{3} \\ 0 & 0 \end{pmatrix} = H_1 H_2 \begin{pmatrix} 3 & \frac{5}{3} \\ 0 & \frac{\sqrt{2}}{3} \\ 0 & 0 \end{pmatrix}$$

$$
= \begin{pmatrix} \frac{1}{3} & \frac{2}{3} & \frac{2}{3} \\ \frac{2}{3} & \frac{1}{3} & -\frac{2}{3} \\ \frac{2}{3} & -\frac{2}{3} & \frac{1}{3} \end{pmatrix} \begin{pmatrix} 1 & 0 & 0 \\ 0 & \frac{1}{\sqrt{2}} & \frac{1}{\sqrt{2}} \\ 0 & \frac{1}{\sqrt{2}} & -\frac{1}{\sqrt{2}} \end{pmatrix} \begin{pmatrix} 3 & \frac{5}{3} \\ 0 & \frac{\sqrt{2}}{3} \\ 0 & 0 \end{pmatrix}
$$

$$
= \begin{pmatrix} \frac{1}{3} & \frac{4}{3\sqrt{2}} & 0 \\ \frac{2}{3} & -\frac{1}{3\sqrt{2}} & \frac{1}{\sqrt{2}} \\ \frac{2}{3} & -\frac{1}{3\sqrt{2}} & -\frac{1}{\sqrt{2}} \end{pmatrix} \begin{pmatrix} 3 & \frac{5}{3} \\ 0 & \frac{\sqrt{2}}{3} \\ 0 & 0 \end{pmatrix}
$$

Moreover, a QR–factorization of A in which Q is 3×2 can be read directly from this answer, namely

$$
A = \begin{pmatrix} \frac{1}{3} & \frac{4}{3\sqrt{2}} \\ \frac{2}{3} & -\frac{1}{3\sqrt{2}} \\ \frac{2}{3} & -\frac{1}{3\sqrt{2}} \end{pmatrix} \begin{pmatrix} 3 & \frac{5}{3} \\ 0 & \frac{\sqrt{2}}{3} \end{pmatrix}
$$

$\square$

Exercises 5.4

1. Let W be the subspace of $\mathbf{R}^4$ spanned by $v_1 = (1,1,2,0)$; $v_2 = (0,1,1,-1)$; and $v_3 = (1,0,1,1)$. Find the point of W that is closest to $x = (1,0,0,1)$. What is the distance from x to W? (Apply theorems *carefully!*)

2. Let $P = \begin{pmatrix} \frac{2}{3} & -\frac{1}{3} & \frac{1}{3} \\ -\frac{1}{3} & \frac{2}{3} & \frac{1}{3} \\ \frac{1}{3} & \frac{1}{3} & \frac{2}{3} \end{pmatrix}$.

   (a) Show that P is the orthogonal projection of $\mathbf{R}^3$ onto some subspace W.
   (b) Find a basis for the subspace W.
   (c) Let $x = (1,2,0)$. Find w and u so that $x = w + u$ where $w \in W$ and $u \in W^{\perp}$.

3. Let $Q = \begin{pmatrix} -1 & 2 & -2 \\ 0 & 1 & 0 \\ 1 & -1 & 2 \end{pmatrix}$ and let $W = \text{range}(Q)$.

   (a) Show that $Q^2 = Q$.
   (b) Is Q the orthogonal projection onto W? Why or why not?

4. Let W be a subspace of the inner product space V and let P be the orthogonal projection of V onto W.
   Let $Q = I - P$. Show that Q is the orthogonal projection of V onto $W^{\perp}$.

5. Let M be the hyperplane in $\mathbf{R}^4$ with equation $a + b - c + 2d = 0$. Find the point of M that is closest to $x = (1, 1, 1, 1)$. What is the distance from x to M?
   (Hint: you may wish to use Problem 5.4.4. and the fact that $M = \mathcal{W}^\perp$ where $\mathcal{W}$ is span$\{(1, 1, -1, 2)\}$.)

$\Diamond$ 6. A group of spies communicates using a code based on a secret matrix S. To send the message vector m to a colleague, a spy multiplies m by S and then adds a random vector r that is orthogonal to the range of S. (The vector r is chosen by the spy and not known by anyone else!) The result is the code vector c which is then sent to the spy's colleague. That is, if the message vector is m, then the vector that the colleague receives is $c = Sm + r$ where r is orthogonal to the range of S. Unfortunately for the spies, you know that the secret matrix is

$$S = \begin{pmatrix} 1 & 1 & 1 & 1 \\ 0 & 1 & 1 & 1 \\ 1 & 0 & 1 & 1 \\ 0 & 1 & 0 & 1 \\ 1 & 1 & 1 & 0 \\ 0 & 0 & 1 & 1 \end{pmatrix}$$

You have intercepted a coded message that you believe will tell the time of the next shipment of missile parts: $c = (31, 75, 65, 2, 15, 0)$. Find the message vector m that the spy sent. (Note: the message words are converted into vectors, 4 letters at a time, using the simple substitution $A \leftrightarrow 1$, $B \leftrightarrow 2$, etc.)

7. The matrix A has QR–factorization

$$A = \begin{pmatrix} 1 & 1 & -3 \\ 1 & 1 & -2 \\ -1 & 0 & 0 \\ -1 & 0 & 1 \end{pmatrix} = \begin{pmatrix} \frac{1}{2} & \frac{1}{2} & -\frac{1}{2} \\ \frac{1}{2} & \frac{1}{2} & \frac{1}{2} \\ -\frac{1}{2} & \frac{1}{2} & -\frac{1}{2} \\ -\frac{1}{2} & \frac{1}{2} & \frac{1}{2} \end{pmatrix} \begin{pmatrix} 2 & 1 & -3 \\ 0 & 1 & -2 \\ 0 & 0 & 1 \end{pmatrix}$$

Use this factorization to find the least squares solution to the system

$$\begin{cases} x & + & y & - & 3z & = & 1 \\ x & + & y & - & 2z & = & 2 \\ -x & & & & & = & 3 \\ -x & & & + & z & = & 0 \end{cases}$$

8.(a) Find the QR factorization for the matrix

$$A = \begin{pmatrix} 1 & -1 & 0 \\ 1 & 2 & 1 \\ 0 & -1 & 1 \\ 0 & 0 & -1 \end{pmatrix}$$

(b) Use the result of part (a) and Theorem 5.15 to find the least squares solution of the following inconsistent system:

$$\begin{cases} x - y = -1 \\ x + 2y + z = 2 \\ - y + z = 1 \\ - z = -1 \end{cases}$$

9. Let A be an $m \times n$ matrix A that has rank n and let let $A = QR$ be a QR–factorization in which Q is $m \times n$ and R is $n \times n$. Prove that QQ' is the orthogonal projection of $\mathbf{R}^m$ onto the range of A.

10. Let U and V be unitary matrices.
    Prove that UV is also a unitary matrix.

11. Let P be the orthogonal projection onto the subspace $\mathcal{W}$ and let $R = I - 2P$.
    (a) Show that R is unitary.
    (b) For u in $\mathcal{W}$ and v in $\mathcal{W}^\perp$, calculate Ru, Rv, and $R(2u + 3v)$.
    (c) Sketch $\mathcal{W}^\perp$, $2u + 3v$, and $R(2u + 3v)$ and explain how R is a "reflection".

12. Find a Householder transformation H that takes the vector $x = (2, 1, 3, -2)$ onto a multiple of the vector $z = (1, 0, 0, 0)$, then compute Hx.

13. Let B be the matrix

$$B = \begin{pmatrix} 2 & 3 \\ -1 & 0 \\ 2 & -1 \end{pmatrix}.$$

Find the QR factorization for B by finding two appropriate Householder transformations. (You may leave Q in the factored form, $Q = H_1 H_2$.)

14. Use the result of exercise 5.4.13. to find the least squares solution of the following inconsistent system:

$$\begin{cases} 2x + 3y = -1 \\ -x = 2 \\ 2x - y = 1 \end{cases}$$

15. Rework problem 5.4.8. using Householder transformations.
    (You may leave Q in factored form.)

5.5   A Project on Circles in Space

In modern manufacturing plants, much of the work is done by robots. While robots are very efficient, they are less cognizant than humans of errors that have been made. To combat this, as part of the quality control process, measurements are taken to compare the actual outcome with the desired outcome. In this instance, the coordinates of ten points on the finished edge of the fuel tank filler tube (that is, a circular cylinder) have been measured. If the part is well made and properly installed, these ten points will lie on a circle whose radius is the radius of the cylinder. If the part has been cut crookedly, these points are on an ellipse and if the piece has been bent, the points are on a bent circle. The question, therefore, is to determine whether ten points whose coordinates are given lie on a circle or whether they do not.[4] These measurements are made after the fuel tank and the attached filler tube have been installed on the vehicle, but before the side panels of the vehicle have been attached. If the end of the filler tube is not circular, the assembly will be removed and replaced before the side panels are installed.

The measured coordinates are the coordinates in three–space of the circle (we hope) with respect to the robot's coordinate system; the filler tube cannot be assumed to have some special relationship to this coordinate system, e.g. the circular edge is unlikely to be horizontal or vertical.

The data points are measured to four digit precision. Thus, it is quite unlikely that they exactly lie on a circle, but that is, in any case, unnecessary: what is necessary is for the cap to fit well enough to prevent leaks. The goal, then, is to determine whether the data points "nearly" lie on a circle or whether they are "far" from being on a circle. We will not formally address the issue of "tolerance" to ask, for example, if the points are within .001 " of being on a circle, although, you may wish to consider such a question.

Exercises 5.5

1. What is a "circle" in three dimensional space? How can a circle be described?
2. Three points determine a circle. Explain why, if we are trying to find a circle that (nearly) passes through ten or more points, it might be inappropriate to choose three of the points in space and find the circle through these three. (Hint: one reason might be analogous to the situation pictured in Figure 5.1.)

---

[4]Thanks are due Professor Michael A. Lachance, University of Michigan at Dearborn, and Dr. Edward Moylan, Ford Motor Company, for their assistance in preparing this material and bringing it to the attention of the author.

3. Decide if the following points in space lie on a circle. If they do, find the center, the radius, and describe the location of the circle in space. If they do not, describe their configuration. In any case, provide some supporting calculations and explain how your calculations justify your answer.

Data Set 1			
	x	y	z
point 1	4.3729	−1.6616	1.5295
point 2	2.0486	−1.7919	4.0798
point 3	2.1505	4.0611	2.1235
point 4	2.1448	4.0583	2.1305
point 5	0.9450	−0.1201	4.7440
point 6	5.2728	1.1598	−0.3320
point 7	4.3190	3.5298	−0.0499
point 8	0.7992	1.9328	4.2537
point 9	1.8906	−1.6627	4.2097
point 10	2.1323	−1.8517	4.0084

4. Decide if the following points in space lie on a circle. If they do, find the center, the radius, and describe the location of the circle in space. If they do not, describe their configuration. In any case, provide some supporting calculations and explain how your calculations justify your answer.

Data Set 2			
	x	y	z
point 1	0.4155	1.4804	−4.1936
point 2	2.5749	7.7438	2.6115
point 3	−0.8669	7.1880	1.0171
point 4	5.3092	6.7052	2.4255
point 5	1.4171	1.0525	−4.3084
point 6	6.4762	5.6856	1.7804
point 7	4.7269	7.0463	2.5828
point 8	−1.7171	6.4185	0.0029
point 9	1.7591	0.9468	−4.3078
point 10	6.4560	5.7093	1.7975

5. Decide if the following points in space lie on a circle. If they do, find the center, the radius, and describe the location of the circle in space. If they do not, describe their configuration. In any case, provide some supporting calculations and explain how your calculations justify your answer.

Data Set 3			
	x	y	z
point 1	0.5040	1.1207	0.4701
point 2	6.3771	2.1594	4.7143
point 3	5.4575	6.9082	0.9256
point 4	6.7997	5.8453	2.7227
point 5	0.9260	4.1247	−1.0883
point 6	2.2393	5.8237	−1.0722
point 7	7.0842	4.6441	3.7244
point 8	0.6881	0.6785	0.9051
point 9	6.0103	6.6889	1.5268
point 10	6.3577	2.1266	4.7189

6. Decide if the following points in space lie on a circle. If they do, find the center, the radius, and describe the location of the circle in space. If they do not, describe their configuration. In any case, provide some supporting calculations and explain how your calculations justify your answer.

Data Set 4			
	x	y	z
point 1	1.7111	0.9193	7.0628
point 2	5.7343	6.0145	−0.2558
point 3	−0.9765	1.1465	1.2708
point 4	5.3876	3.4867	6.5194
point 5	3.7665	2.2732	6.9986
point 6	3.7440	5.1722	−1.5722
point 7	6.0606	6.0993	0.1171
point 8	7.1174	4.9961	5.3769
point 9	−0.7617	1.5549	0.4788
point 10	−0.9473	0.5215	3.1650

6

Eigenvalues and Eigenvectors

6.1   A Dynamics Example

Differential equations are frequently used to describe a physical system. The differential equations arise because the variables of interest are functions of time and their interactions are governed by physical laws that relate the variables and their time derivatives. If the relevant physical principles are linear, such as Hooke's Law concerning springs, Ohm's Law concerning resistance, Faraday's Law concerning inductors, the differential equations that result from describing the system are frequently a system of linear differential equations. Engineers and scientists are concerning with predicting the behavior, over time, of the physical system, given the starting configuration. For example, suppose a physical system is described by the system of differential equations

$$\begin{cases} 4\dfrac{d^2x_1}{dt^2} & = & -12x_1 + 4x_2 \\[2mm] 2\dfrac{d^2x_2}{dt^2} & = & 4x_1 - 4x_2 \end{cases}$$

with initial conditions

$$\begin{cases} x_1(0) & = & 2 \\[1mm] x_2(0) & = & 1 \\[2mm] \dfrac{dx_1}{dt}(0) & = & 0 \\[2mm] \dfrac{dx_2}{dt}(0) & = & 0 \end{cases}$$

The problem is to find functions $x_1(t)$ and $x_2(t)$ so that their values at zero are as specified and the functions and their derivatives satisfy the given differential equations.

In this chapter, we will see that this differential equations problem can be converted to a linear algebra problem, namely, the problem of finding the *eigenvalues* and *eigenvectors* of a matrix and then converting the answer to the linear algebra problem into the solution of the differential equations problem.

6.2   Eigenvalues and Eigenvectors

In this chapter, we will work with some of the most important algebraic and geometric aspects of matrices and linear transformations. We will investigate the existence of special bases for a vector space so that the matrix for a particular transformation will be especially simple and use such bases, when they exist, to solve a variety of problems, for example, systems of ordinary differential equations. Recall that $\mathbf{C}^n$ denotes the set of n-tuples of complex numbers, that is, v is a vector in $\mathbf{C}^n$ if $v = (v_1, v_2, \cdots, v_n)$ for complex numbers $v_1, \cdots, v_n$. We will see shortly why we must work in $\mathbf{C}^n$, not just $\mathbf{R}^n$ in this chapter.

DEFINITION   If A is an $n \times n$ matrix, v is a non–zero vector in $\mathbf{C}^n$, and λ is a number, we say v is an eigenvector *for A with eigenvalue λ if $Av = \lambda v$.*

This definition calls attention to a special relationship between the the matrix A, the vector v, and the number λ. In this chapter we will explore this special relationship. Notice that in the definition, v is *NOT* permitted to be zero, while λ can be any number. This is because the equation $A0 = \lambda 0$ is true for all matrices and all scalars, so the equation does not give any special relationship between A, λ, and 0, but for v not zero, the relationship is special and says something about A, λ, and v, even in the case λ is zero.

Geometrically, an eigenvector is a vector whose image under A is pointing in the same direction as the vector, if the eigenvalue is positive, or the opposite direction, if the eigenvalue is negative.

As an example, suppose A is the matrix

$$A = \begin{pmatrix} 1 & -1 \\ 2 & 4 \end{pmatrix} \tag{6.2.1}$$

Then $u = (1, -1)$ is an eigenvector for A with eigenvalue $\lambda = 2$. Indeed, we can check the definition:

$$Au = \begin{pmatrix} 1 & -1 \\ 2 & 4 \end{pmatrix} \begin{pmatrix} 1 \\ -1 \end{pmatrix} = \begin{pmatrix} 2 \\ -2 \end{pmatrix} = 2u$$

If we are given only the eigenvalue, it is an easy matter to find an eigenvector. To find an eigenvector for the eigenvalue 3 of A, we need to solve the system $Av = 3v$. Writing $v = (x, y)$, $Av = 3v$ is

$$\begin{cases} x & - & y & = & 3x \\ 2x & + & 4y & = & 3y \end{cases}$$

This is the same as $Av - 3v = 0$ or $(A - 3I)v = 0$ which is

$$\begin{cases} -2x & - & y & = & 0 \\ 2x & + & y & = & 0 \end{cases}$$

This system is easily seen to have infinitely many solutions: $y = -2x$ and x is arbitrary. In particular, $v = (1, -2)$ is an eigenvector for the eigenvalue 3 and all other eigenvectors for the eigenvalue 3 are multiples of $(1, -2)$.

On the other hand, if we try to find eigenvectors corresponding to $\lambda = 4$ for the matrix A, we are led to solve the system $(A - 4I)v = 0$ which is

$$\begin{cases} -3x & - & y & = & 0 \\ 2x & & & = & 0 \end{cases}$$

Since this system has only the trivial solution $x = 0$, $y = 0$, we find that 4 is not an eigenvalue, because there are no non–zero vectors v that satisfy the equation $Av = 4v$.

To see the geometry, Figure 6.1 shows unit vectors in the direction of u, v and $e = (1, 0)$ and their images under multiplication by A. It illustrates that Au is pointing in the same direction as u, but is 2 times as long. Similarly, Av points in the same direction as v but is 3 times as long. On the other hand, e is not an eigenvector because Ae is not pointing in the same (or opposite) direction as e

This example and others like it suggest the following theorem.

THEOREM 6.1

If A is an $n \times n$ matrix, then λ is an eigenvalue of A if and only if $A - \lambda I$ is not invertible. Moreover, if λ is an eigenvalue of A, a non–zero vector v is an eigenvector for A with eigenvalue λ if and only if v is in the nullspace of $A - \lambda I$.

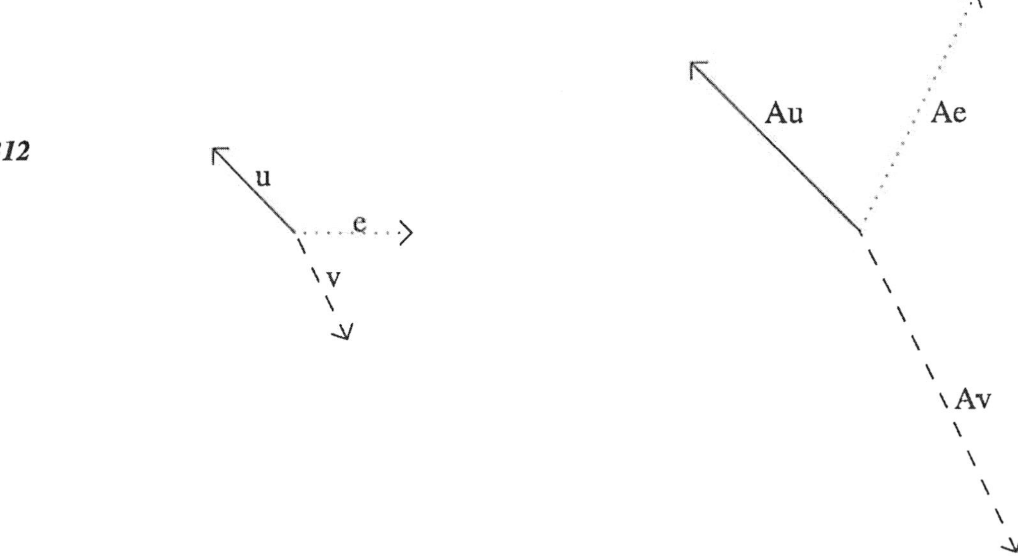

FIGURE 6.1

Vectors and their images under A

PROOF   If v is an eigenvector for A with eigenvalue λ, that is, if $Av = \lambda v$ and v is non-zero, then $Av - \lambda v = 0$ which means $(A - \lambda I)v = 0$. This is the definition of v being in the nullspace of $A - \lambda I$ and by Theorem 3.63, the fact that the homogeneous system $(A - \lambda I)X = 0$ has a non-zero solution is equivalent to the matrix $A - \lambda I$ being not invertible.

Conversely, if $A - \lambda I$ is not invertible, Theorem 3.63 implies the system $(A - \lambda I)X = 0$ has a non-zero solution, say v is such a solution. Then $(A - \lambda I)v = 0$ which means $Av - \lambda v = 0$ or $Av = \lambda v$ and v is an eigenvector for A with eigenvalue λ.   ∎

DEFINITION   If A is an $n \times n$ matrix and λ is an eigenvalue of A, the eigenspace of A corresponding to λ is the nullspace of $A - \lambda I$.

Notice that the eigenspaces are subspaces and consist of eigenvectors together with the vector 0. In the example above, the eigenspace corresponding to $\lambda = 2$ is the subspace with basis $\{(1, -1)\}$ and the eigenspace corresponding to $\lambda = 3$ is the subspace with basis $\{(1, -2)\}$. When answering the question "What are the eigenvectors of A?", what is usually wanted (and *always* in this book) is a basis for each of the eigenspaces for A.

We have seen that not every number is an eigenvalue for every matrix, so the question becomes, "How do we find the eigenvalues of a matrix?" The traditional approach is with the use of determinants, although in practice, this is not always possible.

Since the matrix $A - \lambda I$ is invertible if and only if $\det(A - \lambda I)$ is non–zero, the eigenvalues of A are those numbers λ for which $\det(A - \lambda I) = 0$. In our example,

$$\det(A - \lambda I) = \det \begin{pmatrix} 1 - \lambda & -1 \\ 2 & 4 - \lambda \end{pmatrix} = (1 - \lambda)(4 - \lambda) + 2$$

$$- \lambda^2 - 5\lambda + 6 = (\lambda - 2)(\lambda - 3)$$

Thus, λ is an eigenvalue of A if and only if $\det(A - \lambda I) = 0$ which is if and only if $(\lambda - 2)(\lambda - 3) = 0$, so the eigenvalues of A are 2 and 3.

Just as in this example, $\det(A - \lambda I)$ is a polynomial for every matrix A. Indeed, the Laplace expansion formula for the determinant (Definition 2.6) shows that the determinant of any matrix is a sum of products of entries from the matrix. Since each entry of the matrix $A - \lambda I$ is either a number or a first degree polynomial in λ, a sum of products of entries of this matrix must be a polynomial. Moreover, since there are exactly n entries containing λ in $A - \lambda I$ and no entries are multiplied by themselves in expanding the determinant, no term in the expansion has degree more than n. In fact, the only term to have degree n is the product of the diagonal entries, so this polynomial has degree n.

DEFINITION *If A is an $n \times n$ matrix, the* characteristic polynomial *of A is the polynomial* $\det(A - \lambda I)$.

This definition, together with the observation that a matrix is invertible if and only if its determinant is non-zero, results in the following theorem.

THEOREM 6.2

For A an $n \times n$ matrix, λ is an eigenvalue of A if and only if λ is a root of the characteristic polynomial.

Now since the characteristic polynomial of an $n \times n$ matrix is a polynomial of degree n and a polynomial of degree n has at most n roots whether we work with the real or complex numbers we conclude that an $n \times n$ matrix has at most n eigenvalues. Moreover, since every polynomial has at least one complex root, every matrix has a least one, possibly complex, eigenvalue.

The development so far allows us to identify a strategy for finding the eigenvectors and eigenvalues of a matrix. This strategy is the one typically employed in finding the eigenvalues and eigenvectors of a small matrix by hand. At the end of this section, we will discuss difficulties with this approach when applied to solving large problems by machine and mention more practical ways to find the eigenvalues and eigenvectors of a matrix, although we will not be able to discuss these in detail.

Strategy for finding eigenvalues and eigenvectors

1. Find the characteristic polynomial of A by computing $\det(A - \lambda I)$.

2. Find the roots of the characteristic polynomial; these are the eigenvalues of A.

3. For each eigenvalue λ_j of A, find a basis for the nullspace of $A - \lambda_j I$; these are the eigenvectors of A.

EXAMPLE 6.3

Find the eigenvalues and bases for the eigenspaces of the matrix

$$A = \begin{pmatrix} 1 & -1 & 0 \\ -1 & 2 & -1 \\ 0 & -1 & 1 \end{pmatrix}$$

SOLUTION The characteristic polynomial for the matrix A is

$$\det(A - \lambda I) = \det \begin{pmatrix} 1-\lambda & -1 & 0 \\ -1 & 2-\lambda & -1 \\ 0 & -1 & 1-\lambda \end{pmatrix}$$

$$= (1-\lambda)(2-\lambda)(1-\lambda) + 0 + 0 + 0 - (1-\lambda) - (1-\lambda) = -\lambda^3 + 4\lambda^2 - 3\lambda$$

$$= -\lambda(\lambda - 3)(\lambda - 1)$$

so the eigenvalues of A are 0, 1, and 3. For each of these eigenvalues, we need to find a basis for the corresponding eigenspace.

• For the eigenvalue $\lambda_1 = 0$:

We need to solve the system $(A - 0I)X = 0$. Writing $X = (x, y, z)$, this is

$$\begin{cases} x & - & y & & & = & 0 \\ -x & + & 2y & - & z & = & 0 \\ & & -y & + & z & = & 0 \end{cases}$$

Replacing the second equation by the second plus the first gives the system

$$\begin{cases} x & - & y & & & = & 0 \\ & & y & - & z & = & 0 \\ & & -y & + & z & = & 0 \end{cases}$$

We see the last two equations are equivalent, so we can solve for x and z in terms of y to get the eigenvectors of A corresponding to the eigenvalue $\lambda_1 = 0$. These are (x, y, z) such that $x = y$, $z = y$, and y is arbitrary. In other words, $u = (1, 1, 1)$ is a basis for the eigenspace corresponding to 0.

• For the eigenvalue $\lambda_2 = 1$.
We need to solve the system $(A - 1I)X = 0$. Writing $X = (x, y, z)$, this is

$$\begin{cases} & - y & & = 0 \\ -x & + y & - z & = 0 \\ & - y & & = 0 \end{cases}$$

The first and third equations give $y = 0$, and replacing y by 0 in the second equation, we can solve for x in terms of z to get the eigenvectors of A corresponding to the eigenvalue $\lambda_2 = 1$. We get (x, y, z) such that $x = -z$, $y = 0$, and z is arbitrary so $v = (-1, 0, 1)$ is a basis for the eigenspace corresponding to 1.

• For the eigenvalue $\lambda_3 = 3$.
Finally, we need to solve the system $(A - 3I)X = 0$. Writing $X = (x, y, z)$, this is

$$\begin{cases} -2x & - y & & = 0 \\ -x & - y & - z & = 0 \\ & - y & - 2z & = 0 \end{cases}$$

Solving for x and z in terms of y gives (x, y, z) such that $x = z = -y/2$ and y is arbitrary so $w = (-1, 2, -1)$ is a basis for the eigenspace corresponding to 3. $\quad\Box$

MATLAB finds eigenvectors and eigenvalues with the command `[v,d]=eig(A)` in which v is an $n \times n$ matrix whose columns are eigenvectors of A and d is a diagonal $n \times n$ matrix whose diagonal entries are the eigenvalues of A. The command `eig(A)` produces a list of the eigenvalues. We can redo the example above with MATLAB.

```
> A=[1 -1 0; -1 2 -1; 0 -1 1]

A =
      1    -1     0
     -1     2    -1
      0    -1     1
```

```
> eig(A)

ans =
    0.0000
    1.0000
    3.0000
```

Thus, the eigenvalues of A are 0, 1, and 3 as we found above.

```
> [v d]=eig(A)

v =
    0.5774      0.7071      0.4082
    0.5774     -0.0000     -0.8165
    0.5774     -0.7071      0.4082

d =
    0.0000          0           0
         0     1.0000           0
         0          0      3.0000
```

In this calculation, the columns of v are the eigenvectors of A corresponding to the eigenvalues which occur on the diagonal of d, with the first column of v corresponding to the first diagonal entry of d and so on. Notice that the actual basis for the eigenspace found by MATLAB is different than what we found, but the answers are equivalent. MATLAB always gives eigenvectors with length 1.

The following example illustrates the fact that real matrices do not necessarily have real eigenvalues. Of course, since real numbers are also complex numbers, every real matrix must have a complex eigenvalue. Moreover, since the characteristic polynomial of a real matrix has real coefficients and since the non–real roots of a polynomial with real coefficients occur in complex conjugate pairs, the eigenvalues of a real matrix are either real or occur in complex conjugate pairs.

EXAMPLE 6.4

Find the eigenvalues and bases for the eigenspaces of the matrix

$$B = \begin{pmatrix} 0 & 1 \\ -1 & 0 \end{pmatrix}$$

SOLUTION The characteristic polynomial is

$$\det(B - \lambda I) = \det \begin{pmatrix} -\lambda & 1 \\ -1 & -\lambda \end{pmatrix} = \lambda^2 + 1$$

In particular, we see that *NO* real numbers are roots of the characteristic polynomial, so B has no *REAL* eigenvalues.

On the other hand, if we considered B as a matrix on $\mathbf{C}^2$, i and $-i$ are roots of the characteristic polynomial.

• For the eigenvalue $\lambda_1 = i$.

We need to solve the system $(B - iI)X = 0$. Writing $X = (x, y)$, this is

$$\begin{cases} -ix & + & y & = & 0 \\ -x & - & iy & = & 0 \end{cases}$$

Multiplying the second equation by i gives $-ix - i^2 y = 0$, or, since $i^2 = -1$, $-ix + y = 0$ and we see that the two equations are multiples of each other. Solving for y in terms of x gives (x, y) such that $y = ix$ and x is arbitrary so $u = (1, i)$ is a basis for the eigenspace corresponding to $\lambda_1 = i$.

(As a check, we can multiply B times u to get

$$Bu = \begin{pmatrix} 0 & 1 \\ -1 & 0 \end{pmatrix} \begin{pmatrix} 1 \\ i \end{pmatrix} = \begin{pmatrix} i \\ -1 \end{pmatrix} = i \begin{pmatrix} 1 \\ i \end{pmatrix}$$

as it is supposed to be.)

• For the eigenvalue $\lambda_2 = -i$.

We need to solve the system $(B - (-i)I)X = 0$. Writing $X = (x, y)$, this is

$$\begin{cases} ix & + & y & = & 0 \\ -x & + & iy & = & 0 \end{cases}$$

Again we see that the second equation is i times the first, so they are equivalent. Solving for y in terms of x gives $y = -ix$ and x is arbitrary so $v = (1, -i)$ is a basis for the eigenspace corresponding to $\lambda_2 = -i$.

(This should not be a surprise! Since the matrix B has real entries and the eigenvalue $\lambda_2 = -i$ is the complex conjugate of $\lambda_1 = i$, the eigenvectors corresponding to λ_2 should be complex conjugates of the eigenvectors corresponding to λ_1.)

MATLAB has no difficulty in adjusting to this situation.

```
> B=[0 1; -1 0]

B =
        0        1
       -1        0
```

```
> [vb db]=eig(B)

vb =
   0.7071               0.7071
        0 + 0.7071i          0 - 0.7071i


db =
        0 + 1.0000i          0
        0                    0 - 1.0000i
```

which is, up to normalization, the answer we obtained above. $\square$

We have noted that an $n \times n$ matrix cannot have more than n eigenvalues, but it can have fewer.

EXAMPLE 6.5

Find the eigenvalues and bases for the eigenspaces of the matrix

$$C = \begin{pmatrix} 4 & -8 & 10 \\ -3 & 9 & -10 \\ -3 & 8 & -9 \end{pmatrix}$$

The characteristic polynomial is

$$\det(C - \lambda I) = \det \begin{pmatrix} 4-\lambda & -8 & 10 \\ -3 & 9-\lambda & -1 \\ -3 & 8 & -9-\lambda \end{pmatrix}$$

$$= -(\lambda^3 - 4\lambda^2 + 5\lambda - 2) = -(\lambda - 1)^2(\lambda - 2)$$

Thus, the eigenvalues of C are 1 and 2.

To find the eigenvectors associated with the eigenvalue 1, we need to solve the system $(C - I)X = 0$, or writing $X = (x, y, z)$, we must solve

$$\begin{cases} 3x & - & 8y & + & 10z & = & 0 \\ -3x & + & 8y & - & 10z & = & 0 \\ -3x & + & 8y & - & 10z & = & 0 \end{cases}$$

Clearly, the solution of this system is $x = (8y - 10z)/3$ and y and z arbitrary. If we choose $y = 3$, $z = 0$, and $y = 0$, $z = 3$, we get $(8, 3, 0)$ and $(-10, 0, 3)$ as a basis for the eigenspace of C corresponding to the eigenvalue 1, that is, we have a two dimensional eigenspace.

Similarly, to find the eigenvectors for the eigenvalue 2, we must solve the system $(C - 2I)X = 0$. Solving this system, we get $(1, -1, -1)$ is a basis for the eigenspace corresponding to the eigenvalue 2.

Let's see how MATLAB handles this case.

```
> C=[4 -8 10; -3 9 -10; -3 8 -9]

C =
       4      -8      10
      -3       9     -10
      -3       8      -9

> [vc dc]=eig(C)

vc =
     0.5774     0.4264    -0.8658
    -0.5774    -0.6396    -0.4839
    -0.5774    -0.6396    -0.1273


dc =
     2.0000          0          0
          0     1.0000          0
          0          0     1.0000
```

MATLAB found, as we did, that the eigenvalues of C are 1 and 2, and that, up to normalization, the basis for the eigenspace corresponding to 2 is the same as we found. It also found that the eigenspace corresponding to $\lambda = 1$ is two dimensional. It is less clear that our answers agree, but, as the following calculation with the eigenvectors MATLAB found and the eigenvectors we found, they do agree:

```
> rank([vc(:,2:3) [8;3;0] [-10;0;3]])

ans =
     2
```

[]

However, the situation is not always quite this straightforward. Sometimes, the number of eigenvectors is also smaller than we would hope for.

EXAMPLE 6.6

Find the eigenvalues and bases for the eigenspaces of the matrix

$$D = \begin{pmatrix} -9 & 5 & -1 \\ -17 & 10 & -1 \\ 13 & -6 & 3 \end{pmatrix}$$

The characteristic polynomial is

$$\det(D - \lambda I) = \det \begin{pmatrix} -9-\lambda & 5 & -1 \\ -17 & 10-\lambda & -1 \\ 13 & -6 & 3-\lambda \end{pmatrix}$$

$$= (-9-\lambda)(10-\lambda)(3-\lambda) - 65 + 221 - 6(-9-\lambda) + 13(10-\lambda) + 85(3-\lambda)$$

$$= 2 - 5\lambda + 4\lambda^2 - \lambda^3 = (2-\lambda)(1-\lambda)^2$$

So the eigenvalues of the matrix D are 1 and 2. Solving the system $(D - I)v = 0$ we see that a basis for the eigenspace corresponding to $\lambda = 1$ is $w_1 = (4, 7, -5)$ and a basis for the eigenspace corresponding to $\lambda = 2$ is $(1, 2, -1)$.

MATLAB has much more trouble with this situation.

```
> D=[ -9 5 -1; -17 10 -1; 13 -6 3]

D =
     -9       5      -1
    -17      10      -1
     13      -6       3

> [vd dd]=eig(D)

vd =
    0.4216    -0.4216     0.4082
    0.7379    -0.7379     0.8165
   -0.5270     0.5270    -0.4082

dd =
```

```
 1.0000          0              0
      0     1.0000              0
      0          0         2.0000
```

In this case, MATLAB has identified the eigenvalues of D as 1 and 2, but it has associated the eigenvalue 1 with two columns of vd. In general, these would be linearly independent eigenvectors for the (repeated) eigenvalue 1, however, in this case, the eigenvectors in the first and second columns appear to be negatives of each other, that is, they appear to be dependent. Of course, having worked the problem by hand, we know that there is only a one dimensional eigenspace associated with the eigenvalue 1. Because the eigenvalue 1 is a double root of the characteristic polynomial but the eigenspace is only one dimensional, numerical solution of this eigenvalue problem is unstable and the results tend to be somewhat confusing. In fact, if we check to see if the first two columns of vd are linearly dependent, we find that MATLAB thinks they are not:

```
> vd(:,1)+vd(:,2)

ans =
   1.0e-08 *

   0.9248
  -0.7825
  -0.3557
```

Even though their sum is *small*, it is non-zero. This happened because in this problem, the errors in the numerical solution are more severe than usual. In situations where there may be double roots of the characteristic polynomial, care must be used in interpreting the results of any numerical algorithm.   ▯

There is terminology to describe this situation:

DEFINITION *If A is an $n \times n$ matrix with characteristic polynomial*

$$p(\lambda) = \pm(\lambda - \lambda_1)^{r_1}(\lambda - \lambda_2)^{r_2} \cdots (\lambda - \lambda_k)^{r_k}$$

where the eigenvalues λ_1, λ_2, $\cdots$, λ_k are distinct, then r_j is called the algebraic multiplicity *of the eigenvalue λ_j. The* geometric multiplicity *of the eigenvalue λ_j is the dimension of the eigenspace corresponding to λ_j, that is the dimension of the nullspace of $A - \lambda_j I$.*

In Example 6.5, we found that the algebraic multiplicity and geometric multiplicity of the eigenvalue $\lambda_1 = 1$ for C are both 2 and the algebraic and geometric multiplicities of the eigenvalue $\lambda_2 = 2$ for C are both 1. However, in Example 6.6, we found that the algebraic multiplicity of the eigenvalue $\lambda_1 = 1$ for D is 2 but the geometric multiplicity of this eigenvalue is only 1. The algebraic and geometric multiplicities of the eigenvalue $\lambda_2 = 2$ for D are both 1.

Although its proof is beyond the scope of this book, the following theorem gives the relationship of the algebraic and geometric multiplicities of the eigenvalues of a matrix.

THEOREM 6.7

If A is an $n \times n$ matrix and μ is an eigenvalue of A, then the geometric multiplicity of μ is less than or equal to the algebraic multiplicity of μ.

The examples above have shown that both equality and inequality can occur in this theorem. In order for the number μ to be an eigenvalue at all, the algebraic and geometric multiplicities must be at least 1. Moreover, since the degree of the characteristic polynomial of an $n \times n$ matrix is n, the sum of the algebraic multiplicities of the eigenvalues of an $n \times n$ matrix is exactly n.

The difficulty of Example 6.6 occurs because the geometric multiplicity of $\lambda_1 = 1$ for D is less and not equal to the algebraic multiplicity of $\lambda_1 = 1$. For the same reason, MATLAB found it difficult to deal with D. Fortunately, this phenomenon is rare: almost all matrices have the algebraic multiplicities equal to the geometric multiplicities for all the eigenvalues. We will investigate this situation somewhat more in Sections 6.5 and 6.7, but we will not be able to explore it completely in this text.

The strategy we employ in this book to find eigenvectors and eigenvalues, find the characteristic polynomial, find its roots to find the eigenvalues, and for each eigenvalue solve the eigenvalue equation for the associated eigenvectors, is not practical for most problems of importance for engineering and science. Indeed, only the last step can be carried out easily! Finding the characteristic polynomial by means of expanding a determinant in which some of the entries are symbolic expressions is relatively difficult, although there are indirect methods for finding the coefficients of the characteristic polynomial that will avoid this. The biggest roadblock in this strategy is that finding the roots of a polynomial with large degree is very difficult. Problems to be done by hand in this or other textbooks are either 2×2 matrices whose eigenvalues can be found from the quadratic formula or they have been specially constructed so that the roots are easily guessed. In fact, practice is quite the opposite. One popular method for finding the roots of a polynomial is to construct a matrix whose characteristic polynomial is the given polynomial and finding the eigenvalues of the matrix — exactly reversing the process we are using! So how do programs like MATLAB find eigenvalues? The eigenvectors and eigenvalues are found simultaneously by an iterative (recursive) algorithm, typically a modification of the QR–algorithm (which is beyond the scope of this book). That is, approximations to the eigenvectors and eigenvalues are found, then at the next stage, better approximations are found, and so on, until the answers are considered to be acceptable. This is analogous to Newton's method for finding roots of polynomials.

Exercises 6.2

It would be most helpful to work each of the Exercises 6.2.7., 6.2.10., 6.2.11., and 6.2.12. both by hand and by machine to be able to compare the forms of the answers in simple situations. In each of Exercises 6.2.7., 6.2.10., 6.2.11., 6.2.12., and 6.2.19., find the algebraic and geometric multiplicities of all the eigenvalues.

1. Let $u = (1, 1, 1, 1, 1)$, $v = (0, 1, 0, 1, 0)$, $w = (1, 0, 1, 0, 1)$, $x = (1, 1, 0, 0, 0)$, and $y = (0, 0, 1, 1, 1)$. Some of these vectors are eigenvectors for the matrix

$$E = \begin{pmatrix} 6 & -3 & -1 & 4 & -2 \\ 1 & 2 & -1 & 2 & 0 \\ 3 & -3 & 2 & 4 & -2 \\ 1 & -1 & -1 & 5 & 0 \\ 2 & -2 & -1 & 3 & 2 \end{pmatrix}$$

For each vector, decide if it is an eigenvector of E or not, and if it is, find the corresponding eigenvalue.

2. Let $u = (1, 1, 1, 1, 1)$, $v = (0, 1, 0, 1, 0)$, $w = (1, 1, 0, 1, 1)$, $x = (1, 0, 0, 0, 1)$, and $y = (0, 0, 1, 1, 1)$. Some of these vectors are eigenvectors for the matrix

$$A = \begin{pmatrix} 0 & -2 & 4 & 2 & 2 \\ 1 & -3 & 2 & 7 & -1 \\ 4 & -4 & 6 & 4 & -4 \\ 1 & -1 & 2 & 5 & -1 \\ 2 & -2 & 4 & 2 & 0 \end{pmatrix}$$

For each vector, decide if it is an eigenvector of A or not, and if it is, find the corresponding eigenvalue.

3. The eigenvalues of

$$P = \begin{pmatrix} 2 & 4 \\ 1 & -1 \end{pmatrix}$$

are 3 and -2. Find an eigenvector for each eigenvalue of P.

4. The eigenvalues of

$$T = \begin{pmatrix} 17 & -9 \\ 30 & -16 \end{pmatrix}$$

are -1 and 2. Find a basis for each eigenspace of T.

5. The eigenvalues of

$$Q = \begin{pmatrix} -4 & -12 & -6 \\ 3 & 8 & 3 \\ -3 & -6 & -1 \end{pmatrix}$$

are -1 and 2. Find a basis for each eigenspace of Q.

6. The eigenvalues of

$$R = \begin{pmatrix} 6 & 3 & 1 \\ -3 & 0 & -1 \\ -6 & -6 & 1 \end{pmatrix}$$

are 3 and 1. Find a basis for each eigenspace of R.

Find eigenvalues and bases for the eigenspaces for each of the following matrices.

7. $\begin{pmatrix} 2 & 3 \\ -1 & 6 \end{pmatrix}$

8. $\begin{pmatrix} 5 & -6 \\ 3 & -4 \end{pmatrix}$

9. $\begin{pmatrix} 2 & -1 \\ 1 & 2 \end{pmatrix}$

10. $\begin{pmatrix} 0 & 2 & -4 \\ 2 & -3 & -2 \\ -4 & -2 & 0 \end{pmatrix}$

11. $\begin{pmatrix} 2 & 2 & -1 \\ -1 & 0 & 1 \\ 0 & 1 & 1 \end{pmatrix}$

12. $\begin{pmatrix} 2 & 1 & 0 \\ -1 & 1 & 1 \\ -3 & -4 & 0 \end{pmatrix}$

13. $\begin{pmatrix} 8 & -10 \\ 5 & -7 \end{pmatrix}$

14. $\begin{pmatrix} 1 & -3 \\ 3 & -5 \end{pmatrix}$

15. $\begin{pmatrix} 2 & -1 & 1 \\ 0 & 3 & 0 \\ 1 & 1 & 2 \end{pmatrix}$

16. $\begin{pmatrix} -8 & 14 & 22 \\ -4 & 4 & 8 \\ -1 & 4 & 5 \end{pmatrix}$

17. $\begin{pmatrix} 2 & -5 & -4 \\ -2 & 3 & 3 \\ 4 & -8 & -7 \end{pmatrix}$

18. $\begin{pmatrix} -2 & 10 & -1 \\ -1 & 5 & 0 \\ 3 & -7 & 3 \end{pmatrix}$

◇ 19. Find the eigenvalues and bases for the eigenspaces of the matrix

$$\begin{pmatrix} 36 & -24 & 3 & -11 & 4 \\ 49 & -34 & 1 & -18 & 6 \\ -25 & 15 & -4 & 5 & -1 \\ 13 & -6 & 5 & 1 & -1 \\ 37 & -24 & 1 & -12 & 2 \end{pmatrix}$$

Be careful to interpret the answers given by your machine correctly!

20. Find a 2×2 matrix with eigenvalues -1 and 2 and eigenvectors $(2, 1)$ and $(3, 2)$.

21. Find a 3×3 matrix with eigenvalues 1 with algebraic multiplicity 1 and 3 with algebraic multiplicity 2 and $(1, -1, 2)$, $(-12 - 3)$, and $(1, -2, 2)$.

22.(a) Prove that if E is an $n \times n$ matrix, then the eigenvalues of E^t, the transpose of E, are the same as those of E.

   (b) Give an example of a 2×2 matrix, E, such that the eigenvectors of E^t are different from those of E.

6.3 Systems of Differential Equations

A differential equation is an equation that involves an unknown function and its derivatives. To solve the differential equation means to find all functions that satisfy the equation. It is very common for relationships between physical quantities to be expressed as a differential equation. For example, the fall of an object under the influence of gravity is governed by the differential equation

$$\frac{d^2 x}{dt^2} = -g$$

where x is the height of the object as a function of time and g is the (constant) acceleration due to gravity. The growth of a population that has unlimited resources can be modeled by the equation

$$\frac{dy}{dt} = ay$$

where y is the size of the population at time t and a is a constant associated with the rate at which the population is growing (it is $\ln 2$ divided by the time for the population to double). We learned how to solve both of these differential equations in calculus class! To solve the first, we simply integrated twice to get

$$x(t) = -\frac{g}{2}t^2 + v_0 t + x_0$$

where v_0 is the initial velocity and x_0 is the initial height. To solve the second, we divided by y, integrated, and exponentiated to get

$$y(t) = Ce^{at}$$

where C is the initial population. In both cases, we have found all functions that satisfy the given equation. The population equation $\frac{dy}{dt}(t) = ay(t)$, and its solution $y(t) = Ce^{at}$ for any constant C, is a prototype of a class of differential equations that can be solved using matrix techniques.

The differential equation $2\frac{d^2y}{dt^2} - 5\frac{dy}{dt} + 3y = 0$ is called *linear* because no powers or roots or other complicated functions of y occur in the equation, or better, because if $y(t) = f(t)$ and $y(t) = g(t)$ both solve the equation, so will $f(t) + g(t)$ and so will $\alpha f(t)$ for any constant α. Because 2, -5, and 3 are numbers rather than non-trivial functions of t, the equation $2\frac{d^2y}{dt^2} - 5\frac{dy}{dt} + 3y = 0$ is said to have *constant coefficients*. The second derivative of the unknown function y occurs, but no higher order derivative, so the equation is said to be a second order equation. (It is customary, since we understand that y is a function of t in this situation, to omit the explicit mention of the independent variable t: often one writes $2\frac{d^2y}{dt^2} - 5\frac{dy}{dt} + 3y = 0$ rather than $2\frac{d^2y}{dt^2}(t) - 5\frac{dy}{dt}(t) + 3y(t) = 0$ because the latter equation conveys little more information than the former, but is much more difficult to read or write. We will usually follow this convention, but will make the independent variable explicit when it is helpful to do so.)

More complicated relationships are frequently expressed as a system of differential equations involving several unknown functions and their derivatives. In this section, we will consider solutions of systems of first order linear differential equations with constant coefficients, that is, a collection of equations involving several unknown functions and their first derivatives in which each equation is linear and in which all the coefficients occurring in the equations are constants.

A very easy example of the sort of problem we will consider is the system

$$\begin{cases} \dfrac{dy_1}{dt} = 3y_1 \\ \dfrac{dy_2}{dt} = \quad\quad - 4y_2 \end{cases} \tag{6.3.2}$$

We can see a solution immediately: $y_1(t) = c_1 e^{3t}$ and $y_2(t) = c_2 e^{-4t}$. Indeed, this hardly deserves to be called a system of differential equations because the variables are *uncoupled*, that is, they do not interact with each other.

A somewhat harder example of the kind of problem we want to study is the system

$$\begin{cases} \dfrac{dy_1}{dt} = y_1 - y_2 \\[2mm] \dfrac{dy_2}{dt} = 2y_1 + 4y_2 \end{cases} \tag{6.3.3}$$

in which the unknown functions are coupled. The theory we develop is analogous the easy case above, $\dfrac{dy}{dt} = ay$, but to see this, we must develop the analogy by changing notation. If we write the vector function $Y = (y_1, y_2)$ (short for $Y(t) = (y_1(t), y_2(t))$) in which the two unknown functions are the coordinates, then the rules for taking derivatives of vector functions specify that we should take the derivatives componentwise, so that $\dfrac{dY}{dt} = (\dfrac{dy_1}{dt}, \dfrac{dy_2}{dt})$. In this notation, the system above becomes

$$\frac{dY}{dt} = AY$$

where

$$A = \begin{pmatrix} 1 & -1 \\ 2 & 4 \end{pmatrix}$$

Indeed, for this coefficient matrix A,

$$AY(t) = \begin{pmatrix} 1 & -1 \\ 2 & 4 \end{pmatrix} \begin{pmatrix} y_1(t) \\ y_2(t) \end{pmatrix} = \begin{pmatrix} y_1(t) - y_2(t) \\ 2y_1(t) + 4y_2(t) \end{pmatrix}$$

so the equation $\dfrac{dY}{dt} = AY$ is Equation (6.3.3).

The solutions of the equation $\dfrac{dy}{dt} = ay$ are a constant times some exponential function; we will seek the same sort of solutions to the system represented as $\dfrac{dY}{dt} = AY$, except that because Y is a vector function, we will look for solutions that are an exponential function times some constant *vector*! The solutions of the system of Equation (6.3.2) can be written this way; one solution was

$$Y(t) = e^{3t} \begin{pmatrix} 1 \\ 0 \end{pmatrix}$$

and another solution was

$$Y(t) = e^{-4t} \begin{pmatrix} 0 \\ 1 \end{pmatrix}$$

If $Y(t) = e^{\lambda t}v$ is a solution of the differential equation $\dfrac{dY}{dt} = AY$ for some number λ and some constant vector v, then

$$\frac{dY}{dt} = \frac{d}{dt}\left(e^{\lambda t}v\right) = \left(\frac{d}{dt}e^{\lambda t}\right)v = \lambda e^{\lambda t}v$$

On the other hand,

$$\frac{dY}{dt} = AY(t) = A\left(e^{\lambda t}v\right) = e^{\lambda t}(Av)$$

so we must have

$$e^{\lambda t}(Av) = \lambda e^{\lambda t}v$$

for all values of t. Since $e^{\lambda t}$ is never zero, we can divide by it and conclude that

$$Av = \lambda v$$

If v is not the zero vector, which would imply Y is the zero function, this means that v is an eigenvector for A with eigenvalue λ!

Moreover, the steps can be reversed: if v is an eigenvector for A with eigenvalue λ and $Y(t) = e^{\lambda t}v$, then

$$\frac{dY}{dt} = \lambda e^{\lambda t}v = e^{\lambda t}(\lambda v) = e^{\lambda t}(Av) = A\left(e^{\lambda t}v\right) = AY(t)$$

so this vector function is a solution of the system $\dfrac{dY}{dt} = AY$.

In Section 6.2, we found that A has eigenvalues 2 and 3 associated with the eigenvectors $(1, -1)$ and $(1, -2)$. The above calculations show that

$$U(t) = e^{2t} \begin{pmatrix} 1 \\ -1 \end{pmatrix} = \begin{pmatrix} e^{2t} \\ -e^{2t} \end{pmatrix}$$

is a solution of the system and so is

$$V(t) = e^{3t} \begin{pmatrix} 1 \\ -2 \end{pmatrix} = \begin{pmatrix} e^{3t} \\ -2e^{3t} \end{pmatrix}$$

We can check these in Equation (6.3.3) directly: $U(t) = (u_1(t), u_2(t))$ where $u_1(t) = e^{2t}$ and $u_2(t) = -e^{2t}$. So

$$\frac{du_1}{dt} = 2e^{2t} = e^{2t} - (-e^{2t}) = u_1(t) - u_2(t)$$

and

$$\frac{du_2}{dt} = -2e^{2t} = 2e^{2t} + 4(-e^{2t}) = 2u_1(t) + 4u_2(t)$$

so U satisfies the system of Equation (6.3.3). Similarly,

$$\frac{dv_1}{dt} = 3e^{3t} = e^{3t} - (-2e^{3t}) = v_1(t) - v_2(t)$$

and

$$\frac{dv_2}{dt} = -6e^{3t} = 2e^{3t} + 4(-2e^{3t}) = 2v_1(t) + 4v_2(t)$$

so $V(t)$ satisfies the system of Equation (6.3.3).

Now, as we noted earlier, the $\dfrac{dY}{dt} = AY$ is a linear system; the property of the theorem below, called the *Principle of Superposition*, justifies the name. Another way to state this theorem is to say that the set of solutions of the system $\dfrac{dY}{dt} = AY$ is a subspace.

THEOREM 6.8

If A is an $n \times n$ matrix and U and V are solutions of the system $\dfrac{dY}{dt} = AY$, then for all numbers α and β, the function $W = \alpha U + \beta V$ is also a solution.

PROOF   If $W(t) = \alpha U(t) + \beta V(t)$, then $\dfrac{dW}{dt} = \alpha \dfrac{dU}{dt} + \beta \dfrac{dV}{dt}$. Since U and V are solutions of $\dfrac{dY}{dt} = AY$, the terms in this expression can be replaced by AU and AV respectively. Thus,

$$\frac{dW}{dt} = \alpha \frac{dU}{dt} + \beta \frac{dV}{dt} = \alpha AU(t) + \beta AV(t) = A(\alpha U(t) + \beta V(t)) = AW(t)$$

which is the conclusion.   ∎

To continue the example above, since

$$U(t) = e^{2t} \begin{pmatrix} 1 \\ -1 \end{pmatrix} \text{ and } V(t) = e^{3t} \begin{pmatrix} 1 \\ -2 \end{pmatrix}$$

are solutions, taking $\alpha = -4$ and $\beta = 5$, the theorem says

$$W(t) = -4U(t) + 5V(t) = -4e^{2t} \begin{pmatrix} 1 \\ -1 \end{pmatrix} + 5e^{3t} \begin{pmatrix} 1 \\ -2 \end{pmatrix}$$

$$= \begin{pmatrix} -4e^{2t} + 5e^{3t} \\ 4e^{2t} - 10e^{3t} \end{pmatrix}$$

is a solution. To check this,

$$\frac{dw_1}{dt} = -8e^{2t} + 15e^{3t}$$

and

$$w_1(t) - w_2(t) = (-4e^{2t} + 5e^{3t}) - (4e^{2t} - 10e^{3t}) = -8e^{2t} + 15e^{3t}$$

so the first equation is satisfied. Also

$$\frac{dw_2}{dt} = 8e^{2t} - 30e^{3t}$$

and

$$2w_1(t) + 4w_2(t) = 2(-4e^{2t} + 5e^{3t}) + 4(4e^{2t} - 10e^{3t}) = 8e^{2t} - 30e^{3t}$$

so the second equation of the system holds.

Because the derivative of a constant function is zero, some information about a function is lost in taking the derivative, and many differential equations have an infinite number of solutions. For example, there are infinitely many solutions of the equation $\dfrac{dy}{dt} = 2y$: in this case $y(t) = ce^{2t}$ is a solution of the differential equation for any constant c. In cases where there are an infinite number of solutions, sometimes some extra information is given to specify which solution is wanted, for example, the value of the function at some time might be given. If we want the function y such that $\dfrac{dy}{dt} = 2y$ and $y(0) = 5$, we know from calculus that the only solution is $y(t) = 5e^{2t}$. The extra condition $y(0) = 5$ is called an initial condition and a problem in which a system of differential equations is given together with the values of the unknown functions at some time is called an *initial value problem*.

In the problem we considered above, we can add an initial condition for each function and seek the functions y_1 and y_2 such that

$$\begin{cases} \dfrac{dy_1}{dt} = y_1 - y_2 \\ \dfrac{dy_2}{dt} = 2y_1 + 4y_2 \end{cases} \qquad \begin{cases} y_1(0) = 1 \\ y_2(0) = -3 \end{cases} \qquad (6.3.4)$$

In vector notation, this initial value problem becomes

$$\frac{dY}{dt} = AY \quad \text{and} \quad Y(0) = C$$

where

$$A = \begin{pmatrix} 1 & -1 \\ 2 & 4 \end{pmatrix} \quad \text{and} \quad C = \begin{pmatrix} 1 \\ -3 \end{pmatrix}$$

We know many solutions of the system $\dfrac{dY}{dt} = AY$: for every number α and β, we know $\alpha U + \beta V$ is a solution of the system. Can we find α and β so that $\alpha U(0) + \beta V(0) = C$? Since $U(0) = (1, -1)$ and $V(0) = (1, -2)$, we are looking for α and β so that

$$C = \begin{pmatrix} 1 \\ -3 \end{pmatrix} = \alpha \begin{pmatrix} 1 \\ -1 \end{pmatrix} + \beta \begin{pmatrix} 1 \\ -2 \end{pmatrix} = \begin{pmatrix} \alpha + \beta \\ -\alpha - 2\beta \end{pmatrix}$$

which is

$$\begin{cases} \alpha + \beta &= 1 \\ -\alpha - 2\beta &= -3 \end{cases}$$

This system has solution $\alpha = -1$ and $\beta = 2$, so the initial value problem $\dfrac{dY}{dt} = AY$ and $Y(0) = C$ has solution $Y = -U + 2V$, that is, $y_1(t) = -e^{2t} + 2e^{3t}$ and $y_2(t) = e^{2t} - 4e^{3t}$. We will see that something like this process always works, although it is sometimes a bit more complicated.

The fundamental theorem on which our work is based, although we will not prove it, is the following *existence and uniqueness theorem*. If A is an $n \times n$ matrix, a set of n linearly independent solutions of $\dfrac{dY}{dt} = AY$ is called a *fundamental set of solutions* of the system.

THEOREM 6.9

If A is an $n \times n$ matrix whose entries are real or complex numbers, then there are n linearly independent solutions of the system $\dfrac{dY}{dt} = AY$. Moreover, if a fundamental set of solutions of $\dfrac{dY}{dt} = AY$ is given, then, for any C in $\mathbf{C}^n$, there is exactly one solution of the initial value problem $\dfrac{dY}{dt} = AY$ and $Y(0) = C$, and that solution is a linear combination of the functions that form the fundamental set of solutions.

Finally, the condition of linear independence of the fundamental set of solutions, which might sound formidable, can be checked at any single point!

THEOREM 6.10

Let A be an $n \times n$ matrix whose entries are real or complex numbers and let t_0 be a real number. If $U^{(1)}$, $U^{(2)}$, $\cdots$, $U^{(k)}$ are solutions of $\dfrac{dY}{dt} = AY$, then $U^{(1)}$, $U^{(2)}$, $\cdots$, $U^{(k)}$ are linearly independent vector functions if and only if the vectors $U^{(1)}(t_0)$, $U^{(2)}(t_0)$, $\cdots$, $U^{(k)}(t_0)$ are linearly independent.

In the example above, A is a 2×2 matrix and the solutions $U(t) = (e^{2t}, -e^{2t})$ and $V(t) = (e^{3t}, -2e^{3t})$ are a fundamental set of solutions of the system. Indeed, since these vector functions satisfy the system of differential equations, Theorem 6.10 says it is enough to check the values at $t_0 = 0$: $U(0) = (1, -1)$ and $V(0) = (1, -2)$. Since $(1, -1)$ and $(1, -2)$ are linearly independent, $U(t)$ and $V(t)$ are linearly independent vector functions. In the second part of the example, we found a solution of the initial value problem as a linear combination of these solutions, and Theorem 6.9 says that the solution we found is the only solution of the initial value problem.

We can summarize the strategy outlined above in the following corollary: it is the principle of superposition in action.

COROLLARY 6.11

Let A be an $n \times n$ matrix with real or complex entries and suppose v_1, v_2, $\cdots$, v_k are eigenvectors of A corresponding to eigenvalues λ_1, λ_2, $\cdots$, λ_k. If C is a vector in $\mathbf{C}^n$ with

$$C = \alpha_1 v_1 + \alpha_2 v_2 + \cdots + \alpha_k v_k$$

then the initial value problem $\dfrac{dY}{dt} = AY$ and $Y(0) = C$ has exactly one solution, namely,

$$Y(t) = \alpha_1 e^{\lambda_1 t} v_1 + \alpha_2 e^{\lambda_2 t} v_2 + \cdots + \alpha_k e^{\lambda_k t} v_k$$

PROOF   This function satisfies the initial condition:

$$Y(0) = \alpha_1 e^{\lambda_1 0} v_1 + \alpha_2 e^{\lambda_2 0} v_2 + \cdots + \alpha_k e^{\lambda_k 0} v_k$$
$$= \alpha_1 v_1 + \alpha_2 v_2 + \cdots + \alpha_k v_k = C$$

Also,

$$\frac{dY}{dt} = \alpha_1\lambda_1 e^{\lambda_1 t}v_1 + \alpha_2\lambda_2 e^{\lambda_2 t}v_2 + \cdots + \alpha_k\lambda_k e^{\lambda_k t}v_k$$

and

$$AY(t) = A\left(\alpha_1 e^{\lambda_1 t}v_1 + \alpha_2 e^{\lambda_2 t}v_2 + \cdots + \alpha_k e^{\lambda_k t}v_k\right)$$

$$= \alpha_1 e^{\lambda_1 t}Av_1 + \alpha_2 e^{\lambda_2 t}Av_2 + \cdots + \alpha_k e^{\lambda_k t}Av_k$$

$$= \alpha_1\lambda_1 e^{\lambda_1 t}v_1 + \alpha_2\lambda_2 e^{\lambda_2 t}v_2 + \cdots + \alpha_k\lambda_k e^{\lambda_k t}v_k$$

so $\dfrac{dY}{dt} = AY$. Since Theorem 6.9 guarantees that there is exactly one solution of the initial value problem, this is the only solution there is. ∎

We can solve the system for every initial condition that can be written as a linear combination of eigenvectors of A. Naturally, if there is a basis for $\mathbf{C}^n$ consisting of eigenvectors for A, then *every* vector C can be written as a linear combination of eigenvectors of A and we can solve *every* initial value problem. This was the case with the system in the example above, and is the case for most systems.

EXAMPLE 6.12

Solve the initial value problem

$$\begin{cases} \dfrac{dy_1}{dt} = 3y_1 + 3y_2 \\ \dfrac{dy_2}{dt} = 2y_1 - 2y_2 \end{cases} \qquad \begin{cases} y_1(0) = 1 \\ y_2(0) = 5 \end{cases}$$

SOLUTION To begin, we rewrite the system of equations in matrix form: $\dfrac{dY}{dt} = AY$ and $Y(0) = C$, where

$$A = \begin{pmatrix} 3 & 3 \\ 2 & -2 \end{pmatrix} \quad \text{and} \quad C = \begin{pmatrix} 1 \\ 5 \end{pmatrix}$$

We will look for the eigenvectors and eigenvalues of A. If we can find a pair of linearly independent vectors that are eigenvectors of A, we can apply the results of the corollary to this problem.

$$\det(A - \lambda I) = \det\begin{pmatrix} 3-\lambda & 3 \\ 2 & -2-\lambda \end{pmatrix} = \lambda^2 - \lambda - 12 = (\lambda - 4)(\lambda + 3)$$

so the eigenvalues of A are 4 and -3. Now, finding eigenvectors in the usual way, we find that $(3, 1)$ is a basis for the eigenspace corresponding to $\lambda = 4$ and $(-1, 2)$ is a basis for the eigenspace corresponding to $\lambda = -3$. Since $(3, 1)$ and $(-1, 2)$ are a basis for $\mathbf{C}^2$, Corollary 6.11 will apply. To write C as a linear combination of these vectors, we must find α and β so that

$$C = \alpha \begin{pmatrix} 3 \\ 1 \end{pmatrix} + \beta \begin{pmatrix} -1 \\ 2 \end{pmatrix}$$

Solving the resulting system, we find $\alpha = 1$ and $\beta = 2$ so

$$C = \begin{pmatrix} 1 \\ 5 \end{pmatrix} = \begin{pmatrix} 3 \\ 1 \end{pmatrix} + 2 \begin{pmatrix} -1 \\ 2 \end{pmatrix}$$

Thus,

$$Y(t) = e^{4t} \begin{pmatrix} 3 \\ 1 \end{pmatrix} + 2e^{-3t} \begin{pmatrix} -1 \\ 2 \end{pmatrix} = \begin{pmatrix} 3e^{4t} - 2e^{-3t} \\ e^{4t} + 4e^{-3t} \end{pmatrix}$$

In other words, $y_1(t) = 3e^{4t} - 2e^{-3t}$ and $y_2(t) = e^{4t} + 4e^{-3t}$ are the solutions to the initial value problem above. It is easily checked that $y_1(0) = 1$ and $y_2(0) = 5$. We leave it to the reader to check that y_1 and y_2 satisfy the system of differential equations. $\square$

Many differential equations that arise in engineering and science involve more than just the first derivative; we would like to be able to solve higher order initial value problems. There is a standard technique that converts this kind of problem into a problem that we can solve.

EXAMPLE 6.13

Solve the differential equation $\dfrac{d^3y}{dt^3} - 2\dfrac{d^2y}{dt^2} - \dfrac{dy}{dt} + 2y = 0$ subject to the initial conditions $y(0) = -1$, $\dfrac{dy}{dt}(0) = 2$, and $\dfrac{d^2y}{dt^2}(0) = 1$.

SOLUTION The idea for solving this problem as a linear algebra problem is to substitute for the unknown function and its derivatives. For example, if we take $y_1(t) = y(t)$, $y_2(t) = \dfrac{dy_1}{dt} = \dfrac{dy}{dt}$, and $y_3(t) = \dfrac{dy_2}{dt} = \dfrac{d^2y}{dt^2}$. This means the differential equation becomes $\dfrac{dy_3}{dt} = 2y_3 + y_2 - 2y_1$ and the initial conditions become $y_1(0) = -1$, $y_2(0) = \dfrac{dy}{dt}(0) = 2$, and $y_3(0) = \dfrac{d^2y}{dt^2}(0) = 1$. The other two equations of the system are two of the equations we used to substitute. Putting this all together, we must solve the initial value problem

$$\begin{cases} \dfrac{dy_1}{dt} & = & & & y_2 & \\[2mm] \dfrac{dy_2}{dt} & = & & & & y_3 \\[2mm] \dfrac{dy_3}{dt} & = & -2y_1 & + & y_2 & + & 2y_3 \end{cases} \qquad \begin{cases} y_1(0) & = & -1 \\ y_2(0) & = & 2 \\ y_3(0) & = & 1 \end{cases}$$

Now we can rewrite the system of equations in matrix form: $\dfrac{dY}{dt} = AY$ and $Y(0) = C$, where

$$A = \begin{pmatrix} 0 & 1 & 0 \\ 0 & 0 & 1 \\ -2 & 1 & 2 \end{pmatrix} \quad \text{and} \quad C = \begin{pmatrix} -1 \\ 2 \\ 1 \end{pmatrix}$$

Finding the eigenvectors and eigenvalues of A in the usual way, we find that $(1, 1, 1)$ is a basis for the eigenspace corresponding to $\lambda = 1$, $(1, -1, 1)$ is a basis for the eigenspace corresponding to $\lambda = -1$, and $(1, 2, 4)$ is a basis for the eigenspace corresponding to $\lambda = 2$. Solving the appropriate system, we find

$$C = \begin{pmatrix} -1 \\ 2 \\ 1 \end{pmatrix} = -\frac{1}{2}\begin{pmatrix} 1 \\ 1 \\ 1 \end{pmatrix} - \frac{7}{6}\begin{pmatrix} 1 \\ -1 \\ 1 \end{pmatrix} + \frac{2}{3}\begin{pmatrix} 1 \\ 2 \\ 4 \end{pmatrix}$$

so the solution of our system is

$$-\frac{1}{2}e^t\begin{pmatrix} 1 \\ 1 \\ 1 \end{pmatrix} - \frac{7}{6}e^{-t}\begin{pmatrix} 1 \\ -1 \\ 1 \end{pmatrix} + \frac{2}{3}e^{2t}\begin{pmatrix} 1 \\ 2 \\ 4 \end{pmatrix}$$

Rewriting these functions individually, this is

$$y_1(t) = -\frac{1}{2}e^t - \frac{7}{6}e^{-t} + \frac{2}{3}e^{2t}$$

$$y_2(t) = -\frac{1}{2}e^t + \frac{7}{6}e^{-t} + \frac{4}{3}e^{2t}$$

$$y_3(t) = -\frac{1}{2}e^t - \frac{7}{6}e^{-t} + \frac{8}{3}e^{2t}$$

But in the original problem, we only wanted

$$y(t) = y_1(t) = -\frac{1}{2}e^t - \frac{7}{6}e^{-t} + \frac{2}{3}e^{2t}$$

and the other functions were the derivatives of y. We can check that this satisfies the initial value problem we were given. □

You may have noticed in the example above that the matrix A that arose from the technique had an unusual form. Indeed, this is always the case for problems in which a higher order equation in one unknown function is replaced by a system of first order equations by choosing the functions as we did. Specifically, the equation

$$\frac{d^n y}{dt^n} + \alpha_{n-1}\frac{d^{n-1}y}{dt^{n-1}} + \cdots + \alpha_1\frac{dy}{dt} + \alpha_0 y = 0$$

becomes a system with coefficient matrix

$$A = \begin{pmatrix} 0 & 1 & 0 & \cdots & 0 & 0 \\ 0 & 0 & 1 & & 0 & 0 \\ & \vdots & & \ddots & \vdots & \\ 0 & 0 & 0 & & 0 & 1 \\ -\alpha_0 & -\alpha_1 & -\alpha_2 & \cdots & -\alpha_{n-2} & -\alpha_{n-1} \end{pmatrix}$$

A matrix of this form is called a *companion matrix*.

The overall plan, as outlined in Corollary 6.11, for solving a system of linear differential equations with constant coefficients given initial values is to rewrite the system as a matrix differential equation $\dfrac{dY}{dt} = AY$, find a basis for $\mathbf{C}^n$ consisting of eigenvectors of A, writing the initial condition as a linear combination of the eigenvectors, and writing the solution of the differential equation as a linear combination of exponentials. In this section, we have dealt with systems of differential equations in which the coefficient matrix was real, the eigenvalues were real, and the $\mathbf{R}^n$ had a basis consisting of eigenvectors. However, some real matrices have complex eigenvalues; in Section 6.4, we deal with systems of differential equations for which the eigenvalues of the coefficient matrix are complex and extend the method above to this case. Moreover, as we saw in Example 6.6, there are matrices for which there is no basis of $\mathbf{C}^n$ consisting of eigenvectors for the matrix. For these matrices, the method given above does not apply to all initial value problems. The solution of these more difficult problems is deferred until Section 6.7.

Exercises 6.3

1. Let A be the matrix
$$A = \begin{pmatrix} 12 & 10 \\ -15 & -13 \end{pmatrix}$$

   (a) Show that $(1, -1)$ is an eigenvector for A with eigenvalue 2.
   (b) Show that $Y(t) = e^{2t}(1, -1)$, that is, $y_1(t) = e^{2t}$ and $y_2(t) = -e^{2t}$, is a solution of the system
$$\begin{cases} \dfrac{dy_1}{dt} = 12y_1 + 10y_2 \\[2mm] \dfrac{dy_2}{dt} = -15y_1 - 13y_2 \end{cases} \tag{6.3.5}$$

   (c) Show that $(2, -3)$ is also an eigenvector for A and find a solution of the system in Equation (6.3.5) that is a multiple of $(2, -3)$.
   (d) Find the solution of the system in Equation (6.3.5) that satisfies $y_1(0) = 1$ and $y_2(0) = -2$.

2. Let A be the matrix
$$A = \begin{pmatrix} 1 & 4 \\ -2 & 7 \end{pmatrix}$$

   (a) Find the eigenvalues and a basis for each of the eigenspaces of A.
   (b) Solve the initial value problem
$$\begin{cases} \dfrac{dy_1}{dt} = y_1 + 4y_2 \\[2mm] \dfrac{dy_2}{dt} = -2y_1 + 7y_2 \end{cases} \quad \text{and} \quad \begin{cases} y_1(0) = 1 \\ y_2(0) = 2 \end{cases}$$

3.
$$\text{Let } F = \begin{pmatrix} 0 & -1 \\ -2 & 1 \end{pmatrix}.$$

   (a) Find the eigenvectors and eigenvalues of F.

   (b) Solve the initial value problem
$$\begin{cases} \dfrac{du}{dt} = -v \\[2mm] \dfrac{dv}{dt} = -2u + v \end{cases} \quad \text{where} \quad \begin{cases} u(0) = -3 \\ v(0) = 5 \end{cases}$$

4. Solve the initial value problem
$$\begin{cases} \dfrac{du}{dt} = 7u - 4v \\[2mm] \dfrac{dv}{dt} = 8u - 5v \end{cases} \quad \text{where} \quad \begin{cases} u(0) = 3 \\ v(0) = -4 \end{cases}$$

5. Solve the initial value problem

$$\begin{cases} \dfrac{dy_1}{dt} = 4y_1 + 9y_2 \\[2mm] \dfrac{dy_2}{dt} = -6y_2 - 11y_2 \end{cases} \qquad \text{where} \qquad \begin{cases} y_1(0) = 1 \\[1mm] y_2(0) = 5 \end{cases}$$

6. Solve the initial value problem

$$\begin{cases} \dfrac{dy_1}{dt} = 8y_1 - 6y_2 \\[2mm] \dfrac{dy_2}{dt} = 9y_2 - 13y_2 \end{cases} \qquad \text{where} \qquad \begin{cases} y_1(0) = 1 \\[1mm] y_2(0) = -4 \end{cases}$$

7. Solve the initial value problem

$$\begin{cases} \dfrac{du}{dt} = -4u - 5v - 5w \\[2mm] \dfrac{dv}{dt} = 7u + 8v + 7w \\[2mm] \dfrac{dw}{dt} = -3u - 3v - 2w \end{cases} \qquad \text{where} \qquad \begin{cases} u(0) = 3 \\[1mm] v(0) = -4 \\[1mm] w(0) = 2 \end{cases}$$

8. Solve the initial value problem

$$\begin{cases} \dfrac{du}{dt} = -5u + 12v - 12w \\[2mm] \dfrac{dv}{dt} = 2u - 4v + 6w \\[2mm] \dfrac{dw}{dt} = 4u - 9v + 11w \end{cases} \qquad \text{where} \qquad \begin{cases} u(0) = 1 \\[1mm] v(0) = 3 \\[1mm] w(0) = 1 \end{cases}$$

9. Solve the initial value problem

$$\begin{cases} \dfrac{du}{dt} = - v + w \\[2mm] \dfrac{dv}{dt} = 4u + 4v - 2w \\[2mm] \dfrac{dw}{dt} = -2u - v + 3w \end{cases} \qquad \text{where} \qquad \begin{cases} u(0) = 1 \\[1mm] v(0) = 3 \\[1mm] w(0) = 4 \end{cases}$$

10. Solve the initial value problem $\dfrac{d^2y}{dt^2} + 8\dfrac{dy}{dt} + 15y = 0$ where $y(0) = 1$ and $\dfrac{dy}{dt}(0) = 4$ using the substitution technique of Example 6.13.

11. Solve the initial value problem $\dfrac{d^2y}{dt^2} - \dfrac{dy}{dt} - 20y = 0$ where $y(0) = -3$ and $\dfrac{dy}{dt}(0) = 1$ using the substitution technique of Example 6.13.

12. Solve the initial value problem $\dfrac{d^3y}{dt^3} + \dfrac{d^2y}{dt^2} - 4\dfrac{dy}{dt} - 4y = 0$ and $y(0) = 3$, $\dfrac{dy}{dt}(0) = -1$, and $\dfrac{d^2y}{dt^2}(0) = 1$ using the substitution technique of Example 6.13.

13.(a) Find the characteristic polynomial of the companion matrix

$$A_2 = \begin{pmatrix} 0 & 1 \\ -\alpha_0 & -\alpha_1 \end{pmatrix}$$

(b) Find the characteristic polynomial of the companion matrix

$$A_3 = \begin{pmatrix} 0 & 1 & 0 \\ 0 & 0 & 1 \\ -\alpha_0 & -\alpha_1 & -\alpha_2 \end{pmatrix}$$

(c) Find the characteristic polynomial of the companion matrix

$$A_4 = \begin{pmatrix} 0 & 1 & 0 & 0 \\ 0 & 0 & 1 & 0 \\ 0 & 0 & 0 & 1 \\ -\alpha_0 & -\alpha_1 & -\alpha_2 & -\alpha_3 \end{pmatrix}$$

(d) Make a conjecture for the characteristic polynomial of a companion matrix of order n. Can you prove your conjecture?

14. The matrix $F = \begin{pmatrix} 2 & 6 & -3 & 1 \\ 0 & -1 & 1 & 0 \\ 3 & 7 & -4 & 1 \\ 4 & 8 & -7 & 3 \end{pmatrix}$ has two positive and two negative eigenvalues.

Let M be the subspace spanned by the eigenvectors corresponding to the negative eigenvalues of F.

(We will call the subspace M the *stable manifold* of F because solutions with initial conditions in this manifold move toward zero over time whereas solutions whose initial conditions have components with positive eigenvalues, i.e. exponentials with positive exponent, move away from zero.)

(a) Find the matrix for the orthogonal projection of $\mathbf{R}^4$ onto M.

(b) Find the point of the stable manifold M that is closest to $(1, -1, 1, 0)$.

15. Show that $U(t) = (1, t)$ and $V(t) = (1, t^2)$ are linearly independent vector functions. Nevertheless $U(0) = (1, 0) = V(0)$. Explain why this does not contradict Theorem 6.10.

6.4   Real Differential Equations with Complex Eigenvalues

If an initial value problem has coefficients and data that are real numbers, we would expect that the problem has real–valued solutions. Although the same in principle as the examples in the previous section if we are dealing with $\mathbf{C}^n$ and not $\mathbf{R}^n$, there are special techniques for getting real–valued solutions in a situation in which the system of differential equations has real coefficients but the eigenvectors and eigenvalues of the coefficient matrix are complex. More information about the complex numbers themselves is contained in Appendix A.

One of the issues that we must deal with is the meaning of the exponential function, $e^{\lambda t}$, when λ is not real. Euler's formula, which can be justified by power series calculations, says that the complex exponential function $e^{(a+bi)t}$ can be rewritten as

$$e^{(a+bi)t} = e^{at}\left(\cos(bt) + i\sin(bt)\right) \tag{6.4.6}$$

To see why Equation (6.4.6) is correct, first notice that

$$e^{(a+bi)t} = e^{at+ibt} = e^{at}e^{ibt}$$

by the properties of exponents; e^{at} is an ordinary real exponential function, so we can deal with e^{ibt}, an exponential function with purely imaginary exponent, separately. We can write out the power series for e^{ibt}. Recalling from calculus that

$$e^x = 1 + x + \frac{1}{2!}x^2 + \frac{1}{3!}x^3 + \cdots$$

we get

$$e^{ibt} = 1 + ibt + \frac{1}{2!}(ibt)^2 + \frac{1}{3!}(ibt)^3 + \frac{1}{4!}(ibt)^4 + \frac{1}{5!}(ibt)^5 + \cdots$$

$$= 1 + ibt + \frac{1}{2!}i^2(bt)^2 + \frac{1}{3!}i^3(bt)^3 + \frac{1}{4!}i^4(bt)^4 + \frac{1}{5!}i^5(bt)^5 + \cdots$$

$$= 1 + ibt - \frac{1}{2!}(bt)^2 - i\frac{1}{3!}(bt)^3 + \frac{1}{4!}(bt)^4 + i\frac{1}{5!}(bt)^5 + \cdots$$

where, in the last line, we have used $i^2 = -1$, $i^3 = -i$, $i^4 = 1$, etc. Now, collecting the real and imaginary terms of this series, we get

$$e^{ibt} = 1 + ibt - \frac{1}{2!}(bt)^2 - i\frac{1}{3!}(bt)^3 + \frac{1}{4!}(bt)^4 + i\frac{1}{5!}(bt)^5 - \cdots$$

$$= 1 - \frac{1}{2!}(bt)^2 + \frac{1}{4!}(bt)^4 - \frac{1}{6!}(bt)^6 + \cdots$$

$$+ ibt - i\frac{1}{3!}(bt)^3 + i\frac{1}{5!}(bt)^5 - i\frac{1}{7!}(bt)^7 + \cdots$$

$$= 1 - \frac{1}{2!}(bt)^2 + \frac{1}{4!}(bt)^4 - \frac{1}{6!}(bt)^6 + \cdots$$

$$+ i\left(bt - \frac{1}{3!}(bt)^3 + \frac{1}{5!}(bt)^5 - \frac{1}{7!}(bt)^7 + \cdots\right)$$

Also recalling from calculus that

$$\cos(x) = 1 - \frac{1}{2!}x^2 + \frac{1}{4!}x^4 - \frac{1}{6!}x^6 + \cdots$$

and

$$\sin(x) = x - \frac{1}{3!}x^3 + \frac{1}{5!}x^5 - \frac{1}{7!}x^7 + \cdots$$

we see that the final line in the calculation above gives

$$e^{ibt} = \cos(bt) + i\sin(bt)$$

Putting this together with $e^{(a+bi)t} = e^{at}e^{ibt}$ gives Euler's Formula.

EXAMPLE 6.14

Use Euler's Formula to rewrite the complex exponential $e^{(2-3i)t}$ and find its real and imaginary parts.

SOLUTION Equation (6.4.6) says

$$e^{(2-3i)t} = e^{2t-i3t} = e^{2t}e^{-i3t} = e^{2t}\left(\cos(-3t) + i\sin(-3t)\right)$$

Since $\cos(-3t) = \cos(3t)$ and $\sin(-3t) = -\sin(3t)$, we get

$$e^{(2-3i)t} = e^{2t}\left(\cos(3t) - i\sin(3t)\right)$$

Now multiplying this out, we get

$$e^{(2-3i)t} = e^{2t}\cos(3t) - ie^{2t}\sin(3t)$$

so the real and imaginary parts are

$$\mathrm{Re}(e^{(2-3i)t}) = e^{2t}\cos(3t)$$

and

$$\mathrm{Im}(e^{(2-3i)t}) = -e^{2t}\sin(3t)$$

$\Box$

Let's investigate an example of system of differential equations in which the coefficient matrix is real, but the eigenvalues are complex, to see what happens and see how to handle the difficulty.

EXAMPLE 6.15

Solve the initial value problem

$$\begin{cases} \dfrac{dy_1}{dt} = y_1 - 2y_2 \\ \dfrac{dy_2}{dt} = 2y_1 + y_2 \end{cases} \qquad \begin{cases} y_1(0) = 3 \\ y_2(0) = -2 \end{cases}$$

SOLUTION To begin, we rewrite the system of equations in matrix form: $\dfrac{dY}{dt} = AY$ and $Y(0) = C$, where

$$A = \begin{pmatrix} 1 & -2 \\ 2 & 1 \end{pmatrix} \quad \text{and} \quad C = \begin{pmatrix} 3 \\ -2 \end{pmatrix}$$

As before, we will look for the eigenvectors and eigenvalues of A and try to apply the results of Corollary 6.11 to this problem.

$$\det(A - \lambda I) = \det \begin{pmatrix} 1 - \lambda & -2 \\ 2 & 1 - \lambda \end{pmatrix} = \lambda^2 - 2\lambda + 5$$

Using the quadratic formula to solve $\lambda^2 - 2\lambda + 5 = 0$, we find $\lambda = 1 + 2i$ and $\lambda = 1 - 2i$ and the eigenvalues of A are $1 + 2i$ and $1 - 2i$. Now, finding eigenvectors in the usual way, we must solve the system $(A - (1 + 2i)I)v = 0$. Writing $v = (x, y)$, this is

$$\begin{pmatrix} 1 - (1 + 2i) & -2 \\ 2 & 1 - (1 + 2i) \end{pmatrix} \begin{pmatrix} x \\ y \end{pmatrix} = \begin{pmatrix} 0 \\ 0 \end{pmatrix}$$

or

$$\begin{cases} -2ix - 2y = 0 \\ 2x - 2iy = 0 \end{cases}$$

Multiplying the second equation by i and adding it to the first, since $i(2x - 2iy) = 2ix - 2i^2y = 2ix + 2y$, we get the equivalent system

$$\begin{cases} 0 = 0 \\ 2x - 2iy = 0 \end{cases}$$

(Of course, we knew this must happen! The fact that $1 + 2i$ is an eigenvalue means the system has infinitely many solutions and, therefore, that the equations are multiples of each other.) Solving the second equation, we find that the solution is $x = iy$ and y is arbitrary, so, for example, $(i, 1)$ is an eigenvector for A with eigenvalue $1 + 2i$. (Check this by multiplying!)

Similarly, solving the system $(A - (1 - 2i)I)v = 0$, we have

$$\begin{pmatrix} 1 - (1 - 2i) & -2 \\ 2 & 1 - (1 - 2i) \end{pmatrix} \begin{pmatrix} x \\ y \end{pmatrix} = \begin{pmatrix} 0 \\ 0 \end{pmatrix}$$

or

$$\begin{cases} 2ix - 2y & = & 0 \\ 2x + 2iy & = & 0 \end{cases}$$

Multiplying the second equation by $-i$ and adding it to the first, since $-i(2x + 2iy) = -2ix - 2i^2 y = -2ix + 2y$, we get the equivalent system

$$\begin{cases} 0 & = & 0 \\ 2x + 2iy & = & 0 \end{cases}$$

This gives $x = -iy$ and y is arbitrary, so for example, $(-i, 1)$ is an eigenvector for A with eigenvalue $1 - 2i$. (Check this, too, by multiplying!)

Now $(i, 1)$ and $(-i, 1)$ are a basis for $\mathbf{C}^2$, so Corollary 6.11 applies. To write C as a linear combination of these vectors, we must find α and β so that

$$\begin{pmatrix} 3 \\ -2 \end{pmatrix} = C = \alpha \begin{pmatrix} i \\ 1 \end{pmatrix} + \beta \begin{pmatrix} -i \\ 1 \end{pmatrix}$$

Solving the resulting system, we find $\alpha = -1 - 1.5i$ and $\beta = -1 + 1.5i$ so

$$C = (-1 - 1.5i) \begin{pmatrix} i \\ 1 \end{pmatrix} + (-1 + 1.5i) \begin{pmatrix} -i \\ 1 \end{pmatrix}$$

Thus,

$$Y(t) = (-1 - 1.5i)e^{(1+2i)t} \begin{pmatrix} i \\ 1 \end{pmatrix} + (-1 + 1.5i)e^{(1-2i)t} \begin{pmatrix} -i \\ 1 \end{pmatrix}$$

$$= \begin{pmatrix} (-i + 1.5)e^{(1+2i)t} + (i + 1.5)e^{(1-2i)t} \\ (-1 - 1.5i)e^{(1+2i)t} + (-1 + 1.5i)e^{(1-2i)t} \end{pmatrix}$$

In other words, $y_1(t) = (-i + 1.5)e^{(1+2i)t} + (i + 1.5)e^{(1-2i)t}$ and $y_2(t) = (-1 - 1.5i)e^{(1+2i)t} + (-1 + 1.5i)e^{(1-2i)t}$ is the solution to the initial value problem above.

Now this answer is correct, as you can check, but not very satisfying! We would hope to have an answer that does not involve $\sqrt{-1}$ since the system of differential equations has real coefficients. As we will see below, this can be done.

We can use Euler's formula to rewrite this solution. Thus

$$
\begin{aligned}
y_1(t) &= (-i + 1.5)e^{(1+2i)t} + (i + 1.5)e^{(1-2i)t} \\
&= (-i + 1.5)e^t\left(\cos(2t) + i\sin(2t)\right) + (i + 1.5)e^t\left(\cos(-2t) + i\sin(-2t)\right) \\
&= (-i + 1.5)e^t\left(\cos(2t) + i\sin(2t)\right) + (i + 1.5)e^t\left(\cos(2t) - i\sin(2t)\right) \\
&= -ie^t\cos(2t) - i^2e^t\sin(2t) + 1.5e^t\cos(2t) + 1.5ie^t\sin(2t) \\
&\qquad +ie^t\cos(2t) - i^2e^t\sin(2t) + 1.5e^t\cos(2t) - 1.5ie^t\sin(2t) \\
&= 2e^t\sin(2t) + 3e^t\cos(2t)
\end{aligned}
$$

Similarly,

$$
\begin{aligned}
y_2(t) &= (-1 - 1.5i)e^{(1+2i)t} + (-1 + 1.5i)e^{(1-2i)t} \\
&= (-1 - 1.5i)e^t\left(\cos(2t) + i\sin(2t)\right) + (-1 + 1.5i)e^t\left(\cos(-2t) + i\sin(-2t)\right) \\
&= (-1 - 1.5i)e^t\left(\cos(2t) + i\sin(2t)\right) + (-1 + 1.5i)e^t\left(\cos(2t) - i\sin(2t)\right) \\
&= -e^t\cos(2t) - ie^t\sin(2t) - 1.5ie^t\cos(2t) - 1.5i^2e^t\sin(2t) \\
&\qquad -e^t\cos(2t) + ie^t\sin(2t) + 1.5ie^t\cos(2t) - 1.5i^2e^t\sin(2t) \\
&= -2e^t\cos(2t) + 3e^t\sin(2t)
\end{aligned}
$$

and this way of writing the answer is more acceptable to us! It is easy to check that $Y(0) = (y_1(0), y_2(0)) = (3, -2)$ and slightly more tedious, but not difficult to check that these functions work in the original system. ☐

In looking back over our work, we might notice that the eigenvalues of A are $1 + 2i$ and $1 - 2i = \overline{1 + 2i}$, that is, the eigenvalues are conjugates of each other, and the corresponding eigenvectors $(i, 1)$ and $(-i, 1) = \overline{(i, 1)}$ are also conjugates of each other. This is not an accident!

PROPOSITION 6.16

Suppose A is an $n \times n$ matrix and all of the entries of A are real numbers. If λ is an eigenvalue of A with eigenvector v, then $\overline{\lambda}$ is also an eigenvalue of A and $\overline{v}$ is an eigenvector corresponding to $\overline{\lambda}$.

PROOF   If λ is an eigenvalue of A with eigenvector v, then we have $Av = \lambda v$. Taking complex conjugates of the whole equation and remembering that the conjugate of a sum or product are the sum or product of the conjugates, we get

$$\overline{Av} = \overline{A}\,\overline{v} = \overline{\lambda v} = \overline{\lambda}\,\overline{v}$$

Since A is real by hypothesis, $\overline{A} = A$ and this becomes

$$A\overline{v} = \overline{\lambda}\,\overline{v}$$

which is the conclusion.   ∎

In fact, we can also replace the fundamental set of complex–valued solutions by a fundamental set of real–valued solutions in Example 6.15.

PROPOSITION 6.17
If A is a real $n \times n$ matrix and $Y(t)$ is a complex–valued solution of the system $\dfrac{dY}{dt} = AY$, then the real and imaginary parts of $Y(t)$ are real–valued solutions of the system $\dfrac{dY}{dt} = AY$.

PROOF   If $\dfrac{dY}{dt} = AY$, then

$$\frac{d\overline{Y}}{dt} = \overline{\frac{dY}{dt}} = \overline{AY} = \overline{A}\,\overline{Y} = A\overline{Y}$$

so $\overline{Y}$ is also a solution of the system. Since $\dfrac{dY}{dt} = AY$ is a linear system, Theorem 6.8 implies $U = \operatorname{Re}(Y) = .5Y + .5\overline{Y}$ and $V = \operatorname{Im}(Y) = -i.5Y + i.5\overline{Y}$ are also solutions of $\dfrac{dY}{dt} = AY$.   ∎

In Example 6.15 above, for instance, the function

$$Y(t) = e^{(1+2i)t} \begin{pmatrix} 2i \\ 1 \end{pmatrix} = \begin{pmatrix} -2e^t \sin(2t) + 2ie^t \cos(2t) \\ e^t \cos(2t) + ie^t \sin(2t) \end{pmatrix}$$

is a solution of the system. By Proposition 6.17, the real and imaginary parts of this solution are also solutions, that is,

$$U(t) = \begin{pmatrix} -2e^t \sin(2t) \\ e^t \cos(2t) \end{pmatrix}$$

and

$$V(t) = \begin{pmatrix} 2e^t \cos(2t) \\ e^t \sin(2t) \end{pmatrix}$$

are solutions. Moreover, since $U(0) = (0, 1)$ and $V(0) = (2, 0)$ are linearly independent, Theorem 6.10 implies these are a fundamental set of solutions for the system. In this case, since $C = (3, -2) = -2U(0) + 1.5V(0)$, we see that the solution of the initial value problem is

$$Y(t) = -2U(t) + 1.5V(t) = \begin{pmatrix} -4e^t \sin(2t) + 3e^t \cos(2t) \\ 2e^t \cos(2t) + 1.5e^t \sin(2t) \end{pmatrix}$$

as we found before.

EXAMPLE 6.18

Solve the initial value problem of Section 6.1:

$$\begin{cases} 4\dfrac{d^2 x_1}{dt^2} &= -12x_1 + 4x_2 \\[2mm] 2\dfrac{d^2 x_2}{dt^2} &= 4x_1 - 4x_2 \end{cases}$$

with initial conditions

$$\begin{cases} x_1(0) &= 1 \\ x_2(0) &= 2 \\ \dfrac{dx_1}{dt}(0) &= 0 \\ \dfrac{dx_2}{dt}(0) &= 0 \end{cases}$$

SOLUTION We first convert this to a first order system using the technique of Example 6.13. We solve the differential equations for the derivatives and introduce the new variables $x_3 = \dfrac{dx_1}{dt}$ and $x_4 = \dfrac{dx_2}{dt}$. Then the initial value problem becomes

$$\begin{cases} \dfrac{dx_1}{dt} &= x_3 \\[2mm] \dfrac{dx_2}{dt} &= x_4 \\[2mm] \dfrac{dx_3}{dt} &= -3x_1 + x_2 \\[2mm] \dfrac{dx_4}{dt} &= 2x_1 - 2x_2 \end{cases}$$

with initial conditions

$$\begin{cases} x_1(0) &= 2 \\ x_2(0) &= 1 \\ x_3(0) &= 0 \\ x_4(0) &= 0 \end{cases}$$

This is $\dfrac{dY}{dt} = AY$ and $Y(0) = C$ where

$$A = \begin{pmatrix} 0 & 0 & 1 & 0 \\ 0 & 0 & 0 & 1 \\ -3 & 1 & 0 & 0 \\ 2 & -2 & 0 & 0 \end{pmatrix}$$

and $C = (2, 1, 0, 0)$.

The first step is to identify the eigenvalues and eigenvectors of the coefficient matrix.

```
> A=[0 0 1 0; 0 0 0 1; -3 1 0 0; 2 -2 0 0]

A =
      0     0     1     0
      0     0     0     1
     -3     1     0     0
      2    -2     0     0

> [w d]=eig(A)

w =
  -0.316 + 0.000i  -0.316 - 0.000i   0.316 - 0.000i   0.316 + 0.000i
   0.316 - 0.000i   0.316 + 0.000i   0.633 - 0.000i   0.633 + 0.000i
   0.000 - 0.633i   0.000 + 0.633i   0.000 + 0.316i   0.000 - 0.316i
  -0.000 + 0.633i  -0.000 - 0.633i   0.000 + 0.633i   0.000 - 0.633i


d =
  -0.00 + 2.00i         0                0                0
        0         -0.00 - 2.00i          0                0
        0                0         -0.00 + 1.00i          0
        0                0                0         -0.00 - 1.00i
```

so the eigenvalues of A are $\pm i$ and $\pm 2i$ and the corresponding eigenvectors are the columns of w.

For hand calculation, it is more convenient to notice that the first column of w is a multiple of the vector $u = (-1, 1, -2i, 2i)$ and that this is therefore also an eigenvector of A corresponding to the eigenvalue $2i$.

```
> u=[-1;1;-2*i;2*i]

u =
  -1.00
   1.00
       0 - 2.00i
       0 + 2.00i

> A*u

ans =
       0 - 2.00i
       0 + 2.00i
   4.00
  -4.00
```

Because all the entries of A are real numbers, Proposition 6.16 implies $\overline{u} = (-1, 1, 2i, -2i)$ is also an eigenvector of A and its eigenvalue is $\overline{2i} = -2i$. This vector corresponds to the second column of w.

Similarly, $v = (1, 2, i, 2i)$ is an eigenvector for A corresponding to the eigenvalue i.

```
> v=[1;2;i;2*i]

v =
   1.00
   2.00
       0 + 1.00i
       0 + 2.00i

> A*v

ans =
       0 + 1.00i
       0 + 2.00i
  -1.00
  -2.00
```

and $\overline{v} = (1, 2, -i, -2i)$ is an eigenvector for A corresponding to the eigenvalue $\overline{i} = -i$.

Now Theorem 6.10 implies $U^{(1)}(t) = e^{2it}u$, $U^{(2)}(t) = e^{-2it}\overline{u}$, $U^{(3)}(t) = e^{it}v$, and $U^{(4)}(t) = e^{-it}\overline{v}$ are a fundamental set of solutions of the system $\dfrac{dY}{dt} = AY$ because $U^{(1)}(0) = u$, $U^{(2)}(0) = \overline{u}$, $U^{(3)}(0) = v$, and $U^{(4)}(0) = 0\overline{v}$ are linearly independent. Writing these solutions out more completely, we see that

$$
U^{(1)}(t) = e^{2it}u = e^{2it}\begin{pmatrix} -1 \\ 1 \\ -2i \\ 2i \end{pmatrix} = (\cos(2t) + i\sin(2t))\begin{pmatrix} -1 \\ 1 \\ -2i \\ 2i \end{pmatrix}
$$

$$
= \begin{pmatrix} -\cos(2t) - i\sin(2t) \\ \cos(2t) + i\sin(2t) \\ 2\sin(2t) - 2i\cos(2t) \\ -2\sin(2t) + 2i\cos(2t) \end{pmatrix}
$$

and similarly

$$
U^{(2)}(t) = e^{-2it}\overline{u} = e^{-2it}\begin{pmatrix} -1 \\ 1 \\ 2i \\ -2i \end{pmatrix} = \begin{pmatrix} -\cos(2t) + i\sin(2t) \\ \cos(2t) - i\sin(2t) \\ 2\sin(2t) + 2i\cos(2t) \\ -2\sin(2t) - 2i\cos(2t) \end{pmatrix}
$$

$$
U^{(3)}(t) = e^{it}v = \begin{pmatrix} \cos(t) + i\sin(t) \\ 2\cos(t) + i2\sin(t) \\ -\sin(t) + i\cos(t) \\ -2\sin(t) + i2\cos(t) \end{pmatrix}
$$

and

$$
U^{(4)}(t) = e^{-it}\overline{v} = \begin{pmatrix} \cos(t) - i\sin(t) \\ 2\cos(t) - i2\sin(t) \\ -\sin(t) - i\cos(t) \\ -2\sin(t) - i2\cos(t) \end{pmatrix}
$$

Proposition 6.17 says that the real and imaginary parts of each of these solutions is a solution also. For example,

$$
V^{(1)}(t) = \operatorname{Re}(U^{(1)}(t)) = \begin{pmatrix} -\cos(2t) \\ \cos(2t) \\ 2\sin(2t) \\ -2\sin(2t) \end{pmatrix}
$$

and

$$V^{(2)}(t) = \text{Im}(U^{(1)}(t)) = \begin{pmatrix} -\sin(2t) \\ \sin(2t) \\ -2\cos(2t) \\ 2\cos(2t) \end{pmatrix}$$

are both solutions. Notice that the real and imaginary parts of $U^{(2)}$ are these same two solutions. Similarly

$$V^{(3)}(t) = \text{Re}(U^{(3)}(t)) = \begin{pmatrix} \cos(t) \\ 2\cos(t) \\ -\sin(t) \\ -2\sin(t) \end{pmatrix}$$

and

$$V^{(4)}(t) = \text{Im}(U^{(3)}(t)) = \begin{pmatrix} \sin(t) \\ 2\sin(t) \\ \cos(t) \\ 2\cos(t) \end{pmatrix}$$

are solutions. Again, the real and imaginary parts of $U^{(4)}$ agree with these latter two solutions. Moreover, because $V^{(1)}(0) = (-1, 1, 0, 0)$, $V^{(2)}(0) = (0, 0, -2, 2)$, $V^{(3)}(0) = (1, 2, 0, 0)$, and $V^{(4)}(0) = (0, 0, 1, 2)$ are linearly independent, $V^{(1)}$, $V^{(2)}$, $V^{(3)}$, and $V^{(4)}$ are a fundamental set of solutions for $\dfrac{dY}{dt} = AY$.

To solve the initial conditions, we need to find constants a_1, a_2, a_3, and a_4 so that

$$C = a_1 V^{(1)}(0) + a_2 V^{(2)}(0) + a_3 V^{(3)}(0) + a_4 V^{(4)}(0)$$

that is, so that

$$\begin{pmatrix} 2 \\ 1 \\ 0 \\ 0 \end{pmatrix} = a_1 \begin{pmatrix} -1 \\ 1 \\ 0 \\ 0 \end{pmatrix} + a_2 \begin{pmatrix} 0 \\ 0 \\ -2 \\ 2 \end{pmatrix} + a_3 \begin{pmatrix} 1 \\ 2 \\ 0 \\ 0 \end{pmatrix} + a_4 \begin{pmatrix} 0 \\ 0 \\ 1 \\ 2 \end{pmatrix}$$

Solving this system gives $a_1 = -1$, $a_2 = 0$, $a_3 = 1$, and $a_4 = 0$, so the solution of the intial value problem is

$$Y(t) = -V^{(1)}(t) + 0V^{(2)}(t) + 1V^{(3)}(t) + 0V^{(4)}(t) = \begin{pmatrix} \cos(2t) + \cos(t) \\ -\cos(2t) + 2\cos(t) \\ -2\sin(2t) - \sin(t) \\ 2\sin(2t) - 2\sin(t) \end{pmatrix}$$

which means $x_1(t) = \cos(2t) + \cos(t)$ and $x_2(t) = -\cos(2t) + 2\cos(t)$. It is not difficult to check that this satisfies the original system of differential equations and the initial conditions. The dynamics of the system, then, is oscillatory, with no damping. ☐

Exercises 6.4

1. Use Euler's equation to rewrite $y(t) = e^{(-2+3i)t}$ and find the real and imaginary parts of y.

2. Use Euler's equation to rewrite $y(t) = e^{(5-2i)t}$ and find the real and imaginary parts of y.

3. Solve the initial value problem

$$\begin{cases} \dfrac{dy_1}{dt} = 2y_1 + 3y_2 \\ \dfrac{dy_2}{dt} = -3y_1 + 2y_2 \end{cases} \quad \text{where} \quad \begin{cases} y_1(0) = -1 \\ y_2(0) = 4 \end{cases}$$

4. Solve the initial value problem

$$\begin{cases} \dfrac{dy_1}{dt} = 4y_1 + 5y_2 \\ \dfrac{dy_2}{dt} = -y_1 + 2y_2 \end{cases} \quad \text{where} \quad \begin{cases} y_1(0) = 1 \\ y_2(0) = -3 \end{cases}$$

5. Solve the initial value problem

$$\begin{cases} \dfrac{dy_1}{dt} = 7y_1 + 20y_2 \\ \dfrac{dy_2}{dt} = -2y_1 - 5y_2 \end{cases} \quad \text{where} \quad \begin{cases} y_1(0) = -2 \\ y_2(0) = 5 \end{cases}$$

6. Solve the initial value problem

$$\begin{cases} \dfrac{dy_1}{dt} = 7y_1 + 17y_2 \\ \dfrac{dy_2}{dt} = y_1 - y_2 \end{cases} \quad \text{where} \quad \begin{cases} y_1(0) = 1 \\ y_2(0) = 0 \end{cases}$$

7. Solve the initial value problem

$$\begin{cases} \dfrac{dy_1}{dt} = 7y_1 + 4y_2 - 8y_3 \\ \dfrac{dy_2}{dt} = \phantom{{}+{}} y_2 + 2y_3 \\ \dfrac{dy_3}{dt} = 4y_1 + 2y_2 - 3y_3 \end{cases} \quad \text{where} \quad \begin{cases} y_1(0) = 1 \\ y_2(0) = -1 \\ y_3(0) = 0 \end{cases}$$

8. Solve the initial value problem $\dfrac{d^2y}{dt^2} + 2\dfrac{dy}{dt} + 5y = 0$ and $y(0) = 2$ and $\dfrac{dy}{dt}(0) = -1$ using the substitution technique of Example 6.13.

9. Solve the initial value problem $\dfrac{d^2y}{dt^2} - 6\dfrac{dy}{dt} + 10y = 0$ and $y(0) = 1$ and $\dfrac{dy}{dt}(0) = 3$ using the substitution technique of Example 6.13.

10. Solve the initial value problem $\dfrac{d^3y}{dt^3} - \dfrac{d^2y}{dt^2} - 4\dfrac{dy}{dt} - 6y = 0$ and $y(0) = 1$, $\dfrac{dy}{dt}(0) = -1$,

    and $\dfrac{d^2y}{dt^2}(0) = 0$

    using the substitution technique of Example 6.13.

6.5   Similarity and Diagonalization

In the previous section, we found that we could solve a system of linear differential equations with constant coefficients if there was a basis for $\mathbf{C}^n$ consisting of eigenvectors of the coefficient matrix. In this section, we want to explore situations under which this occurs and see some of the other conditions that are closely related to this one.

We will begin by developing the idea of a function of a matrix. The simplest functions we know are the polynomials. Although we have already used the idea informally, in writing A^2 for AA, $A^3 = AAA$, for example, it will be helpful to have a formal definition for a polynomial of a matrix.

DEFINITION *If p is the polynomial $p(x) = a_k x^k + \cdots + a_1 x + a_0$ and A is an $n \times n$ matrix, define $p(A)$ by*

$$p(A) = a_k A^k \cdots + a_1 A + a_0 I$$

For example, suppose A is the matrix of Equation (6.2.1), and $p(x) = 2x^2 - 3x - 5$, then $p(A)$ is

$$p(A) = 2A^2 - 3A - 5I$$

$$= 2\begin{pmatrix} 1 & -1 \\ 2 & 4 \end{pmatrix}\begin{pmatrix} 1 & -1 \\ 2 & 4 \end{pmatrix} - 3\begin{pmatrix} 1 & -1 \\ 2 & 4 \end{pmatrix} - 5\begin{pmatrix} 1 & 0 \\ 0 & 1 \end{pmatrix}$$

$$= \begin{pmatrix} -10 & -7 \\ 14 & 11 \end{pmatrix} \tag{6.5.7}$$

Applying functions to matrices allows us to discover the structure of a matrix and to use the structure that is present. One of the keys to using the structure is the *Spectral Mapping Theorem* which relates the eigenvectors and eigenvalues of a function of a matrix to those of the original matrix. Here, we will state the theorem for polynomials, and in the exercises, it is extended to the function $1/x$ and rational functions (that is, quotients of polynomials). The class of functions to which this applies can be further extended to include all "entire functions" and other "analytic functions" from the theory of functions of a complex variable.

THEOREM 6.19   Spectral Mapping Theorem
Let A be an $n \times n$ matrix and let p be a polynomial. If v is an eigenvector of A with eigenvalue λ, then v is an eigenvector of $p(A)$ with eigenvalue $p(\lambda)$.

That is, the spectral mapping theorem says $Av = \lambda v$ implies $p(A)v = p(\lambda)v$.
PROOF   Let $p(x) = a_k x^k + \cdots + a_1 x + a_0$. Since $Av = \lambda v$, we see

$$A^2 v = A(Av) = A(\lambda v) = \lambda A(v) = \lambda^2 v$$

so

$$A^3 v = A(A^2 v) = A(\lambda^2 v) = \lambda^2 Av = \lambda^3 v$$

and similarly, $A^j v = \lambda^j v$ for all positive integers j. Note also that $Iv = v$.
  Now

$$p(A)v = \left(a_k A^k + \cdots + a_1 A + a_0 I \right) v = a_k A^k v + \cdots + a_1 Av + a_0 Iv$$

$$= a_k \lambda^k v + \cdots + a_1 \lambda v + a_0 v = \left(a_k \lambda^k + \cdots + a_1 \lambda + a_0 \right) v = p(\lambda)v$$

∎

We found that $v = (1, -2)$ is an eigenvector for the matrix A of Equation (6.2.1),

$$A = \begin{pmatrix} 1 & -1 \\ 2 & 4 \end{pmatrix}$$

with eigenvalue 3. Letting $p(x) = 2x^2 - 3x - 5$, then recalling the calculation of $p(A)$ in Equation (6.5.7),

$$p(A)v = \begin{pmatrix} -10 & -7 \\ 14 & 11 \end{pmatrix} \begin{pmatrix} 1 \\ -2 \end{pmatrix} = \begin{pmatrix} 4 \\ -8 \end{pmatrix} = 4 \begin{pmatrix} 1 \\ -2 \end{pmatrix}$$

This is the relationship predicted by the Spectral Mapping Theorem since $p(3) = 4$.

The Spectral Mapping Theorem is called that because the the set of eigenvalues of a matrix is called the *spectrum* of the matrix, and the functions "map" the spectrum of the original matrix onto the spectrum of the new matrix.

As an application of the spectral mapping theorem, we want to show that eigenvectors associated with different eigenvalues of a matrix are linearly independent. To do this we first recall a property of polynomials: if y, and x_1, x_2, $\cdots$, x_k are distinct numbers, it is possible to find a polynomial p so that $p(y) \neq 0$ and $p(x_i) = 0$ for $i = 1, \cdots, k$. For example, one convenient polynomial is

$$p(x) = (x - x_1)(x - x_2) \cdots (x - x_k)$$

Since $y \neq x_i$, none of the factors of $p(y)$ is zero and $p(y) \neq 0$.

(Recall that our terminology is that eigenvectors are non–zero!)

THEOREM 6.20

If A is an $n \times n$ matrix with distinct eigenvalues λ_1, λ_2, $\cdots$, λ_k and corresponding eigenvectors v_1, v_2, $\cdots$, v_k, then the eigenvectors v_1, v_2, $\cdots$, v_k are linearly independent.

PROOF   Suppose α_1, α_2, $\cdots$, α_k are scalars so that

$$\alpha_1 v_1 + \alpha_2 v_2 + \cdots + \alpha_k v_k = 0$$

(We want to show that this means all the α's are zero.)

For $1 \leq j \leq k$, let p_j be a polynomial so that $p_j(\lambda_j) \neq 0$ but $p_j(\lambda_i) = 0$ for $i \neq j$. Then,

$$p_j(A)\,(\alpha_1 v_1 + \alpha_2 v_2 + \cdots + \alpha_k v_k) = p_j(A)0$$

or

$$\alpha_1 p_j(A)v_1 + \alpha_2 p_j(A)v_2 + \cdots + \alpha_k p_j(A)v_k = 0$$

and, by the Spectral Mapping Theorem,

$$\alpha_1 p_j(\lambda_1)v_1 + \alpha_2 p_j(\lambda_2)v_2 + \cdots + \alpha_k p_j(\lambda_k)v_k = 0$$

By the choice of p_j, all terms but one are zero; the equation becomes

$$\alpha_j p_j(\lambda_j)v_j = 0$$

Since $p_j(\lambda_j) \neq 0$ and $v_j \neq 0$ (since v_j is an eigenvector), this is only possible when $\alpha_j = 0$. Since this is the case for each j, all the coefficients are zero and the eigenvectors are linearly independent.   ∎

COROLLARY 6.21
*If A is an $n \times n$ matrix with n distinct eigenvalues, then there is a basis for $\mathbf{C}^n$
consisting of eigenvectors for A.*

PROOF Let v_1, v_2, $\cdots$, v_n be eigenvectors associated with the n distinct eigenvalues. By Theorem 6.20, these vectors are linearly independent; since there are n
linearly independent vectors in $\mathbf{C}^n$, they form a basis for $\mathbf{C}^n$. $\blacksquare$

We found that 2 and 3 are eigenvalues for the matrix A of Equation (6.2.1),

$$A = \begin{pmatrix} 1 & -1 \\ 2 & 4 \end{pmatrix}$$

Theorem 6.20 says that the associated eigenvectors $u = (1, -1)$ and $v = (1, -2)$
must be linearly independent, and we can see easily that they are. Of course, we
cannot expect eigenvectors for the same eigenvalue to be independent: $w = (-5, 5)$
and $u = (1, -1)$ are both eigenvectors for the eigenvalue 2 and they are linearly
dependent.

The way we have found the eigenvalues of a matrix, finding the roots of the
characteristic polynomial, makes it seem reasonable that most $n \times n$ matrices have
n distinct eigenvalues: a matrix can have repeated eigenvalues only if the coefficients
of the characteristic polynomial are related to each other in a very special way. If
we change the entries of a matrix very slightly, the coefficients of the characteristic
polynomial change slightly and there is no reason for us to believe they would
continue to be special. Indeed, this thinking can be made precise and we come
to the conclusion that "most" matrices have distinct eigenvalues. An immediate
consequence of Corollary 6.21 is therefore that for "most" $n \times n$ matrices, there is a
basis for $\mathbf{C}^n$ consisting of eigenvectors for the matrix. This has unfortunate results
if we wish to do accurate numerical calculations for a matrix that we expect to have
repeated eigenvalues. Indeed, in programs such as MATLAB that find eigenvectors
and eigenvalues numerically, the program almost always calculates what it expects
is a basis of eigenvectors because after changing the matrix slightly because of
small round-off errors in the calculations, the resulting matrix almost certainly has
distinct eigenvalues and a basis of eigenvectors. Since there is little we can do to
change this situation, we should be aware that this is the case.

In contrasting the example of Equation (6.3.2), in which the unknown functions
were uncoupled, with the examples in the rest of Section 6.3, you may have wondered if the unknown functions in most systems of equations could be decoupled
through some change of variables. In the remainder of the section, we will see that
in every case we were successful in solving the system, the answer is yes and we

will see that the decoupling is equivalent to the solution given in that section. The decoupling is accomplished by a similarity transformation described in the next definition.

DEFINITION *If A and B are $n \times n$ matrices, we say A and B are* similar *if there is an invertible matrix S so that $B = S^{-1}AS$.*

Similar matrices have certain properties in common, but because matrix multiplication is not commutative, they need not be equal. Our goal in using similarity will be to show that a matrix we are interested in is similar to a matrix that is more understandable.

EXAMPLE 6.22

If

$$A = \begin{pmatrix} 1 & -1 \\ 2 & 4 \end{pmatrix} \quad \text{and} \quad S = \begin{pmatrix} 1 & 2 \\ 1 & 3 \end{pmatrix}$$

then

$$B = S^{-1}AS = \begin{pmatrix} 3 & -2 \\ -1 & 1 \end{pmatrix}\begin{pmatrix} 1 & -1 \\ 2 & 4 \end{pmatrix}\begin{pmatrix} 1 & 2 \\ 1 & 3 \end{pmatrix} = \begin{pmatrix} -12 & -35 \\ 6 & 17 \end{pmatrix}$$

The definition says A and B are similar.   ☐

These two matrices do not appear to be very much alike! The next result explains some of the ways in which they are alike.

THEOREM 6.23

If A and B are similar $n \times n$ matrices, the eigenvalues of A and B are the same. Moreover, if v is an eigenvector for $B = S^{-1}AS$, then Sv is an eigenvector for A with the same eigenvalue.

PROOF   Suppose S is an invertible matrix so that $B = S^{-1}AS$. The characteristic polynomial of B is

$$\det(B - \lambda I) = \det(S^{-1}AS - \lambda I)$$

$$= \det(S^{-1}(A - \lambda I)S) = \det(S^{-1})\det(A - \lambda I)\det(S) = \det(A - \lambda I)$$

because the determinants of S and S^{-1} are reciprocals. Thus, A and B have the same characteristic polynomial, hence the same eigenvalues.

Suppose $Bv = \lambda v$ for some non–zero vector v. Since S is invertible, Sv is also non–zero. Now $B = S^{-1}AS$ implies $SB = AS$, so

$$A(Sv) = SBv = S(\lambda v) = \lambda(Sv)$$

and Sv is an eigenvector for A, also with eigenvalue λ. ∎

EXAMPLE 6.22, continued. If

$$A = \begin{pmatrix} 1 & -1 \\ 2 & 4 \end{pmatrix} \quad \text{and} \quad B = \begin{pmatrix} -12 & -35 \\ 6 & 17 \end{pmatrix}$$

then we saw that $B = S^{-1}AS$ where

$$S = \begin{pmatrix} 1 & 2 \\ 1 & 3 \end{pmatrix}$$

We saw in Example (6.2.1) that 2 is an eigenvalue of A. The Theorem above shows that 2 is also an eigenvalue of B: checking we find that $v = (5, -2)$ is an eigenvector for B with eigenvalue 2:

$$\begin{pmatrix} -12 & -35 \\ 6 & 17 \end{pmatrix} \begin{pmatrix} 5 \\ -2 \end{pmatrix} = \begin{pmatrix} 10 \\ -4 \end{pmatrix} = 2 \begin{pmatrix} 5 \\ -2 \end{pmatrix}$$

The theorem says, that under these conditions, Sv is an eigenvector for A. Now,

$$Sv = \begin{pmatrix} 1 & 2 \\ 1 & 3 \end{pmatrix} \begin{pmatrix} 5 \\ -2 \end{pmatrix} = \begin{pmatrix} 1 \\ -1 \end{pmatrix}$$

and it is easily checked that

$$\begin{pmatrix} 1 & -1 \\ 2 & 4 \end{pmatrix} \begin{pmatrix} 1 \\ -1 \end{pmatrix} = \begin{pmatrix} 2 \\ -2 \end{pmatrix}$$

Similarly, 3 is an eigenvalue for B with eigenvector $w = (-7, 3)$ and $Sw = (-1, 2)$ is an eigenvector for A with eigenvalue 3:

$$\begin{pmatrix} 1 & -1 \\ 2 & 4 \end{pmatrix} \begin{pmatrix} -1 \\ 2 \end{pmatrix} = \begin{pmatrix} -3 \\ 6 \end{pmatrix} = 3 \begin{pmatrix} -1 \\ 2 \end{pmatrix}$$

□

The simplest matrices are the diagonal matrices. For example, the matrix form for the system of Equation (6.3.2) is

$$\frac{dY}{dt} = \begin{pmatrix} 3 & 0 \\ 0 & -4 \end{pmatrix} Y$$

in which the coefficient matrix is diagonal; indeed, it is diagonal because the equations are uncoupled! We will see that it is very convenient to work with matrices that are similar to diagonal matrices!

DEFINITION *An $n \times n$ matrix A is called* diagonalizable *if A is similar to a diagonal matrix.*

Suppose A is diagonalizable and S is a matrix so that $D = S^{-1}AS$ is diagonal. It is easy to see that the eigenvalues of a diagonal matrix are just the diagonal entries and the eigenvectors are just the standard basis vectors, that is, $De_j = d_j e_j$ where the diagonal entries of D are $d_1, \cdots, d_n$ and $e_1, \cdots, e_n$ are the standard basis vectors. For example,

$$\begin{pmatrix} -3 & 0 & 0 \\ 0 & 7 & 0 \\ 0 & 0 & 4 \end{pmatrix} \begin{pmatrix} 1 \\ 0 \\ 0 \end{pmatrix} = \begin{pmatrix} -3 \\ 0 \\ 0 \end{pmatrix} = -3 \begin{pmatrix} 1 \\ 0 \\ 0 \end{pmatrix}$$

and

$$\begin{pmatrix} -3 & 0 & 0 \\ 0 & 7 & 0 \\ 0 & 0 & 4 \end{pmatrix} \begin{pmatrix} 0 \\ 1 \\ 0 \end{pmatrix} = \begin{pmatrix} 0 \\ 7 \\ 0 \end{pmatrix} = 7 \begin{pmatrix} 0 \\ 1 \\ 0 \end{pmatrix}$$

Since A and D are similar, they must have the same eigenvalues; in other words, the diagonal matrix D has diagonal entries the eigenvalues of A. Moreover, since the eigenvectors of D are the standard basis vectors, Theorem 6.23 says that Se_1, $Se_2, \cdots$, and Se_n are eigenvectors of A. But Se_j is just the jth column of S! Since S is invertible, its columns are a basis for $\mathbf{C}^n$, so we see that if A is diagonalizable, the columns of S are a basis for $\mathbf{C}^n$ consisting of eigenvectors of A. Conversely, if there is a basis for $\mathbf{C}^n$ consisting of eigenvectors of A, let S be the matrix with these vectors as columns. Since the columns of S are linearly independent, S is invertible and since the columns of S are eigenvectors for A, we see that $D = S^{-1}AS$ is diagonal. We have proved the following theorem.

THEOREM 6.24
An $n \times n$ matrix A is diagonalizable if and only if there is a basis for $\mathbf{C}^n$ consisting of eigenvectors for A.

Although $\mathbf{C}^n$ has been in the background throughout our study of matrix analysis, we have focused $\mathbf{R}^n$ as the more important object and noted occasionally that similar results hold for $\mathbf{C}^n$. In the study of eigenvalues and diagonalization of matrices, however, because matrices with all entries real sometimes have complex eigenvalues, we need to concentrate more on $\mathbf{C}^n$. If A has real entries, has only real eigenvalues, and there is a basis for $\mathbf{C}^n$ consisting of eigenvectors for A, then there is also a basis for $\mathbf{R}^n$ consisting of eigenvectors of A. On the other hand, there are real matrices with complex eigenvalues for which there is a basis for $\mathbf{C}^n$ consisting of eigenvectors, but for which there is no basis for $\mathbf{R}^n$ consisting of eigenvectors.

One important way to understand the matrix S is that it is a change of coordinate matrix: namely it changes coordinates in the eigenvector basis into coordinates in the standard basis.

We have seen examples (see Example 6.6) of matrices for which there is no basis of eigenvectors, so there are matrices that are not diagonalizable. However by Corollary 6.21, since eigenvectors corresponding to distinct eigenvalues are linearly independent, any matrix with distinct eigenvalues is diagonalizable. As noted earlier, "most" matrices have distinct eigenvalues and every matrix is near a matrix with distinct eigenvalues, so most matrices are diagonalizable and if you make a small error in computing the entries of your matrix, you are very likely to end up with a diagonalizable matrix. For this reason, MATLAB or any other numerical software for finding eigenvalues and eigenvectors will usually indicate that your matrix is diagonalizable! In the command `[v,d]=eig(A)`, the matrix v is just the matrix we have called S above, and d is D.

EXAMPLE 6.25
Find an invertible matrix S and a diagonal matrix D so that $D = S^{-1}AS$ where

$$A = \begin{pmatrix} 3 & 3 \\ 2 & -2 \end{pmatrix}$$

SOLUTION In Example 6.12 we found that the eigenvalues of A are 4 and -3 and the corresponding eigenvectors are $(3, 1)$ and $(-1, 2)$. The proof of the Theorem says we construct S by making its columns the basis of eigenvectors we have found, so

$$S = \begin{pmatrix} 3 & -1 \\ 1 & 2 \end{pmatrix}$$

In this case,

$$S^{-1} = \begin{pmatrix} \frac{2}{7} & \frac{1}{7} \\ -\frac{1}{7} & \frac{3}{7} \end{pmatrix}$$

and

$$S^{-1}AS = \begin{pmatrix} \frac{2}{7} & \frac{1}{7} \\ -\frac{1}{7} & \frac{3}{7} \end{pmatrix} \begin{pmatrix} 3 & 3 \\ 2 & -2 \end{pmatrix} \begin{pmatrix} 3 & -1 \\ 1 & 2 \end{pmatrix} = \begin{pmatrix} 4 & 0 \\ 0 & -3 \end{pmatrix}$$

The latter matrix is what we would expect because we know the entries of D along the diagonal, which are the eigenvalues of D, must be the eigenvalues of A. Thus, S diagonalizes A, as we wanted.   ☐

If A is the matrix of Equation (6.2.1), then

$$S = \begin{pmatrix} 1 & 1 \\ -1 & -2 \end{pmatrix}$$

is a matrix whose columns are eigenvectors for A, so

$$D = S^{-1}AS = \begin{pmatrix} 2 & 1 \\ -1 & -1 \end{pmatrix} \begin{pmatrix} 1 & -1 \\ 2 & 4 \end{pmatrix} \begin{pmatrix} 1 & 1 \\ -1 & -2 \end{pmatrix} = \begin{pmatrix} 2 & 0 \\ 0 & 3 \end{pmatrix}$$

is diagonal. Notice that there are many matrices S that diagonalize A because there are infinitely many eigenvectors for each eigenvalue. Moreover, since the diagonal entries of D correspond to the columns of S, the order of the eigenvalues along the diagonal in D can be changed by reordering the columns of S.

It is not a coincidence that the condition for a matrix to be diagonalizable is the same as the condition of Section 6.3 for there to be an easily computable solution to an initial value problem: if A is diagonalizable and $D = S^{-1}AS$, then S^{-1} gives the change of variables that allows the equations to be decoupled. (However, it is usually easier to solve the problem directly in terms of the eigenvectors as in Section 6.3.)

The system of Example 6.12 was

$$\begin{cases} \dfrac{dy_1}{dt} = 3y_1 + 3y_2 \\ \dfrac{dy_2}{dt} = 2y_1 - 2y_2 \end{cases}$$

or $\dfrac{dY}{dt} = AY$. Changing variables by $U = S^{-1}Y$, we get $\dfrac{dU}{dt} = \dfrac{d}{dt}(S^{-1}Y) = S^{-1}\dfrac{dY}{dt}$ because S^{-1} is a constant matrix. Then $\dfrac{dU}{dt} = S^{-1}\dfrac{dY}{dt} = S^{-1}AY =$

$S^{-1}ASU = DU$. In this example, this is

$$\begin{pmatrix} u_1 \\ u_2 \end{pmatrix} = \begin{pmatrix} \frac{2}{7} & \frac{1}{7} \\ -\frac{1}{7} & \frac{3}{7} \end{pmatrix} \begin{pmatrix} y_1 \\ y_2 \end{pmatrix} = \begin{pmatrix} \frac{2}{7}y_1 + \frac{1}{7}y_2 \\ -\frac{1}{7}y_1 + \frac{3}{7}y_2 \end{pmatrix}$$

Since $u_1 = \frac{2}{7}y_1 + \frac{1}{7}y_2$, we get

$$\begin{aligned}
\frac{du_1}{dt} &= \frac{2}{7}\frac{dy_1}{dt} + \frac{1}{7}\frac{dy_2}{dt} \\[2mm]
&= \frac{2}{7}(3y_1 + 3y_2) + \frac{1}{7}(2y_1 - 2y_2) \\[2mm]
&= \frac{8}{7}y_1 + \frac{4}{7}y_2 \\[2mm]
&= 4\left(\frac{2}{7}y_1 + \frac{1}{7}y_2\right) = 4u_1
\end{aligned}$$

Similarly, $u_2 = -\frac{1}{7}y_1 + \frac{3}{7}y_2$

$$\begin{aligned}
\frac{du_2}{dt} &= -\frac{1}{7}\frac{dy_1}{dt} + \frac{3}{7}\frac{dy_2}{dt} \\[2mm]
&= -\frac{1}{7}(3y_1 + 3y_2) + \frac{3}{7}(2y_1 - 2y_2) \\[2mm]
&= \frac{3}{7}y_1 - \frac{9}{7}y_2 \\[2mm]
&= 3\left(\frac{1}{7}y_1 - \frac{3}{7}y_2\right) = 3u_2
\end{aligned}$$

and the equations of the system in the new variables are decoupled.

Exercises 6.5

1. A is a 3×3 matrix;

    the vector u is an eigenvector of A with eigenvalue -1;

    the vector v is an eigenvector of A with eigenvalue 2;

    and the vector w is an eigenvector of A with eigenvalue 5.

    (a) Evaluate: $(A^2 + 4A - I)(u)$.
    (b) Evaluate: $(A^2 + 4A - I)(v)$.
    (c) Evaluate: $(A^2 + 4A - I)(w)$.
    (d) Evaluate: $(A^2 + 4A - I)(3u - 2v + w)$.

2. Let B be an $n \times n$ matrix whose (only) eigenvalues are 1, 2, -3, and 3. Find the four eigenvalues of $B^2 - B + 3I$.

3. Let C be an $n \times n$ matrix that is invertible, let λ be an eigenvalue of C, and let x be an eigenvector of C corresponding to λ. Show that $\lambda \neq 0$ and that λ^{-1} is an eigenvalue of C^{-1} with eigenvector x.

4. Let B be the matrix of Exercise 6.5.2.. If w is an eigenvector of B corresponding to the eigenvalue 2, find $(B - 5I)^{-1}w$.

5. Suppose N is an $n \times n$ matrix such that $N^k = 0$ for some positive integer k. What numbers can be eigenvalues of N.

6. Suppose P is a matrix such that $P = P^2$. What numbers can be eigenvalues of P.

7. Find an invertible matrix S and a diagonal matrix D so that $D = S^{-1}RS$ where

$$R = \begin{pmatrix} 3 & -2 \\ -1 & 4 \end{pmatrix}$$

8. Find an invertible matrix S and a diagonal matrix D so that $D = S^{-1}AS$ where

$$A = \begin{pmatrix} -6 & -4 & 1 \\ 6 & 5 & 0 \\ -8 & -4 & 3 \end{pmatrix}$$

9. Find an invertible matrix S and a diagonal matrix D so that $D = S^{-1}BS$ where

$$B = \begin{pmatrix} 5 & -4 & 4 \\ 2 & -1 & 2 \\ -1 & 1 & 0 \end{pmatrix}$$

10. Do the necessary calculations, then explain why it is not possible to diagonalize the matrix

$$J = \begin{pmatrix} 0 & -3 & -7 \\ 1 & 5 & 10 \\ -1 & -2 & -3 \end{pmatrix}$$

11. Change variables in the system

$$\begin{cases} \dfrac{dy_1}{dt} = y_1 + 4y_2 \\ \dfrac{dy_2}{dt} = -2y_1 + 7y_2 \end{cases}$$

so that in the new variables, the equations are decoupled. (Compare Exercise 6.3.2.)

6.6 The Matrix Exponential and Systems of Differential Equations

Motivated by the single equation case, $\frac{dy}{dt} = ay$, we have written our systems as $\frac{dY}{dt} = AY$ where Y is a vector function and A is a matrix. In this section, we will develop the analogy more deeply. The solution to the initial value problem $\frac{dy}{dt} = ay$ with $y(0) = c$ is $y(t) = ce^{at}$. If it made sense, the solution of the initial value problem $\frac{dY}{dt} = AY$ with $Y(0) = C$ should involve $Y(t) = e^{At}$ if such a thing made sense. We will show that we can make an appropriate definition of matrix exponential, and that with this definition, the solution of the initial value problem is $Y(t) = e^{tA}C$ (where the order must have the constant C last because e^{tA} is defined to be a matrix and C is a vector). The definition is exactly analogous to the definition of the exponential function from calculus using infinite series.

DEFINITION If B is an $n \times n$ matrix, the matrix e^B is defined by the series

$$e^B = I + B + \frac{1}{2!}B^2 + \frac{1}{3!}B^3 + \frac{1}{4!}B^4 + \cdots$$

Of course, we should worry about whether the series converges, and so on, but we will instead just state the facts and leave the proofs to more advanced books such as *Linear Algebra for Engineering and Science*, by Carl C. Cowen. The following theorem gives the important information about the matrix exponential function. In particular, the theorem says that it provides the solution to initial value problems of the sort that we are considering. The proof is analogous to the proof in calculus that the series for e^t converges absolutely for every t. Our biggest difficulty is that we do not have a definition for convergence of series of matrices and we have not developed the notion of "norm of a matrix" that makes it possible to define absolute convergence in this context.

THEOREM 6.26

Let A be an $n \times n$ matrix and let C be a vector in $\mathbf{R}^n$ or $\mathbf{C}^n$. Then

- *The series for e^{tA} converges for all t.*

- *The function $y(t) = e^{tA}C$ has derivative $\dfrac{dy}{dt} = Ae^{tA}C$.*

- *The function $y(t) = e^{tA}C$ satisfies the initial contition $y(0) = C$.*

- *Moreover, the function $y(t) = e^{tA}C$ is the only solution of the initial value problem $\dfrac{dY}{dt} = AY$ and $Y(0) = C$.*

Idea of PROOF   The series for $y(t) = e^{tA}C$ is

$$y(t) = C + tAC + \frac{t^2}{2!}A^2C + \frac{t^3}{3!}A^3C + \frac{t^4}{4!}A^4C + \cdots$$

Accepting that the series can be differentiated term by term, we find

$$\frac{dy}{dt} = 0 + AC + \frac{2t}{2!}A^2C + \frac{3t^2}{3!}A^3C + \frac{4t^3}{4!}A^4C + \cdots$$

$$= AC + \frac{t}{1}A^2C + \frac{t^2}{2!}A^3C + \frac{t^3}{3!}A^4C + \cdots$$

$$= A\left(I + \frac{t}{1}A + \frac{t^2}{2!}A^2 + \frac{t^3}{3!}A^3 + \cdots\right)C = Ae^{tA}C$$

and

$$y(0) = C + 0 + 0 + \cdots = C$$

The remainder of the proof will be omitted.   ∎

Although the theorem gives the solution of the initial value problem $\dfrac{dY}{dt} = AY$ and $Y(0) = C$, it gives it as the sum of an infinite series! We can handle this for a few simple examples, but we need a better way to deal with the problem.

EXAMPLE 6.27

If D is a diagonal matrix, then e^{tD} is a diagonal matrix whose diagonal entries are the exponentials of the diagonal entries of D.

Indeed, suppose

$$D = \begin{pmatrix} d_1 & 0 & \cdots & 0 \\ 0 & d_2 & & 0 \\ \vdots & & \ddots & \vdots \\ 0 & 0 & \cdots & d_n \end{pmatrix}$$

Then

$$e^{tD} = I + tD + \frac{t^2}{2!}D^2 + \cdots$$

$$= \begin{pmatrix} 1 \cdots 0 \\ \ddots \\ 0 \cdots 1 \end{pmatrix} + t \begin{pmatrix} d_1 \cdots 0 \\ \ddots \\ 0 \cdots d_n \end{pmatrix} + \frac{t^2}{2!} \begin{pmatrix} d_1^2 \cdots 0 \\ \ddots \\ 0 \cdots d_n^2 \end{pmatrix} + \cdots$$

$$= \begin{pmatrix} 1 \cdots 0 \\ \ddots \\ 0 \cdots 1 \end{pmatrix} + \begin{pmatrix} td_1 \cdots 0 \\ \ddots \\ 0 \cdots td_n \end{pmatrix} + \begin{pmatrix} \frac{(td_1)^2}{2!} \cdots 0 \\ \ddots \\ 0 \cdots \frac{(td_n)^2}{2!} \end{pmatrix} + \cdots$$

$$= \begin{pmatrix} 1 + td_1 + \frac{(td_1)^2}{2!} + \cdots\cdots & 0 \\ & \ddots \\ 0 & \cdots 1 + td_n + \frac{(td_n)^2}{2!} + \cdots \end{pmatrix}$$

$$= \begin{pmatrix} e^{td_1} & 0 & \cdots & 0 \\ 0 & e^{td_2} & & 0 \\ \vdots & & \ddots & \vdots \\ 0 & 0 & \cdots & e^{td_n} \end{pmatrix}$$

$\square$

However, most matrices are much harder to exponentiate!

EXAMPLE 6.28

Let

$$B = \begin{pmatrix} 1 & 1 \\ 0 & 1 \end{pmatrix}$$

To find e^{tB}, we need to find the powers of B and sum the infinite series.

$$B^2 = \begin{pmatrix} 1 & 1 \\ 0 & 1 \end{pmatrix} \begin{pmatrix} 1 & 1 \\ 0 & 1 \end{pmatrix} = \begin{pmatrix} 1 & 2 \\ 0 & 1 \end{pmatrix}$$

$$B^3 = BB^2 = \begin{pmatrix} 1 & 1 \\ 0 & 1 \end{pmatrix} \begin{pmatrix} 1 & 2 \\ 0 & 1 \end{pmatrix} = \begin{pmatrix} 1 & 3 \\ 0 & 1 \end{pmatrix}$$

$$B^4 = BB^3 = \begin{pmatrix} 1 & 1 \\ 0 & 1 \end{pmatrix} \begin{pmatrix} 1 & 3 \\ 0 & 1 \end{pmatrix} = \begin{pmatrix} 1 & 4 \\ 0 & 1 \end{pmatrix}$$

indeed, it is not difficult to prove that, for all positive integers k,

$$B^k = \begin{pmatrix} 1 & k \\ 0 & 1 \end{pmatrix}$$

Therefore,

$$e^{tB} = I + tB + \frac{t^2}{2!}B^2 + \frac{t^3}{3!}B^3 + \cdots$$

$$= \begin{pmatrix} 1 & 0 \\ 0 & 1 \end{pmatrix} + t\begin{pmatrix} 1 & 1 \\ 0 & 1 \end{pmatrix} + \frac{t^2}{2!}\begin{pmatrix} 1 & 2 \\ 0 & 1 \end{pmatrix} + \frac{t^3}{3!}\begin{pmatrix} 1 & 3 \\ 0 & 1 \end{pmatrix} + \cdots$$

$$= \begin{pmatrix} 1 & 0 \\ 0 & 1 \end{pmatrix} + t\begin{pmatrix} t & t \\ 0 & t \end{pmatrix} + \begin{pmatrix} \frac{t^2}{2!} & \frac{2t^2}{2!} \\ 0 & \frac{t^2}{2!} \end{pmatrix} + \begin{pmatrix} \frac{t^3}{3!} & \frac{3t^3}{3!} \\ 0 & \frac{t^3}{3!} \end{pmatrix} + \cdots$$

$$= \begin{pmatrix} 1 + t + \frac{t^2}{2!} + \frac{t^3}{3!} + \cdots & 0 + t + t^2 + \frac{t^3}{2!} + \cdots \\ 0 & 1 + t + \frac{t^2}{2!} + \frac{t^3}{3!} + \cdots \end{pmatrix}$$

$$= \begin{pmatrix} e^t & te^t \\ 0 & e^t \end{pmatrix}$$

$\square$

·Notice that in order to compute the matrix exponential from the definition, we must be able to find all the powers of the matrix and then recognize the sum of an infinite series. This is usually difficult to do. In particular, it looks too difficult to carry out in the case of the initial value problem of Equation (6.3.4). We find a way around this difficulty by proving the spectral mapping theorem for the matrix exponential!

THEOREM 6.29   *Spectral Mapping Theorem for the Matrix Exponential*
If A is an $n \times n$ matrix with eigenvector v with eigenvalue λ, then v is an eigenvector of the matrix e^{tA} with eigenvalue $e^{\lambda t}$.

PROOF   We have

$$e^{tA}v = \left(I + tA + \frac{t^2}{2!}A^2 + \frac{t^3}{3!}A^3 + \cdots \right)v$$

$$= Iv + tAv + \frac{t^2}{2!}A^2v + \frac{t^3}{3!}A^3v + \cdots$$

$$= v + \lambda t v + \frac{(\lambda t)^2}{2!} v + \frac{(\lambda t)^3}{3!} v + \cdots$$

$$= \left(1 + \lambda t + \frac{(\lambda t)^2}{2!} + \frac{(\lambda t)^3}{3!} + \cdots \right) v$$

$$= e^{\lambda t} v$$

∎

Applying the spectral mapping theorem to several eigenvectors simultaneously, we get the solution to our problem in many cases. This is the same result as Corollary 6.11 from Section 6.3 where we approached the problem somewhat differently.

COROLLARY 6.30

Let A be an $n \times n$ matrix with eigenvectors v_1, v_2, $\cdots$, v_k corresponding to the eigenvalues λ_1, λ_2, $\cdots$, λ_k. If $C = \alpha_1 v_1 + \alpha_2 v_2 + \cdots + \alpha_k v_k$ then the unique solution of the initial value problem $\dfrac{dY}{dt} = AY$ and $Y(0) = C$ is

$$Y(t) = e^{tA} C = \alpha_1 e^{\lambda_1 t} v_1 + \alpha_2 e^{\lambda_2 t} v_2 + \cdots + \alpha_k e^{\lambda_k t} v_k$$

In particular, if there is a basis for $\mathbf{C}^n$ consisting of eigenvectors for A, then every vector C can be written as a linear combination of the eigenvectors and this theorem gives the solution of initial value problems! The interpretation of this corollary is that even though we do not have a good global formula for e^{tA}, we have a formula that is good on each eigenspace and we can use the prinicple of superposition to put them together to compute $e^{tA} C$ for any vector that can be written as a linear combination of eigenvectors.

EXAMPLE 6.31

(Example 6.12 revisited.) Solve the initial value problem

$$\begin{cases} \dfrac{dy_1}{dt} = 3y_1 + 3y_2 \\ \dfrac{dy_2}{dt} = 2y_1 - 2y_2 \end{cases} \qquad \begin{cases} y_1(0) = 1 \\ y_2(0) = 5 \end{cases}$$

SOLUTION To begin, we rewrite the system of equations in matrix form: $\dfrac{dY}{dt} = AY$ and $Y(0) = C$, where

$$A = \begin{pmatrix} 3 & 3 \\ 2 & -2 \end{pmatrix} \quad \text{and} \quad C = \begin{pmatrix} 1 \\ 5 \end{pmatrix}$$

Now, the solution is $y(t) = e^{tA}C$, but this is not explicit since we have trouble calculating the matrix exponential. To get around this difficulty, we will use Theorem 6.29. Before we can do this, though, we need to find the eigenvectors and eigenvalues for A, then, if we get a basis of eigenvectors, we will write C as a linear combination of them, and then use Theorem 6.29 to write down the answer.

As was calculated earlier, the eigenvalues of A are 4 and -3 and $(3, 1)$ is a basis for the eigenspace corresponding to $\lambda = 4$ and $(-1, 2)$ is a basis for the eigenspace corresponding to $\lambda = -3$. Solving an appropriate system to write C as a linear combination of the eigenvectors, we find

$$C = \begin{pmatrix} 1 \\ 5 \end{pmatrix} = \begin{pmatrix} 3 \\ 1 \end{pmatrix} + 2 \begin{pmatrix} -1 \\ 2 \end{pmatrix}$$

Theorem 6.29 tells us how to write the solution down:

$$y(t) = e^{tA}C = e^{4t} \begin{pmatrix} 3 \\ 1 \end{pmatrix} + 2e^{-3t} \begin{pmatrix} -1 \\ 2 \end{pmatrix}$$

or to put it in the form the problem was stated in, $y_1(t) = 3e^{4t} - 2e^{-3t}$ and $y_2(t) = e^{4t} + 4e^{-3t}$. This answer can be checked in the original system for accuracy. ▯

As we saw in Example 6.6, there are matrices for which there is no basis of $\mathbf{C}^n$ consisting of eigenvectors for the matrix. For these matrices, the solution of an initial value problem is still $e^{tA}C$, but Corollary 6.30 does not give a formula that can be used to compute this explicitly. The solution of these more difficult problems is deferred until Section 6.7.

It is not a coincidence that the condition for a matrix to be diagonalizable (Theorem 6.24) is the same as the condition implicit in Corollary 6.30 for there to be an easily computable solution to an initial value problem: if A is diagonalizable, we can easily compute the matrix exponential. (However, if one is only solving an initial value problem, it is probably easier to find the solution using the eigenvector expansion rather than finding e^{tA}.)

Suppose $D = S^{-1}AS$. Then $D^2 = S^{-1}ASS^{-1}AS = S^{-1}A^2S$, so D^2 and A^2 are similar as well. In the same way, $D^j = S^{-1}A^jS$, so

$$e^{tD} = I + tD + \frac{t^2}{2!}D^2 + \frac{t^3}{3!}D^3 + \cdots$$

$$= I + tS^{-1}AS + \frac{t^2}{2!}S^{-1}A^2S + \frac{t^3}{3!}S^{-1}A^3S + \cdots$$

$$= S^{-1}\left(I + tA + \frac{t^2}{2!}A^2 + \frac{t^3}{3!}A^3 + \cdots\right)S = S^{-1}e^{tA}S$$

It follows that $e^{tA} = Se^{tD}S^{-1}$. Since e^{tD} is easily computable as in Example 6.27, we can find e^{tA}.

For the matrix of Equation (6.2.1), we have

$$e^{tA} = Se^{tD}S^{-1} = \begin{pmatrix} 1 & 1 \\ -1 & -2 \end{pmatrix}\begin{pmatrix} e^{2t} & 0 \\ 0 & e^{3t} \end{pmatrix}\begin{pmatrix} 2 & 1 \\ -1 & -1 \end{pmatrix}$$

$$= \begin{pmatrix} 2e^{2t} - e^{3t} & e^{2t} - e^{3t} \\ -2e^{2t} + 2e^{3t} & -e^{2t} + 2e^{2t} \end{pmatrix}$$

Exercises 6.6

1.(a)   Let

$$N = \begin{pmatrix} 0 & 1 & 0 & 0 \\ 0 & 0 & 1 & 0 \\ 0 & 0 & 0 & 1 \\ 0 & 0 & 0 & 0 \end{pmatrix}$$

Find the eigenvectors and eigenvalues of N.

(b)   In spite of your answer to part (a), Theorem 6.26 is still usable! Use the definition of the matrix exponential function as an infinite series to compute e^{tN}.

2.   Use your solution to Exercise 6.6.1. to solve the initial value problem Solve the initial value problem

$$\begin{cases} \dfrac{dy_1}{dt} = y_2 \\[2mm] \dfrac{dy_2}{dt} = y_3 \\[2mm] \dfrac{dy_3}{dt} = y_4 \\[2mm] \dfrac{dy_4}{dt} = 0 \end{cases} \quad \text{and} \quad \begin{cases} y_1(0) = 2 \\ y_2(0) = 3 \\ y_3(0) = -1 \\ y_4(0) = 1 \end{cases}$$

Check your answer by relating this problem the technique of Example 6.13.

3.(a)   Let

$$A = \begin{pmatrix} -6 & -5 & -2 & 3 & 1 \\ 10 & 9 & 3 & -5 & -2 \\ -9 & -8 & -4 & 4 & 1 \\ -6 & -4 & -5 & 2 & -1 \\ 19 & 16 & 10 & -8 & -1 \end{pmatrix}$$

Use the definition of the matrix exponential function as an infinite series to compute e^{tA}.

(b)   Use your answer to solve the initial value problem $\dfrac{dY}{dt} = AY$ and $Y(0) = (1, 0, -1, 1, 2)$.

4.   Let

$$P = \begin{pmatrix} 2 & 4 \\ 1 & -1 \end{pmatrix}$$

Calculate e^{tP}. (Compare with Exercise 6.2.3.)

5.   Let

$$R = \begin{pmatrix} 3 & -2 \\ -1 & 4 \end{pmatrix}$$

Calculate e^{tR}. (Compare with Exercise 6.5.7.)

6.   Let

$$B = \begin{pmatrix} 5 & -4 & 4 \\ 2 & -1 & 2 \\ -1 & 1 & 0 \end{pmatrix}$$

Calculate e^{tB}. (Compare with Exercise 6.5.9.)

6.7   More on Differential Equations

In this section, we will address the problem of solving a system of differential equations when the characteristic polynomial of the coefficient matrix has multiple roots. The first step is Taylor's theorem for the matrix exponential function.

PROPOSITION 6.32
If A is an $n \times n$ matrix and α is any number, then

$$e^{tA} = e^{\alpha t}\left[I + t(A - \alpha I) + \frac{t^2}{2!}(A - \alpha I)^2 + \frac{t^3}{3!}(A - \alpha I)^3 + \cdots \right]$$

We have seen that we can solve systems of differential equations as long as there is a basis of eigenvectors for the coefficient matrix. The following result gives a replacement for the basis of eigenvectors when there is none.

THEOREM 6.33
Let A be an $n \times n$ matrix with characteristic polynomial

$$p(\lambda) = \pm(\lambda - \lambda_1)^{r_1}(\lambda - \lambda_2)^{r_2} \cdots (\lambda - \lambda_k)^{r_k}$$

where the λ_i are distinct, and let M_i denote the nullspace of $(A - \lambda_i I)^{r_i}$ for $i = 1, 2, \ldots, k$. Then

(i) The dimension of M_i is r_i.

For each $i = 1, 2, \ldots, k$, let $v_{i,1}, v_{i,2}, \ldots, v_{i,r_i}$ be a basis for M_i. Then

(ii) $\{v_{i,j} : i = 1, \ldots, k \text{ and } j = 1, \ldots, r_i\}$ is a basis for $\mathbf{C}^n$.

DEFINITION *The subspace M_i is called the* generalized eigenspace *of A associated with λ_i and the non-zero vectors in M_i are called* generalized eigenvectors *corresponding to λ_i.*

EXAMPLE 6.34

For the matrix

$$A = \begin{pmatrix} 2 & -1 & 1 \\ 2 & 1 & -2 \\ 1 & -3 & 2 \end{pmatrix}$$

we will determine bases for the generalized eigenspaces M_i. The characteristic polynomial of A is

$$\det(A - \lambda I) = -\lambda^3 + 5\lambda^2 - 3\lambda - 9 = -(\lambda - 3)^2(\lambda + 1)$$

In the notation of the theorem, $\lambda_1 = 3$ and $\lambda_2 = -1$ are the eigenvalues and $r_1 = 2$ and $r_2 = 1$.

Now M_1 is the nullspace of

$$(A - 3I)^2 = \begin{pmatrix} -1 & -1 & 1 \\ 2 & -2 & -2 \\ 1 & -3 & -1 \end{pmatrix}^2 = \begin{pmatrix} 0 & 0 & 0 \\ -8 & 8 & 8 \\ -8 & 8 & 8 \end{pmatrix}.$$

An easy calculation shows that $v_{1,1} = (1, 1, 0)$ and $v_{1,2} = (0, -1, 1)$ form a basis for M_1. Similarly, M_2 is the nullspace of $(A + I) = \begin{pmatrix} 3 & -1 & 1 \\ 2 & 2 & -2 \\ 1 & -3 & 3 \end{pmatrix}$, and $v_{2,1} = (0, 1, 1)$ is a basis for M_2.

Theorem 6.33 says that $\{v_{1,1}, v_{1,2}, v_{2,1}\}$ is a basis for $\mathbf{C}^3$; it is easily checked that this is so. $\square$

Next, we see how this result can be used in the solution of differential equations.

COROLLARY 6.35

Let A be an $n \times n$ matrix with characteristic polynomial

$$p(\lambda) = \pm(\lambda - \lambda_1)^{r_1}(\lambda - \lambda_2)^{r_2} \cdots (\lambda - \lambda_k)^{r_k}$$

where the λ_i are distinct, and let M_i denote the nullspace of $(A - \lambda_i I)^{r_i}$ for $i = 1, 2, \ldots, k$. If $C = w_1 + w_2 + \cdots + w_k$ where the w_j are generalized eigenvectors for A, then the unique solution of the system $\dfrac{dY}{dt} = AY$ and $Y(0) = C$ is

$$Y(t) = e^{\lambda_1 t} \left[w_1 + t(A - \lambda_1 I)w_1 + \frac{t^2}{2!}(A - \lambda_1 I)^2 w_1 + \cdots + \frac{t^{r_1}}{r_1!}(A - \lambda_1 I)^{r_1} w_1 \right]$$

$$+ e^{\lambda_2 t} \left[w_2 + t(A - \lambda_2 I)w_2 + \frac{t^2}{2!}(A - \lambda_2 I)^2 w_2 + \cdots + \frac{t^{r_2}}{r_2!}(A - \lambda_2 I)^{r_2} w_2 \right]$$

$$\cdots + e^{\lambda_k t} \left[w_k + t(A - \lambda_k I)w_k + \frac{t^2}{2!}(A - \lambda_k I)^2 w_k + \cdots + \frac{t^{r_k}}{r_k!}(A - \lambda_k I)^{r_k} w_k \right]$$

This is an extension of Corollary 6.11. The essence of Corollary 6.11 is that although there is not (usually) a convenient formula for e^{tA} valid for all vectors in the space, there is a convenient formula for eigenvectors and the principle of superposition allows us to use it to give answers for vectors that are linear combinations of eigenvectors. Corollary 6.35 gives a formula for e^{tA} that is valid for generalized eigenvectors and the principle of superposition allows us to use this formula for vectors that are linear combinations of generalized eigenvectors, that is, for every vector.

EXAMPLE 6.36

Solve the initial value problem

$$\begin{cases} \dfrac{dy_1}{dt} = 2y_1 - y_2 + y_3 \\[2mm] \dfrac{dy_2}{dt} = 2y_1 + y_2 - 2y_3 \\[2mm] \dfrac{dy_3}{dt} = y_1 - 3y_2 + 2y_3 \end{cases} \quad \text{and} \quad \begin{cases} y_1(0) = 1 \\ y_2(0) = 2 \\ y_3(0) = -3. \end{cases}$$

This is just $\dfrac{dY}{dt} = AY$ and $Y(0) = C$ where the matrix A is the matrix of Example 6.34 and $C = (1, 2, -3)$. The solution is $Y(t) = e^{tA}C$, but this is not explicit. Corollary 6.35 above says that the calculation will be easy if we first write $C = w_1 + w_2$ where w_1 and w_2 are in M_1 and M_2 respectively. We express C in the basis $\{v_{1,1}, v_{1,2}, v_{2,1}\}$, group to form w_1, and then do the computation: $(1, 2, -3) = 1v_{1,1} - 2v_{1,2} - v_{2,1} = w_1 + w_2$ where $w_1 = 1v_{1,1} - 2v_{1,2} = (1, 3, -2)$ and $w_2 = -v_{2,1} = (0, -1, -1)$. Thus,

$$Y(t) = e^{tA}(w_1 + w_2) = e^{tA}w_1 + e^{tA}w_2.$$

Since w_1 is in M_1 and w_2 is in M_2, we use the Taylor expansion for the exponential with $\alpha = \lambda_1 = 3$ for w_1 and the Taylor expansion for the exponential with $\alpha = \lambda_2 = -1$ for w_2. Since w_1 is in the nullspace of $(A - 3I)^2$, all terms in the corresponding Taylor expansion with powers 2 or higher give zero. As in the corollary, we have

$$Y(t) = e^{3t}[w_1 + t(A - 3I)w_1] + e^{-t}[w_2] = e^{3t}w_1 + te^{3t}(A - 3I)w_1 + e^{-t}w_2$$

Since

$$(A - 3I)w_1 = \begin{pmatrix} -1 & -1 & 1 \\ 2 & -2 & -2 \\ 1 & -3 & -1 \end{pmatrix} \begin{pmatrix} 1 \\ 3 \\ -2 \end{pmatrix} = \begin{pmatrix} -6 \\ 0 \\ -6 \end{pmatrix},$$

we find that
$$\begin{cases} y_1(t) = e^{3t} - 6te^{3t} \\ y_2(t) = 3e^{3t} - e^{-t} \\ y_3(t) = -2e^{3t} - 6te^{3t} - e^{-t}. \end{cases} \qquad \square$$

Exercises 6.7

1.(a) Let

$$B = \begin{pmatrix} -3 & -5 & -2 & 3 & 1 \\ 10 & 12 & 3 & -5 & -2 \\ -9 & -8 & -1 & 4 & 1 \\ -6 & -4 & -5 & 5 & -1 \\ 19 & 16 & 10 & -8 & 2 \end{pmatrix}$$

Show that the generalized eigenspace corresponding to the eigenvalue 3 is all of $\mathbf{R}^5$. (See Exercise 6.6.3.)

(b) Solve the initial value problem $\dfrac{dY}{dt} = BY$ and $Y(0) = (1, 0, -1, 1, 2)$.

2. Solve the initial value problem

$$\begin{cases} \dfrac{du}{dt} = 3u + v \\ \dfrac{dv}{dt} = -u + v \end{cases} \quad \text{where} \quad \begin{cases} u(0) = 2 \\ v(0) = -1 \end{cases}$$

3. Solve the initial value problem

$$\begin{cases} \dfrac{du}{dt} = u - 4v \\ \dfrac{dv}{dt} = u + 5v \end{cases} \quad \text{where} \quad \begin{cases} u(0) = 2 \\ v(0) = -1 \end{cases}$$

4. Solve the initial value problem

$$\begin{cases} \dfrac{du}{dt} = 4u \qquad - 2w \\ \dfrac{dv}{dt} = u - v + 5w \\ \dfrac{dw}{dt} = u - v + 3w \end{cases} \quad \text{where} \quad \begin{cases} u(0) = 0 \\ v(0) = 3 \\ w(0) = -2 \end{cases}$$

5. Solve the initial value problem

$$\begin{cases} \dfrac{du}{dt} = -u + 3v + 5w \\ \dfrac{dv}{dt} = 2u + v - 2w \\ \dfrac{dw}{dt} = u + 2v - w \end{cases} \quad \text{where} \quad \begin{cases} u(0) = 1 \\ v(0) = -1 \\ w(0) = 2 \end{cases}$$

6. Solve the initial value problem

$$\begin{cases} \dfrac{dy_1}{dt} = 3y_1 + y_2 - 6y_3 + 4y_4 \\ \dfrac{dy_2}{dt} = 8y_1 + y_2 - 12y_3 + 8y_4 \\ \dfrac{dy_3}{dt} = 4y_1 + 2y_2 - 9y_3 + 5y_4 \\ \dfrac{dy_4}{dt} = \qquad y_2 - 2y_3 \end{cases} \quad \text{and} \quad \begin{cases} y_1(0) = 2 \\ y_2(0) = -6 \\ y_3(0) = -6 \\ y_4(0) = 1. \end{cases}$$

NOTE: the characteristic polynomial of the matrix $\begin{pmatrix} 3 & 1 & -6 & 4 \\ 8 & 1 & -12 & 8 \\ 4 & 2 & -9 & 5 \\ 0 & 1 & -2 & 0 \end{pmatrix}$

is $(\lambda + 1)^3(\lambda + 2)$.

7
Hermitian Matrices

7.1 Diagonalization of Hermitian Matrices

Diagonalizable matrices are more convenient to work with than non–diagonalizable ones, and if a matrix is unitarily diagonalizable, important structural features of the matrix are preserved in the diagonalization because unitary matrices represent rigid motions of space. In this section, we will work with the class of Hermitian matrices, the most important class of unitarily diagonalizable matrices. These matrices are so important because real symmetric matrices, which are Hermitian, occur in applications so frequently. Sometimes real symmetric matrices occur because of some physical symmetry in the objects being described, but more often, they occur as a consequence of some physical law, for example in circuit theory, Ohm's and Kirchhoff's Laws imply certain matrices must be Hermitian.

We will begin by reviewing the definition of unitary matrices, which have been mentioned earlier but which we have not studied in a systematic way.

DEFINITION *An $n \times n$ matrix U is called* unitary *if $U' = U^{-1}$.*

If U is a real unitary matrix, $U^{-1} = U' = U^t$, we also say U is an orthogonal *matrix.*

THEOREM 7.1
Let U be an $n \times n$ matrix. The following conditions on U are equivalent.
- *(i) U is unitary.*
- *(ii)   For all vectors v, $\|Uv\| = \|v\|$.*
- *(iii) For all vectors x, y, $\langle Ux, Uy \rangle = \langle x, y \rangle$.*

Remember that the phrase "the conditions are equivalent" means that if any one of them is true all the rest are true, and if any one is false, all the rest are false. Thus, this theorem explains more fully than the definition what unitary matrices are. The second condition says that U preserves lengths: if we imagine U as transforming vectors, it says that the length of the vector and the transformed vector are the same. The third condition says that U preserves inner products. Since the size of the angle between two vectors can be found using inner products, this means that U preserves angles: the angle between two vectors is the same as the angle between the transformed vectors. These two conditions, preserving lengths and angles, mean that a unitary matrix describes a rigid motion of space: transforming an object with a unitary matrix does not change it's shape. Rotations, reflections, and translations are rigid motions that are familiar from geometry, but a unitary matrix cannot describe a translation because multiplication by a unitary matrix fixes the origin, $U0 = 0$, and no point is fixed by a translation. However, both reflections and rotations are given by unitary matrices.

PROOF   (of Theorem 7.1)

(i) $\Rightarrow$(ii): Since $U'U = I$, for all vectors v,

$$\|Uv\|^2 = \langle Uv, Uv \rangle = \langle U'Uv, v \rangle = \langle v, v \rangle = \|v\|^2$$

(ii) $\Rightarrow$(iii): The *polarization identity*

$$\langle x, y \rangle = \frac{1}{4}\left(\|x + y\|^2 - \|x - y\|^2 - i\|x + iy\|^2 + i\|x - iy\|^2\right)$$

expresses the inner product in terms of the norm. The identity is proved by writing out all the norms on the right-hand side as inner products and collecting terms. Using the polarization identity and assumption (ii) applied to $v = x \pm y$ and $v = x \pm iy$ gives

$$\langle Ux, Uy \rangle = \frac{1}{4}\left(\|Ux + Uy)\|^2 - \|Ux - Uy)\|^2 - i\|Ux + iUy)\|^2 + i\|Ux - iUy)\|^2\right)$$

$$= \frac{1}{4}\left(\|U(x + y)\|^2 - \|U(x - y)\|^2 - i\|U(x + iy)\|^2 + i\|U(x - iy)\|^2\right)$$

$$= \frac{1}{4}\left(\|x + y\|^2 - \|x - y\|^2 - i\|x + iy\|^2 + i\|x - iy\|^2\right)$$

$$. = \langle x, y \rangle.$$

(iii) $\Rightarrow$(i): By (iii), for all vectors x, y, we have

$$\langle U'Ux, y \rangle = \langle Ux, Uy \rangle = \langle x, y \rangle$$

which gives

$$\langle (U'Ux - x), y \rangle = 0$$

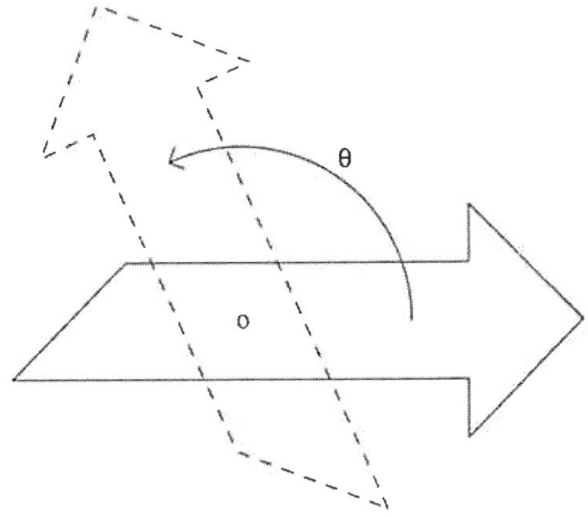

FIGURE 7.1

Rotation by angle θ

Since this holds for all y, we conclude $U'Ux - x = 0$ for all x. That is, $U'Ux = x$ for all x so $U'U = I$ and $U' = U^{-1}$. ∎

For example, the matrix

$$U = \begin{pmatrix} \cos\theta & -\sin\theta \\ \sin\theta & \cos\theta \end{pmatrix}$$

is unitary since

$$U'U = \begin{pmatrix} \cos\theta & \sin\theta \\ -\sin\theta & \cos\theta \end{pmatrix} \begin{pmatrix} \cos\theta & -\sin\theta \\ \sin\theta & \cos\theta \end{pmatrix}$$

$$= \begin{pmatrix} (\cos\theta)^2 + (\sin\theta)^2 & 0 \\ 0 & (\cos\theta)^2 + (\sin\theta)^2 \end{pmatrix} = I$$

where the final equality uses the familiar trigonometric identity $(\cos\theta)^2 + (\sin\theta)^2 = 1$. This unitary matrix represents the rigid motion of rotating the plane through the angle θ (see Figure 7.1).

The matrix

$$R = \begin{pmatrix} \cos^2\theta - \sin^2\theta & 2\sin\theta\cos\theta \\ 2\sin\theta\cos\theta & \sin^2\theta - \cos^2\theta \end{pmatrix} \tag{7.1.1}$$

is also unitary and represents reflection in the line $y = (\tan\theta)x$ (see Figure 7.2 and Exercise 7.1.1.).

One can also recognize unitary matrices by their columns.

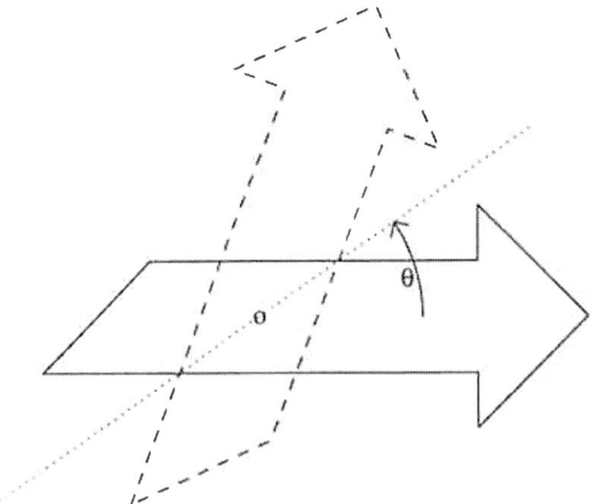

FIGURE 7.2

Reflection in $y = (\tan\theta)x$

PROPOSITION 7.2

An $n \times n$ matrix U is unitary if and only if the columns of U are an orthonormal basis for $\mathbf{R}^n$ ($\mathbf{C}^n$).

PROOF   The corollary follows immediately from Theorem 4.24 by noting that for square matrices, the condition $U'U = I$ is equivalent to $U' = U^{-1}$, and that in an n-dimensional space an orthonormal set with n vectors must be a basis.   ∎

This proposition is closely related to another way to understand unitary matrices: a unitary matrix changes one system of orthogonal coordinates into another (see Figure 7.1). If U is unitary, multiplication of a vector by U' gives the coordinates of the vector in the orthonormal basis consisting of the columns of U. That is, it changes coordinates from the standard (orthonormal) basis into the new basis consisting of the columns of U. For example, the formula given in many calculus books for changing variables by a rotation is more easily understood as multiplication by the unitary matrix

$$U' = \begin{pmatrix} \cos\theta & \sin\theta \\ -\sin\theta & \cos\theta \end{pmatrix}$$

We would like to use this change of orthogonal coordinates to make transformation by some matrix easier to understand. In particular, we will apply unitary similarity to understand Hermitian matrices better.

DEFINITION We say matrices A and B are unitarily similar (or unitarily equivalent) if

$$B = U'AU$$

for some unitary matrix U.

The following easy calculation shows that unitary similarity preserves the adjoint operation. If U is unitary, then

$$(U^{-1}AU)' = (U'AU)' = U'A'(U')' = U^{-1}A'U$$

On the other hand, if two matrices are similar but not unitarily similar, say $B = S^{-1}AS$ where S is not unitary, then, although A' and B' are similar, the similarities between A and B and A' and B' are implemented by the two different matrices S and $(S^{-1})'$. For this reason, in problems for which adjoints, inner products, or orthogonality are important, unitary similarity is a more useful tool than plain similarity. The following proposition is an example of this usefulness.

PROPOSITION 7.3
Suppose A and B are unitarily similar $n \times n$ matrices. Then $A = A'$ if and only if $B = B'$.

PROOF   Exercise 7.1.9.   ∎

We recall that a matrix that satisfies $A' = A$ is called *Hermitian* or *self–adjoint*. The class of self–adjoint matrices includes the very important class of real symmetric matrices. We will begin by proving some of the easier facts about eigenvectors and eigenvalues of Hermitian matrices and finish by stating the main theorem of the section, that Hermitian matrices are unitarily diagonalizable.

First, we show that the eigenvalues of a Hermitian matrix are special.

THEOREM 7.4
If A is Hermitian, then all eigenvalues of A are real.

PROOF   Notice that the proof needs to be done in $\mathbf{C}^n$ because we cannot prove the eigenvalues are all real unless we know about other complex numbers. Let A be Hermitian and let v is an eigenvector for A with eigenvalue λ. On the one hand,

$$\langle Av, v \rangle = \langle \lambda v, v \rangle = \overline{\lambda}\langle v, v \rangle = \overline{\lambda}\|v\|^2$$

On the other hand, because $A' = A$

$$\langle Av, v \rangle = \langle v, A'v \rangle = \langle v, Av \rangle = \langle v, \lambda v \rangle = \lambda\langle v, v \rangle = \lambda\|v\|^2$$

Thus, $\lambda\|v\|^2 = \overline{\lambda}\|v\|^2$. Since v is an eigenvector for A, we know $\|v\|^2 \neq 0$ so $\lambda = \overline{\lambda}$ and λ is real. ∎

Recall that for general matrices, eigenvectors associated with distinct eigenvalues are linearly independent. The following corollary says that eigenvectors for Hermitian matrices are even more special.

COROLLARY 7.5

If A is Hermitian, then eigenvectors of A corresponding to distinct eigenvalues are orthogonal to each other.

PROOF   Let A be Hermitian, that is, $A = A'$, and let v and w be eigenvectors for A corresponding to the distinct eigenvalues λ and μ; in other words, $Av = \lambda v$ and $Aw = \mu w$ and $\mu \neq \lambda$. By Theorem 7.4, both λ and μ are real. On the one hand,

$$\langle Av, w \rangle = \langle \lambda v, w \rangle = \overline{\lambda}\langle v, v \rangle = \lambda\langle v, w \rangle$$

On the other hand, because $A' = A$

$$\langle Av, w \rangle = \langle v, A'w \rangle = \langle v, Aw \rangle = \langle v, \mu w \rangle = \mu\langle v, w \rangle$$

Thus, $\lambda\langle v, w \rangle = \mu\langle v, w \rangle$ or $(\lambda - \mu)\langle v, w \rangle = 0$. Since $\lambda \neq \mu$, we have $\lambda - \mu \neq 0$ and we conclude $\langle v, w \rangle = 0$. Thus, v and w are orthogonal as asserted. ∎

Theorem 7.6 is the main result of this section because it is the foundation for much of the application of Hermitian matrices to science and engineering. The corollary that follows it is just another way of stating the first part of the theorem.

THEOREM 7.6   *Spectral Theorem for Hermitian Matrices*

If A is Hermitian, then A is unitarily diagonalizable. Conversely, if B is unitarily diagonalizable, there is a Hermitian matrix A and a polynomial p so that $B = p(A)$.

PROOF   The proof of the important part of the theorem, that Hermitian matrices are unitarily diagonalizable, is beyond the scope of this book.

The converse is more accessible: if B is unitarily diagonalizable, say U is a unitary matrix and $D = U'BU$ is the diagonal matrix whose diagonal entries are $\lambda_1, \lambda_2, \cdots, \lambda_n$, let p be a polynomial with $p(j) = \lambda_j$ for $j = 1, \cdots, n$. Let C be the diagonal matrix whose diagonal entries are $1, 2, \cdots, n$ and let $A = UCU'$. Notice that $C = C'$, so by Proposition 7.3, A is Hermitian. Now, since $U' = U^{-1}$,

$$A^2 = AA = UCU'UCU' = UCCU' = UC^2U'$$

and

$$A^3 = AA^2 = UCU'UC^2U' = UCC^2U' = UC^3U'$$

etc., so we see $p(A) = Up(C)U'$. We have chosen C and p so that $p(C) = D$, so

$$p(A) = Up(C)U' = UDU' = UU'BUU' = B$$

∎

COROLLARY 7.7
If $A = A'$, there is an orthonormal basis for $\mathbf{C}^n$ consisting of eigenvectors of A.

PROOF The corollary follows from Theorem 7.6 because we have seen that if U diagonalizes A, then the columns of U are eigenvectors of A. Since the columns of a unitary matrix are an orthonormal basis for $\mathbf{C}^n$, the corollary follows. ∎

We have mostly been working in $\mathbf{R}^n$ and stating the theorems for $\mathbf{C}^n$ and/or $\mathbf{R}^n$ without too much fanfare. However, in dealing with eigenvalues, we have seen that it is possible to begin with a real matrix and end up with complex eigenvalues and eigenvectors. It is sometimes important to know that there are real number solutions to problems of this sort. If A is a real symmetric matrix, then $A = A'$ and Theorem 7.4 implies the eigenvalues of A are real. To find the eigenvectors of A, we must solve the system $(A - \lambda I)X = 0$. Since the coefficients of this system are all real numbers, we can find solutions of the system in $\mathbf{R}^n$ and these are real eigenvectors for A. Since we need not introduce complex numbers in the process of normalizing (and possibly orthogonalizing), we can unitarily diagonalize A using only real numbers. We record this fact as a separate corollary.

COROLLARY 7.8
If A is real and symmetric, there is an orthonormal basis for $\mathbf{R}^n$ consisting of eigenvectors of A.

Let's do some computations that illustrate these theorems.

EXAMPLE 7.9
For the matrix

$$A = \begin{pmatrix} 6 & 2 \\ 2 & 3 \end{pmatrix}$$

find a unitary matrix U and a diagonal matrix D so that $D = U^{-1}AU$.

SOLUTION The matrix A is real and symmetric, so $A = A'$ and Theorem 7.6 guarantees that such a unitary exists. Since the columns of U will be eigenvectors for A and the diagonal entries of D will be the eigenvalues of A, we need to find the eigenvalues and eigenvectors of A.

$$\det \begin{pmatrix} 6 - \lambda & 2 \\ 2 & 3 - \lambda \end{pmatrix} = 18 - 9\lambda + \lambda^2 - 4 = \lambda^2 - 9\lambda + 14 = (\lambda - 7)(\lambda - 2)$$

so the eigenvalues of A are 2 and 7. As Theorem 7.4 asserted, the eigenvalues of A are real.

To find the eigenvectors associated with $\lambda = 2$, we must find the nullspace of $(A - 2I)$, that is, we must solve $(A - 2I)X = 0$. This is

$$\begin{pmatrix} 4 & 2 \\ 2 & 1 \end{pmatrix} \begin{pmatrix} x \\ y \end{pmatrix} = \begin{pmatrix} 0 \\ 0 \end{pmatrix}$$

The two equations are equivalent, so we need to solve $2x + y = 0$. As is easily seen, there are infinitely many solutions, $y = -2x$ and x arbitrary. Choosing $x = 1$, $y = -2$ and we find $(1, -2)$ is a basis for the eigenspace of A correspoinding to $\lambda = 2$.

To find the eigenvectors associated with $\lambda = 7$, we must find the nullspace of $(A - 7I)$. Solving

$$\begin{pmatrix} -1 & 2 \\ 2 & -4 \end{pmatrix} \begin{pmatrix} x \\ y \end{pmatrix} = \begin{pmatrix} 0 \\ 0 \end{pmatrix}$$

we get $x - 2y = 0$, so there are infinitely many solutions, $x = 2y$ and y arbitrary. Choosing $y = 1$, $x = 2$ and we find $(2, 1)$ is a basis for the eigenspace of A correspoinding to $\lambda = 7$.

As Theorem 7.5 asserted (since $2 \neq 7$), the eigenvectors $(1, -2)$ and $(2, 1)$ are perpendicular. Using these vectors as the columns of a matrix S, we have (as in Section 6.5) $S^{-1}AS$ is diagonal, but this does not solve our problem because we need to find a unitary matrix that does this and S is not unitary. A matrix is unitary if and only if its columns are an orthonormal basis. Since $(1, -2)$ and $(2, 1)$ are orthogonal, we need only normalize them to get the required unitary matrix:

$$\frac{1}{\left\| \begin{pmatrix} 1 \\ -2 \end{pmatrix} \right\|} \begin{pmatrix} 1 \\ -2 \end{pmatrix} = \frac{1}{\sqrt{5}} \begin{pmatrix} 1 \\ -2 \end{pmatrix} = \begin{pmatrix} \frac{1}{\sqrt{5}} \\ -\frac{2}{\sqrt{5}} \end{pmatrix}$$

and

$$\frac{1}{\left\| \begin{pmatrix} 2 \\ 1 \end{pmatrix} \right\|} \begin{pmatrix} 2 \\ 1 \end{pmatrix} = \frac{1}{\sqrt{5}} \begin{pmatrix} 2 \\ 1 \end{pmatrix} = \begin{pmatrix} \frac{2}{\sqrt{5}} \\ \frac{1}{\sqrt{5}} \end{pmatrix}$$

These vectors are an orthonormal basis for $\mathbf{R}^2$, so the unitary matrix is

$$U = \begin{pmatrix} \frac{1}{\sqrt{5}} & \frac{2}{\sqrt{5}} \\ -\frac{2}{\sqrt{5}} & \frac{1}{\sqrt{5}} \end{pmatrix}$$

Since U is unitary, $U^{-1} = U'$ and computing (or better, just writing down the answer), we get

$$D = U'AU = \begin{pmatrix} \frac{1}{\sqrt{5}} & -\frac{2}{\sqrt{5}} \\ \frac{2}{\sqrt{5}} & \frac{1}{\sqrt{5}} \end{pmatrix} \begin{pmatrix} 6 & 2 \\ 2 & 3 \end{pmatrix} \begin{pmatrix} \frac{1}{\sqrt{5}} & \frac{2}{\sqrt{5}} \\ -\frac{2}{\sqrt{5}} & \frac{1}{\sqrt{5}} \end{pmatrix} = \begin{pmatrix} 2 & 0 \\ 0 & 7 \end{pmatrix}$$

which is a unitary diagonalization of A as required.

The algorithms of MATLAB will yield a unitary matrix that diagonalizes A when one exists. In this case, we get

```
> a=[6 2 ; 2 3]

a =
        6        2
        2        3

> [v d]=eig(a)

v =
     -0.8944      0.4472
     -0.4472     -0.8944


d =
        7        0
        0        2

> v'*v

ans =
     1.0000      0.0000
     0.0000      1.0000
```

where the last calculation verifies that the diagonalizing matrix v is unitary. Note that, except for the order of the columns and the signs, this is the same solution we found by hand. ☐

Not all problems are quite this straightforward by hand. Although distinct eigenvalues must be associated with orthogonal eigenvectors, the eigenvectors associated with repeated eigenvalues need not be orthogonal, so we must work a little harder in this case. Note, however, that unlike the problems of Chapter 6.2 associated with repeated eigenvalues, Theorem 7.6 guarantees that self–adjoint matrices are always diagonalizable.

EXAMPLE 7.10

For the matrix

$$B = \begin{pmatrix} 0 & -2 & 1 \\ -2 & 3 & -2 \\ 1 & -2 & 0 \end{pmatrix}$$

find a unitary matrix U and a diagonal matrix D so that $D = U^{-1}BU$.

SOLUTION As before, we begin by finding the eigenvectors and eigenvalues of B.

$$\det \begin{pmatrix} 0-\lambda & -2 & 1 \\ -2 & 3-\lambda & -2 \\ 1 & -2 & 0-\lambda \end{pmatrix} = -(\lambda^3 - 3\lambda^2 - 9\lambda - 5) = -(\lambda - 5)(\lambda + 1)^2$$

so the eigenvalues of B are 5 and -1 (and -1 has multiplicity 2).

To find the eigenvectors associated with $\lambda = 5$, we need to solve the system $(B - 5I)X = 0$ which is

$$\begin{pmatrix} -5 & -2 & 1 \\ -2 & -2 & -2 \\ 1 & -2 & -5 \end{pmatrix} \begin{pmatrix} x \\ y \\ z \end{pmatrix} = \begin{pmatrix} 0 \\ 0 \\ 0 \end{pmatrix}$$

Dividing the second equation by -2 and adding 5 times it to the first and subtracting from the last equation, we get $x + y + z = 0$ and $-3y - 6z = 0$. This gives $x+y+z = 0$ and $y+2z = 0$, so the solution is $y = -2z$, $x = z$, and z is arbitrary. For example, $v = (1, -2, 1)$ is a basis for the eigenspace corresponding to $\lambda = 5$.

To find the eigenvectors associated with $\lambda = -1$, we need to solve the system $(B + I)X = 0$. This means

$$\begin{pmatrix} 1 & -2 & 1 \\ -2 & 4 & -2 \\ 1 & -2 & 1 \end{pmatrix} \begin{pmatrix} x \\ y \\ z \end{pmatrix} = \begin{pmatrix} 0 \\ 0 \\ 0 \end{pmatrix}$$

which gives $x - 2y + z = 0$, or $x = 2y - z$ and y and z are arbitrary. This means that $(2, 1, 0)$ and $(-1, 0, 1)$ form a basis for the eigenspace corresponding to $\lambda = -1$.

Now the theory guarantees $(1, -2, 1)$ is perpendicular to both $(2, 1, 0)$ and $(-1, 0, 1)$, but $(2, 1, 0)$ and $(-1, 0, 1)$ are not perpendicular to each other, nor need they be. There are infinitely many different bases for the eigenspace corresponding to -1, and our basis is more or less chosen at random from this space, so we really have no right to expect they should be perpendicular. We need an orthonormal basis for this eigenspace and we have little choice but use the Gram–Schmidt procedure to find it. Take $w_1 = (2, 1, 0)$ and

$$w_2 = \begin{pmatrix} -1 \\ 0 \\ 1 \end{pmatrix} - \frac{\left\langle \begin{pmatrix} 2 \\ 1 \\ 0 \end{pmatrix}, \begin{pmatrix} -1 \\ 0 \\ 1 \end{pmatrix} \right\rangle}{\left\langle \begin{pmatrix} 2 \\ 1 \\ 0 \end{pmatrix}, \begin{pmatrix} 2 \\ 1 \\ 0 \end{pmatrix} \right\rangle} \begin{pmatrix} 2 \\ 1 \\ 0 \end{pmatrix}$$

$$= \begin{pmatrix} -1 \\ 0 \\ 1 \end{pmatrix} - \frac{-2}{5} \begin{pmatrix} 2 \\ 1 \\ 0 \end{pmatrix} = \begin{pmatrix} -.2 \\ .4 \\ 1 \end{pmatrix}$$

So

$$u_1 = \frac{1}{\|w_1\|} w_1 = \begin{pmatrix} \frac{2}{\sqrt{5}} \\ \frac{1}{\sqrt{5}} \\ 0 \end{pmatrix} \quad \text{and} \quad u_2 = \frac{1}{\|w_2\|} w_2 = \begin{pmatrix} -\frac{1}{\sqrt{30}} \\ \frac{2}{\sqrt{30}} \\ \frac{5}{\sqrt{30}} \end{pmatrix}$$

are an orthonormal basis for the eigenspace corresponding to -1 and

$$u_3 = \frac{1}{\|v\|} v = \begin{pmatrix} \frac{1}{\sqrt{6}} \\ -\frac{2}{\sqrt{6}} \\ \frac{1}{\sqrt{6}} \end{pmatrix}$$

is an orthonormal basis for the eigenspace corresponding to 5.

Putting these together, we get

$$U = \begin{pmatrix} \frac{2}{\sqrt{5}} & -\frac{1}{\sqrt{30}} & \frac{1}{\sqrt{6}} \\ \frac{1}{\sqrt{5}} & \frac{2}{\sqrt{30}} & -\frac{2}{\sqrt{6}} \\ 0 & \frac{5}{\sqrt{30}} & \frac{1}{\sqrt{6}} \end{pmatrix}$$

is a unitary matrix whose columns are eigenvectors of B. Thus,

$$D = U^{-1}BU$$

$$= \begin{pmatrix} \frac{2}{\sqrt{5}} & \frac{1}{\sqrt{5}} & 0 \\ -\frac{1}{\sqrt{30}} & \frac{2}{\sqrt{30}} & \frac{5}{\sqrt{30}} \\ \frac{1}{\sqrt{6}} & -\frac{2}{\sqrt{6}} & \frac{1}{\sqrt{6}} \end{pmatrix} \begin{pmatrix} 0 & -2 & 1 \\ -2 & 3 & -2 \\ 1 & -2 & 0 \end{pmatrix} \begin{pmatrix} \frac{2}{\sqrt{5}} & -\frac{1}{\sqrt{30}} & \frac{1}{\sqrt{6}} \\ \frac{1}{\sqrt{5}} & \frac{2}{\sqrt{30}} & -\frac{2}{\sqrt{6}} \\ 0 & \frac{5}{\sqrt{30}} & \frac{1}{\sqrt{6}} \end{pmatrix}$$

$$= \begin{pmatrix} -1 & 0 & 0 \\ 0 & -1 & 0 \\ 0 & 0 & 5 \end{pmatrix}$$

is a unitary diagonalization of B as was wanted.

MATLAB is also efficient on this problem:

```
> b=[0 -2 1; -2 3 -2; 1 -2 0]

b =
     0    -2     1
    -2     3    -2
     1    -2     0

> [u d]=eig(b)

u =
    -0.9078     0.0957     0.4082
    -0.3163     0.4830    -0.8165
     0.2753     0.8704     0.4082


d =
    -1.0000          0          0
          0    -1.0000          0
          0          0     5.0000

> u'*u

ans =
     1.0000     0.0000     0.0000
     0.0000     1.0000     0.0000
     0.0000     0.0000     1.0000
```

where, again, the last calculation illustrates that the u found by the machine is unitary. This solution is considerably different than that found by hand: because there are infinitely many orthonormal bases for the eigenspace corresponding to -1, we would not expect MATLAB to find the same one we did. $\quad\Box$

Exercises 7.1

1. Let R be the matrix introduced in Equation (7.1.1) on page 377:
$$R = \begin{pmatrix} \cos^2\theta - \sin^2\theta & 2\sin\theta\cos\theta \\ 2\sin\theta\cos\theta & \sin^2\theta - \cos^2\theta \end{pmatrix}$$

   (a) Show that R is unitary.
   (b) Show that if u is a vector in the line $y = (\tan\theta)x$, then $Ru = u$ and that if v is a vector perpendicular to this line, then $Rv = -v$.
   (c) Find the eigenvectors and eigenvalues of R.
   (d) Using part (b), draw the line $y = (\tan\theta)x$, choose a vector u in the line and v perpendicular to the line (with their tails at 0), draw the vector $3u + 2v$ and $R(3u + 2v) = 3u - 2v$. Compare $3u + 2v$ and $R(3u + 2v)$ to conclude that R reflects in the line.

2. Let U and V be unitary matrices.
   Prove that UV is also a unitary matrix.

3. Let U be a real unitary matrix. Prove that the determinant of U is ± 1.

4. Find a *unitary* matrix U and a diagonal matrix D so that so that $D = U^{-1}AU$ for
$$A = \begin{pmatrix} 1 & 2 \\ 2 & -2 \end{pmatrix}$$

5. Find a *unitary* matrix U and a diagonal matrix D so that so that $D = U^{-1}RU$ for
$$R = \begin{pmatrix} 1 & 12 \\ 12 & -6 \end{pmatrix}$$

6. Find a *unitary* matrix U and a diagonal matrix D so that so that $D = U^{-1}TU$ for
$$T = \begin{pmatrix} 0 & 3 & -5 \\ 2 & -5 & 1 \\ -10 & 3 & 5 \end{pmatrix}$$
The characteristic polynomial of T is $p(\lambda) = -(\lambda - 10)(\lambda + 2)(\lambda + 8)$.

7.
$$\text{Let } B = \begin{pmatrix} 1 & 4 & -2 \\ 4 & 1 & 2 \\ -2 & 2 & 4 \end{pmatrix}$$
The characteristic polynomial of B is $p(\lambda) = -(\lambda - 5)^2(\lambda + 4)$. Find a *unitary* matrix U so that $U^{-1}BU = \Lambda$ is diagonal, and find the diagonal matrix Λ.

8. The numbers -3, 1, and 5 are the eigenvalues of the matrix

$$E = \begin{pmatrix} 0 & -1 & -3 & 1 \\ -1 & 0 & 1 & -3 \\ -3 & 1 & 0 & -1 \\ 1 & -3 & -1 & 0 \end{pmatrix}$$

(a) What property of E guarantees the existence of a unitary matrix U so that $U^{-1}EU$ is diagonal?

(b) Find a unitary matrix U as in (a).

(c) Solve the initial value problem: $\dfrac{dY}{dt} = EY$ and $Y(0) = (1, -1, 2, 0)$.

9. Prove that if B is Hermitian and A is unitarily similar to B, then A is Hermitian also.

10. Let F be the matrix

$$F = \begin{pmatrix} 2 & i \\ -i & 2 \end{pmatrix}$$

(a) What property of F guarantees the existence of a unitary matrix U so that $D = U^{-1}FU$ is diagonal?

(b) Find a unitary matrix U and a diagonal matrix D as in (a).

11. The 5×5 matrix S is Hermitian (self–adjoint) and v is an eigenvector for S with eigenvalue -3. The vector w is perpendicular to v. Prove that Sw is also perpendicular to v.

12. Let G be the matrix

$$G = \begin{pmatrix} 2 & 1 \\ -1 & 2 \end{pmatrix}$$

(a) Show that even though $G' \neq G$, there is a unitary matrix U so that $U^{-1}GU$ is diagonal.

(b) Find a Hermitian matrix A and a polynomial p so that $G = p(A)$.

13. Let B be a self-adjoint matrix and let $A = B^2$. Prove that if λ is an eigenvalue of A, then λ is real and $\lambda \geq 0$.

14.(a) Prove the following converse to Exercise 7.1.13.: If A is a Hermitian matrix all of whose eigenvalues are non-negative, then there is a Hermitian matrix B, all of whose eigenvalues are non-negative, such that $B^2 = A$.

(**Hint:** Unitarily diagonalize, find a square root of the diagonal matrix, then undiagonalize.)

(b) The eigenvalues of the matrix $A = \begin{pmatrix} 5 & -4 \\ -4 & 5 \end{pmatrix}$ are 1 and 9.

Find a Hermitian matrix B so that all the eigenvalues are nonnegative and $B^2 = A$.

15. Let P be a self–adjoint matrix all of whose eigenvalues are positive. (Such matrices are called *positive definite*.) Use the Spectral Theorem (Theorem 7.6) to show that $\langle Pv, v \rangle > 0$ for all non–zero vectors v.

A
Complex Numbers

The *complex numbers* arise mathematically as solutions to some algebraic equations that do not have real number solutions. In engineering and science, the complex numbers arise in descriptions of physical phenomena such as electrical currents and air flowing over a wing.

The imaginary unit i is defined to be a number such that $i^2 = -1$. This number is called *imaginary* because this is a property no real number can have. (In some engineering contexts, the imaginary unit is denoted j; MATLAB can use either i or j as input.) Sometimes one sees $i = \sqrt{-1}$, but we also have $(-i)^2 = -1$ so this is somewhat unclear.

The *complex numbers* are those numbers that can be written as

$$a + bi$$

where a and b are real numbers and i is the imaginary unit. In this expression, a is called the real part of the complex number and b (*not bi!*) is called the imaginary part of the complex number. For example, if $z = 3 - 4i$, then the real part of z is 3 and we write $\mathrm{Re}(z) = 3$. The imaginary part of z is -4 and we write $\mathrm{Im}(z) = -4$. Two complex numbers are equal if and only if their real and imaginary parts are equal. MATLAB uses the commands `real` and `imag` for real and imaginary parts of complex numbers (and matrices).

```
> z = 3-4*i

z =
   3.0000 - 4.0000i

> real(z)

ans =
     3
```

```
> imag(z)

ans =
    -4
```

The *complex conjugate* of the number $a + bi$ is the number $a - bi$, written

$$\overline{a + bi} = a - bi$$

Thus, for $z = 3 - 4i$, we have

$$\bar{z} = \overline{3 - 4i} = 3 - (-4)i = 3 + 4i$$

In MATLAB, the command `conj` is used for the complex conjugate.

```
> conj(z)

ans =
   3.0000 + 4.0000i
```

EXAMPLE A.1

Find the real parts, the imaginary parts, and the complex conjugates of the numbers
$z = -4 + 2i$ and $w = 2.6 + i$.
SOLUTION $\operatorname{Re}(z) = \operatorname{Re}(-4 + 2i) = -4$.
$\operatorname{Im}(z) = \operatorname{Im}(-4 + 2i) = 2$.
$\bar{z} = \overline{-4 + 2i} = -4 - 2i$.
   Similarly, since $w = 2.6 + i = 2.6 + 1i$, $\operatorname{Re}(w) = \operatorname{Re}(2.6 + i) = 2.6$.
$\operatorname{Im}(w) = \operatorname{Im}(2.6 + i) = 1$.
$\bar{w} = \overline{2.6 + i} = 2.6 - i$. $\square$

   In the context of complex numbers, every real number is considered to be a
complex number: $5 = 5 + 0i$. So $\operatorname{Re}(5) = \operatorname{Re}(5 + 0i) = 5$ and $\operatorname{Im}(5) = \operatorname{Im}(5 + 0i) = 0$. Real numbers are their own conjugates:

$$\bar{5} = \overline{5 + 0i} = 5 - 0i = 5$$

Moreover, if z is a complex number that is its own conjugate, say $z = a + bi$
where a and b are its real and imaginary parts, then $z = \bar{z}$ means $a + bi = a - bi$
which happens precisely when $a = a$ and $b = -b$, that is, exactly when $b = 0$
and the number z is $z = a + 0i = a$. In other words, the real numbers are
characterized as those complex numbers that are their own conjugates. Numbers

of the form $bi = 0 + bi$ for b real are called *purely imaginary* and are characterized by their conjugates being their negatives. For example, $3i$ is purely imaginary and $\overline{3i} = \overline{0 + 3i} = 0 - 3i = -3i$. There is not a sensible way to order the complex numbers that is compatible with their arithmetic, so the words "postive" and "negative" are reserved for real numbers. To say that a complex number is "positive" means that it is real and positive.

Complex numbers are frequently represented geometrically in the Cartesian plane by identifying the complex number $a + bi$ with the point (a, b) of the plane. In this representation, the x–axis becomes the "real axis" because all real numbers have imaginary part 0 and the y–axis becomes the "imaginary axis" because all the purely imaginary numbers have real part 0. Geometrically, complex conjugation is reflection across the real axis. When representing complex numbers in this way, the plane is called the *complex plane*.

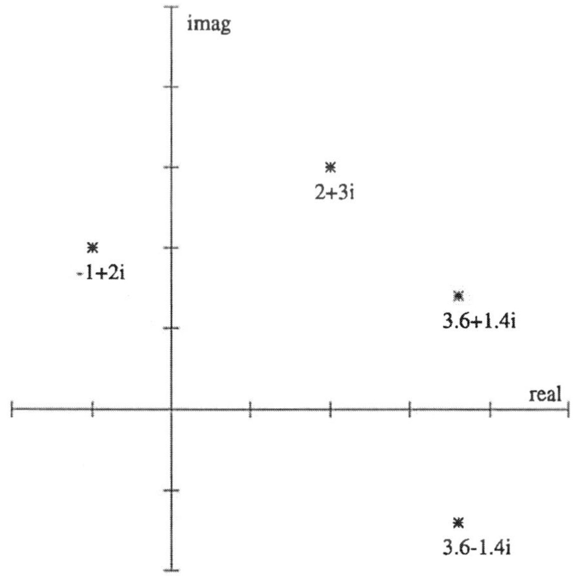

FIGURE A.1

Geometric representation of complex numbers

Arithmetic is defined for the complex numbers in terms of the representation in terms of real and imaginary parts and has the usual formal properties of arithmetic. For example, addition is defined by

$$(a + bi) + (c + di) = (a + c) + (b + d)i$$

For example, if $z = -4 + 2i$ and $w = 2.6 + i$, then

$$z + w = (-4 + 2i) + (2.6 + i) = (-4 + 2.6) + (2 + 1)i = -1.4 + 3i$$

Addition of complex numbers is commutative and associative and the identity for addition is the complex number zero, $0 = 0 + 0i$. Moreover, the additive inverse of $a + bi$ is $-(a + bi) = -a - bi$ which satisfies

$$(a + bi) + (-a + -bi) = (a - a) + (b - b)i = 0 + 0i = 0$$

Moreover, the sum of the conjugates is the conjugate of the sum: $\overline{z + w} = \overline{z} + \overline{w}$. In the example above,

$$\overline{z + w} = \overline{-1.4 + 3i} = -1.4 - 3i$$

and

$$\overline{z} + \overline{w} = \overline{-4 + 2i} + \overline{2.6 + i} = (-4 - 2i) + (2.6 - i) = -1.4 - 3i$$

Multiplication of complex numbers is defined as you would expect from expanding the product of binomials in high school algebra. We would expect

$$(a + bi)(c + di) = ac + adi + bci + bdi^2$$

Since $i^2 = -1$, this becomes $ac + adi + bci - bd$ and this is the basis of the definition of *multiplication* of complex numbers:

$$(a + bi)(c + di) = (ac - bd) + (ad + bc)i$$

For example,

$$(-4 + 5i)(2 - 3i) = ((-4)(2) - (5)(-3)) + ((-4)(-3) + (5)(2))i = 7 + 22i$$

Of course, MATLAB does complex multiplication as well:

```
> (-4+5*i)*(2-3*i)

ans =
   7.0000 +22.0000i
```

Conjugation works as you would hope in complex multiplication: $\overline{zw} = \overline{z}\,\overline{w}$. To see this, simply compute

$$\overline{(a + bi)}\,\overline{(c + di)} = (a - bi)(c - di) = (ac - (-b)(-d)) + (a(-d) + (-b)c)i$$
$$= (ac - bd) - (ad + bc)i$$

and notice that this is indeed $\overline{(a + bi)(c + di)}$.

EXAMPLE A.2

We do some sample computations.

$$6+2i+(3-i)(4+3i) = (6+2i)+(12-4i+9i-3i^2) = (6+2i)+(12+5i) = 21+7i$$

A more complicated computation:

$$\overline{(2.1+1.4i)(2-.9i)} + (1.6-2.6i)(-3.4+2i)$$

$$= \overline{(4.2+2.8i-1.89i-1.26i^2)} + (-5.44+8.84i+3.2i-5.2i^2)$$

$$= \overline{(5.46+.91i)} + (-.24+12.04i)$$

$$= 5.46-.91i-.24+12.04i = 5.22+11.13i$$

Notice that the square of a complex number, such as $2-5i$, need not be a real, positive number.

$$(2-5i)^2 = (2-5i)(2-5i) = 4-10i-10i+25i^2 = -21-20i$$

However,

$$(3+2i)\overline{(3+2i)} = (3+2i)(3-2i) = 9+6i-6i-4i^2 = 9+4 = 13$$

∎

The latter calculation illustrates a general principle, indeed, it is the principle behind the definition of the complex conjugate: the product of a complex number and its conjugate is always non-negative! If a and b are real numbers,

$$(a+bi)\overline{(a+bi)} = (a+bi)(a-bi) = a^2+abi-abi-b^2i^2 = a^2+b^2 \geq 0$$

This allows us to define the *absolute value* or *modulus* of the complex number $z = a+bi$ (where a and b are the real and imaginary parts of z) by

$$|z| = \sqrt{z\bar{z}} = \sqrt{a^2+b^2}$$

Thus, the computation above gives

$$|3+2i| = \sqrt{(3+2i)\overline{(3+2i)}} = \sqrt{(3+2i)(3-2i)} = \sqrt{9+4} = \sqrt{13}$$

MATLAB knows this as well. The MATLAB command for the absolute value of a number is abs.

```
> abs(3+2i)

ans =
    3.6056

> ans^2

ans =
    13.0000
```

so $|3 + 2i| = \sqrt{13}$ in MATLAB as well. Notice that $|z| = 0$ if and only if $z = 0$ (see Exercise A.0.4.).

The familiar formal properties of multiplication for real numbers are also true for multiplication of complex numbers: multiplication of complex numbers is commutative and associative and multiplication distributes over addition: $w(z_1 + z_2) = wz_1 + wz_2$. Moreover, the multiplicative identity for complex multiplication is the real number $1 = 1 + 0i$. Indeed

$$1(a + bi) = (1 + 0i)(a + bi) = 1a + abi + 0ai + 0bi^2 = a + bi$$

Finding multiplicative inverses, and dividing in general, is more tricky. The most efficient way to compute reciprocals or quotients is to use the absolute value!

$$\frac{1}{z} = \frac{\overline{z}}{z\overline{z}} = \frac{\overline{z}}{|z|^2}$$

which is easier because $1/|z|^2$ is a real number and we can distribute the product over the real and imaginary parts. For example,

$$\frac{1}{3 + 4i} = \frac{3 - 4i}{(3 + 4i)(3 - 4i)} = \frac{3 - 4i}{25} = \frac{3}{25} - \frac{4}{25}i = .12 - .16i$$

To check this, we can multiply

$$(3 + 4i)(.12 - .16i) = .36 + .48i - .48i - .64i^2 = .36 + .64 = 1$$

as we expected. For quotients in general, we use the same trick.

$$\frac{w}{z} = \frac{w\overline{z}}{z\overline{z}} = \frac{w\overline{z}}{|z|^2}$$

For example,

$$\frac{4 - 5i}{3 - i} = \frac{(4 - 5i)\overline{(3 - i)}}{(3 - i)\overline{(3 - i)}} = \frac{(4 - 5i)(3 + i)}{(3 - i)(3 + i)} = \frac{17 - 11i}{10} = \frac{17}{10} - \frac{11}{10}i = 1.7 - 1.1i$$

MATLAB gives the same result:

```
> (4-5*i)/(3-i)

ans =
   1.7000 - 1.1000i
```

Absolute value behaves as we hope it would for complex numbers.

PROPOSITION A.3
If w and z are complex numbers and $z \neq 0$, then

$$|wz| = |w||z|$$

and

$$\left|\frac{w}{z}\right| = \frac{|w|}{|z|}$$

PROOF   Exercises A.0.5. and A.0.6.   ∎

The formulas for finding the absolute value of a product or quotient are frequently more efficient than finding the product or quotient first. For example,

$$\left|\frac{4-5i}{3-i}\right| = \frac{|4-5i|}{|3-i|} = \frac{\sqrt{41}}{\sqrt{10}} = \sqrt{4.1}$$

which gives the same answer as $|1.7 - 1.1| = \sqrt{2.89 + 1.21} = \sqrt{4.1}$ without finding the quotient of the complex numbers.

In addition to the usual representation, complex numbers have a polar form. The connection between the representation of a complex number in real and imaginary parts and its representation in polar form corresponds to relationship between the graphical interpretation of $z = a + bi$ as the point (a, b) in the plane and its representation in polar coordinates. Recall that in polar coordinates, the radius is $r = \sqrt{a^2 + b^2} = |z|$ and the angle is $\theta = \arctan b/a$. Since $a/r = \cos(\theta)$ and $b/r = \sin(\theta)$ we have the polar representation

$$z = a + bi = |z|(\cos(\theta) + i\sin(\theta))$$

where θ is any angle (in radians) for which $b/a = \tan(\theta)$. This angle is called the *argument* of z. Just as in analytic geometry in polar coordinates, the argument can have many values, differing by multiples of 2π, and this occasionally causes difficulties. (In MATLAB, the argument of a complex number is given by the command `angle`; this always gives an argument satisfying $-\pi < \text{angle}(t) \leq \pi$.)

This geometric interpretation gives a helpful description of complex multiplication. Using trigonometric identities for the sine and cosine of sums of angles, we see that if $z_1 = |z_1|(\cos(\theta_1) + i\sin(\theta_1))$ and $z_2 = |z_2|(\cos(\theta_2) + i\sin(\theta_2))$ then

$$z_1 z_2 = |z_1||z_2|(\cos(\theta_1 + \theta_2) + i\sin(\theta_1 + \theta_2))$$

For example, $z_1 = 2 + 1.4i$ has polar representation

$$2 + i = \sqrt{5.96}(\cos(\theta_1) + i\sin(\theta_1)) \approx 2.4413(\cos(\theta_1) + i\sin(\theta_1))$$

where $\theta_1 \approx .6107$, and $z_2 = 1 + 2.4i$ has polar representation

$$z_2 = 2.6(\cos(\theta_2) + i\sin(\theta_2))$$

where $\theta_2 \approx 1.1760$. Now $z_1 z_2 = (2 + 1.4i)(1 + 2.4i) = -1.36 + 6.2i$ which has polar representation

$$z_1 z_2 = \sqrt{40.2896}(\cos(\theta) + i\sin(\theta)) \approx 6.3474(\cos(\theta) + i\sin(\theta))$$

where $\theta = \theta_1 + \theta_2 \approx 1.7867$. This calculation is illustrated in Figure A.2.

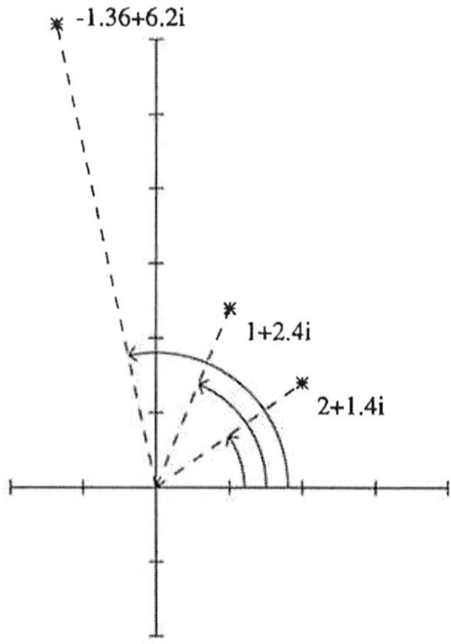

FIGURE A.2

Multiplication of complex numbers

The polar form of complex numbers is a version of *Euler's formula*. Euler's formula, which can be justified by power series calculations, says

$$e^{\alpha + \beta i} = e^{\alpha}e^{\beta i} = e^{\alpha}(\cos(\beta) + i\sin(\beta))$$

Euler's formula is useful in rewriting complex solutions of differential equations in real form (Chapter 6.2) and in finding powers and roots of complex numbers.

We know that polynomials with real coefficients need not have real roots; for example, the polynomial $x^2 + 4$ has roots $2i$ and $-2i$. The Fundamental Theorem of Algebra says that every polynomial of degree n with complex numbers as coefficients has n complex roots (where multiple roots are counted separately if they occur). Complex roots of polynomials with real coefficients must occur in complex conjugate pairs and such polynomials can be factored, in principle, into a product of real linear factors and real quadratic factors. This is helpful in factoring real polynomials of low degree to find their complex roots.

EXAMPLE A.4

Find the four roots of the polynomial $x^4 - 4x^3 - 4x^2 + 36x - 45$.

SOLUTION We first try to guess some small integer roots or plot the graph of the polynomial to try to find (approximately, perhaps) some real roots. This attempt gives 3 and -3 as real roots:

$$3^4 - 4(3^3) - 4(3^2) + 36(3) - 45 = 81 - 108 - 36 + 108 - 45 = 0$$

and

$$(-3)^4 - 4((-3)^3) - 4((-3)^2) + 36(-3) - 45 = 81 + 108 - 36 - 108 - 45 = 0$$

This means that both $x - 3$ and $x - (-3) = x + 3$ are factors of the given polynomial. In other words $x^2 - 9 = (x - 3)(x + 3)$ must divide the polynomial. Carrying out the long division, we find

$$x^4 - 4x^3 - 4x^2 + 36x - 45 = (x^2 - 9)(x^2 - 4x + 5)$$

Now we can use the quadratic formula to find the roots of $x^2 - 4x + 5$:

$$\rho = \frac{-(-4) \pm \sqrt{(-4)^2 - 4(1)(5)}}{2(1)} = \frac{4 \pm \sqrt{-4}}{2} = \frac{4 \pm 2i}{2} = 2 \pm i$$

Thus, the four roots of the original polynomial are $3, -3, 2 + i$, and $2 - i$. $\quad\Box$

Exercises A.0

1. Let $z = 4 - 5i$.
   Find: (a) $\text{Re}(z)$ (b) $\text{Im}(z)$ (c) $|z|$ (d) $\bar{z}$

2. Compute:

   (a) $(3 + 2i)(2 - i) + i(-2 + i)$ (b) $(2 - 3i)^2(4 + 2i)$

   (c) $(2 - i)^2 + (1 + 3i)^2$ (d) $\left(\overline{(2 - i)}\right)^2 + \left(\overline{(1 + 3i)}\right)^2$ *see (c)*

   (e) $\dfrac{1}{3 + 4i}$ (f) $\dfrac{4 - 2i}{1 + i}$

   (g) $\dfrac{2 + 3i}{(2 - i)^2} + \dfrac{i}{1 + i}$ (h) $\left|\dfrac{1 + 3i}{(2 - i)}\right|$

3. Find all (3) roots of the equation

$$z^3 - 3z^2 + 7z - 5 = 0$$

4. Prove that $|z| = 0$ if and only if $z = 0$.
5. Prove that $|wz| = |w||z|$ for all complex numbers w and z.
6. Prove that $\left|\dfrac{w}{z}\right| = \dfrac{|w|}{|z|}$ for all complex numbers w and z where $z \neq 0$.

B

Block Matrices

Many times in applications, the entries of a matrix are grouped in a meaningful way. In this case it is useful to keep track of them in a group and do the matrix arithmetic so as to take advantage of this structure. For example, in design work it might happen that one component of a system is changed while the rest stay the same. If the system is described by a matrix, only the part of the matrix associated with that one component will change. In some cases, recalculations are simplified by taking advantage of the fact that only some of the matrix entries are changed. In this section we introduce blocked (or partitioned) matrices and the arithmetic associated with them. We will see that a recognition of this underlying structure may make solution of problems easier to understand or computationally more efficient.

EXAMPLE B.1

The matrix

$$A = \begin{pmatrix} 1.2 & -.7 & 2.1 & 1.6 & 1.3 \\ -.7 & .9 & -3.2 & 0 & 2.6 \\ 2.1 & -3.2 & 1.8 & -1.4 & .3 \\ 1.6 & 0 & -1.4 & 2.3 & -.2 \\ 1.3 & 2.6 & .3 & -.2 & 1.5 \end{pmatrix}$$

describes a system with two subassemblies and their interconnections. Entries a_{11}, a_{12}, a_{13}, a_{21}, a_{22}, a_{23}, a_{31}, a_{32}, and a_{33} describe the first subassembly internally, entries a_{44}, a_{45}, a_{54}, and a_{55} describe the second subassembly internally, and the remaining entries describe the connections between them.

If we write

$$D = \begin{pmatrix} 1.2 & -.7 & 2.1 \\ -.7 & .9 & -3.2 \\ 2.1 & -3.2 & 1.8 \end{pmatrix} \qquad E = \begin{pmatrix} 1.6 & 1.3 \\ 0 & 2.6 \\ -1.4 & .3 \end{pmatrix}$$

$$F = \begin{pmatrix} 1.6 & 0 & -1.4 \\ 1.3 & 2.6 & .3 \end{pmatrix} \qquad G = \begin{pmatrix} 2.3 & -.2 \\ -.2 & 1.5 \end{pmatrix}$$

then we can think of A as a matrix of matrices

$$A = \begin{pmatrix} D & E \\ F & G \end{pmatrix}$$

MATLAB is set up to use blocked matrices in both input and output. We can enter D, E, F, and G as usual

```
> D=[ 1.2 -.7 2.1;-.7 .9 -3.2; 2.1 -3.2 1.8]

D =
     1.2000    -0.7000     2.1000
    -0.7000     0.9000    -3.2000
     2.1000    -3.2000     1.8000

> E=[1.6 1.3;0 2.6;-1.4 .3]

E =
     1.6000     1.3000
          0     2.6000
    -1.4000     0.3000

> F=[1.6 0 -1.4; 1.3 2.6 .3]

F =
     1.6000          0    -1.4000
     1.3000     2.6000     0.3000

> G=[2.3 -.2 ; -.2 1.5]

G =
     2.3000    -0.2000
    -0.2000     1.5000
```

and build A from the pieces

```
> A=[D E; F G]
```

```
A  =
      1.2000     -0.7000      2.1000      1.6000      1.3000
     -0.7000      0.9000     -3.2000           0      2.6000
      2.1000     -3.2000      1.8000     -1.4000      0.3000
      1.6000           0     -1.4000      2.3000     -0.2000
      1.3000      2.6000      0.3000     -0.2000      1.5000
```

Moreover, submatrices of A can be pulled out directly. For example, we can pull out the block of A in the first four rows and last three columns to use separately.

```
>  P=A(1:4,3:5)

P  =
      2.1000      1.6000      1.3000
     -3.2000           0      2.6000
      1.8000     -1.4000      0.3000
     -1.4000      2.3000     -0.2000
```

Now, if the subassembly corresponding to G is slightly redesigned, we can replace it in the matrix. For example,

```
>  A(4:5,4:5)=[2.4 -.3;-.3 1.5]

A  =
      1.2000     -0.7000      2.1000      1.6000      1.3000
     -0.7000      0.9000     -3.2000           0      2.6000
      2.1000     -3.2000      1.8000     -1.4000      0.3000
      1.6000           0     -1.4000      2.4000     -0.3000
      1.3000      2.6000      0.3000     -0.3000      1.5000
```

replaces the block G by the slightly modified block

$$H = \begin{pmatrix} 2.4 & -.3 \\ -.3 & 1.5 \end{pmatrix}$$

You can see this would be a big savings in effort if A were 1000×1000 and the part of A being changed was only 15×15! □

With this motivation, we make the more formal definitions.

DEFINITION Suppose A is matrix. We say that the matrix B is a submatrix of A if each row and column of B come from a row and column of A. If B is a square submatrix of A and the diagonal entries of B are from the diagonal of A, then B is called a principal submatrix of A.

That is, a submatrix of a matrix is a rectangle of entries from the matrix; it is a principal submatrix if the chosen entries are symmetrically placed with respect to the main diagonal.

EXAMPLE B.2
Let

$$A = \begin{pmatrix} 1 & 2 & 3 & 4 \\ 5 & 6 & 7 & 8 \\ 9 & 10 & 11 & 12 \end{pmatrix}$$

Then

$$B_1 = \begin{pmatrix} 1 & 3 \\ 9 & 11 \end{pmatrix}, \quad B_2 = \begin{pmatrix} 1 & 2 & 3 \\ 5 & 6 & 7 \end{pmatrix}, \text{ and } B_3 = \begin{pmatrix} 2 & 3 \\ 6 & 7 \\ 10 & 11 \end{pmatrix}$$

are submatrices of A and B_1 is principal, whereas

$$B_4 = \begin{pmatrix} 1 & 2 \\ 3 & 4 \end{pmatrix}$$

is not a submatrix of A; for example, the entries 1 and 3 in the first column of B_4 did not come from a single column of A.   ☐

The most important case, and one that will be adequate for almost all of our work, is viewing a matrix as being formed of its columns or rows. Each of the columns of an $m \times n$ matrix is a submatrix, as is each of the rows of the matrix, and the whole matrix is a matrix of these submatrices. As we can see from the example below, MATLAB can easily build a matrix from its columns.

```
> C1=[1;4;7]

C1 =
     1
     4
     7
```

```
> C2=[2;5;8]

C2 =

       2
       5
       8

> C3=[3;6;9]

C3 =

       3
       6
       9

> C=[C1 C2 C3]

C =

       1       2       3
       4       5       6
       7       8       9
```

and similarly with rows.

More generally, we will consider submatrices that arise by drawing vertical lines in the matrix between columns and horizontal lines between rows so that the matrix is broken into a collection of submatrices. The matrix below has some lines drawn in:

$$C = \left(\begin{array}{ccc|c} 1 & 2 & 3 & 4 \\ \hline 5 & 6 & 7 & 8 \\ 9 & 10 & 11 & 12 \end{array}\right)$$

This matrix has been broken into four submatrices

$$C_{11} = \left(\begin{array}{ccc} 1 & 2 & 3 \end{array}\right), \; C_{12} = \left(\begin{array}{c} 4 \end{array}\right), \; C_{21} = \left(\begin{array}{ccc} 5 & 6 & 7 \\ 9 & 10 & 11 \end{array}\right), \; C_{22} = \left(\begin{array}{c} 8 \\ 12 \end{array}\right)$$

and we may write C as a block (or partitioned) matrix

$$C = \left(\begin{array}{cc} C_{11} & C_{12} \\ C_{21} & C_{22} \end{array}\right)$$

We make this idea precise in the following formal definition. While the definition may be difficult to interpret, its goal is to describe carefully the process of subdividing a matrix as above. The numbers i_1, i_2, etc., in the definition indicate where the horizontal lines are placed and j_1, j_2, etc., in the definition indicate where the vertical lines are placed. In the example above, $q = r = 2$ so C was written as a 2×2 block matrix, and since the lines were placed below the first (and third) row and to the right of the third (and fourth) columns, $i_1 = 1$, $i_2 = 3$, $j_1 = 3$, and $j_2 = 4$.

DEFINITION   If A is an $m \times n$ matrix and integers $0 = i_0 < i_1 < i_2 < \cdots < i_q = m$ and $0 = j_0 < j_1 < j_2 < \cdots < j_r = n$ are given, let A_{kl} be the submatrix of A consisting of a_{ij} for $i_{k-1} < i \leq i_k$ and $j_{l-1} < j \leq j_l$ for $k = 1, 2, \ldots, q$ and $l = 1, 2, \ldots, r$. The matrix A is said to be blocked (or partitioned) as a $q \times r$ block matrix and we write

$$A = \begin{pmatrix} A_{11} & A_{12} & \cdots & A_{1r} \\ A_{21} & A_{22} & \cdots & A_{2r} \\ \vdots & \vdots & & \vdots \\ A_{q1} & A_{q2} & \cdots & A_{qr} \end{pmatrix}$$

As noted above, the most important case is obtained by partitioning the matrix into its columns or rows. To partition a matrix into its columns, we would use vertical lines between every pair of columns an no horizontal lines. Thus an $m \times n$ matrix C can be thought of as being built of n column vectors, each in $\mathbf{R}^m$. More precisely, C can be written as a $1 \times n$ block matrix where each block is a column of C. The 3×3 matrix in the example above is $C = (C_1 \ C_2 \ C_3)$ where $C_1 = (1, 4, 7)$, $C_2 = (2, 5, 8)$, and $C_3 = (3, 6, 9)$.

The main idea in this section is that it is possible to do matrix arithmetic with block matrices, that is, with matrices of matrices, completely analogously to the way arithmetic is done with matrices of numbers. For example, it is easy to see that if E and F are $m \times n$ matrices partitioned in the same way that $E + F$ is naturally partitioned and can be computed by adding the corresponding submatrices of E and F. More subtle is the fact that if E is $m \times n$ and F is $n \times p$ where $0 = i_0 < i_1 < i_2 < \cdots < i_q = m$ cut the rows of E, $0 = j_0 < j_1 < j_2 < \cdots < j_r = n$ cut the columns of E and the rows of F, and $0 = k_0 < k_1 < k_2 < \cdots < k_s = p$ cut the columns of F then the product EF is naturally partitioned by cutting its rows $0 = i_0 < i_1 < i_2 < \cdots < i_q = m$ and its columns $0 = k_0 < k_1 < k_2 < \cdots < k_s = p$ and the submatrices of EF are the sums of the products of the corresponding submatrices of E and F. When matrices E and F are partitioned in such a way that EF can be computed as a block matrix, we say they have been partitioned *compatibly* for the product.

EXAMPLE B.3

If C is as above and

$$D = \left(\begin{array}{cc} -1 & -2 \\ -3 & -4 \\ -5 & -6 \\ \hline -7 & -8 \end{array}\right) = \left(\begin{array}{c} D_{11} \\ D_{21} \end{array}\right)$$

then $E = CD$ is partitioned

$$CD = \left(\begin{array}{cc} C_{11} & C_{12} \\ C_{21} & C_{22} \end{array}\right)\left(\begin{array}{c} D_{11} \\ D_{21} \end{array}\right) = \left(\begin{array}{c} C_{11}D_{11} + C_{12}D_{21} \\ C_{21}D_{11} + C_{22}D_{21} \end{array}\right) = \left(\begin{array}{c} E_{11} \\ E_{21} \end{array}\right) = E$$

where

$$E_{11} = C_{11}D_{11} + C_{12}D_{21}$$

$$= \left(\begin{array}{ccc} 1 & 2 & 3 \end{array}\right)\left(\begin{array}{cc} -1 & -2 \\ -3 & -4 \\ -5 & -6 \end{array}\right) + \left(\begin{array}{c} 4 \end{array}\right)\left(\begin{array}{cc} -7 & -8 \end{array}\right) = \left(\begin{array}{cc} -50 & -60 \end{array}\right)$$

and

$$E_{21} = C_{21}D_{11} + C_{22}D_{21}$$

$$= \left(\begin{array}{ccc} 5 & 6 & 7 \\ 9 & 10 & 11 \end{array}\right)\left(\begin{array}{cc} -1 & -2 \\ -3 & -4 \\ -5 & -6 \end{array}\right) + \left(\begin{array}{c} 8 \\ 12 \end{array}\right)\left(\begin{array}{cc} -7 & -8 \end{array}\right) = \left(\begin{array}{cc} -114 & -140 \\ -178 & -220 \end{array}\right)$$

So

$$CD = \left(\begin{array}{cc} -50 & -60 \\ \hline -114 & -140 \\ -178 & -220 \end{array}\right)$$

(Check by multiplying the usual way!) □

The observation of Exercise 1.3.7. can be shown to be generally true. We state the result formally and prove the result with block matrices. One of the goals of considering block matrices is to make this kind of calculation and discovery seem natural.

THEOREM B.4

Suppose A is an $m \times n$ matrix and B is an $n \times p$ matrix. Then the columns of AB are A times the columns of B and the rows of AB are the rows of A times B.

PROOF   We will prove the assertion about columns; the conclusion about rows is Exercise B.0.4.

Partition the $n \times p$ matrix B into its columns, that is,

$$B = \begin{pmatrix} B_1 & B_2 & \cdots & B_p \end{pmatrix}$$

where B_j is the jth column of B. This means that we are thinking of B as a $1 \times p$ block matrix. We regard the $m \times n$ matrix A as being a 1×1 block matrix, so A and B have been partitioned compatibly and AB can be computed as a $1 \times p$ block matrix. To say that AB is a $1 \times p$ block matrix is to say that it is partitioned into columns and

$$AB = A \begin{pmatrix} B_1 & B_2 & \cdots & B_p \end{pmatrix}$$
$$= \begin{pmatrix} AB_1 & AB_2 & \cdots & AB_p \end{pmatrix}$$

In other words, the columns of AB are just A times the columns of B.   ∎

The observation of Exercise 1.3.6. can also be shown to be generally true: If a matrix is multiplied on the left by a diagonal matrix, the jth row of the product is the jth diagonal entry times the the jth row of the original matrix. (Exercise B.0.5. describes the analogous situation for columns.)

EXAMPLE B.5

Suppose D is an $m \times m$ diagonal matrix with diagonal entries $d_1, d_2, \cdots,$ and d_m. If A is an $m \times n$ matrix with rows $A_1, A_2, \cdots,$ and A_m, then DA has rows $d_1 A_1$, $d_2 A_2, \cdots,$ and $d_m A_m$.

Rather than prove this in general, we will write out the details for $m = 4$. The general proof is analogous.

When $m = 4$,

$$D = \begin{pmatrix} d_1 & 0 & 0 & 0 \\ 0 & d_2 & 0 & 0 \\ 0 & 0 & d_3 & 0 \\ 0 & 0 & 0 & d_4 \end{pmatrix}$$

and A, blocked in rows, is

$$A = \begin{pmatrix} A_1 \\ A_2 \\ A_3 \\ A_4 \end{pmatrix}$$

Now D is blocked 4×4 (with each entry a number) and A is blocked 4×1 with each entry a row vector. It follows that D and A are blocked compatibly and that DA is defined and the product can be computed in block form with the result being an $m \times 1$ block matrix with each entry being a sum of objects written as number times row vector, that is, each entry is a row vector. Explicitly, the first row of DA, found by combining the first row of D with the first column of A, is

$$d_1 A_1 + 0A_2 + 0A_3 + 0A_4 = d_1 A_1$$

Similarly, the second row of DA is found by combining the second row of D with the first column of A to get

$$0A_1 + d_2 A_2 + 0A_3 + 0A_4 = d_2 A_2$$

Similarly, the third row of DA is

$$0A_1 + 0A_2 + d_3 A_3 + 0A_4 = d_3 A_3$$

and the fourth row is

$$0A_1 + 0A_2 + 0A_3 + d_4 A_4 = d_4 A_4$$

Putting this all together, we get

$$DA = \begin{pmatrix} d_1 & 0 & 0 & 0 \\ 0 & d_2 & 0 & 0 \\ 0 & 0 & d_3 & 0 \\ 0 & 0 & 0 & d_4 \end{pmatrix} \begin{pmatrix} A_1 \\ A_2 \\ A_3 \\ A_4 \end{pmatrix} = \begin{pmatrix} d_1 A_1 \\ d_2 A_2 \\ d_3 A_3 \\ d_4 A_4 \end{pmatrix}$$

as the statement above asserts. $\quad\square$

EXAMPLE B.6

Show that if A is an $m \times n$ matrix with a row of zeros and B is any $n \times p$ matrix, then AB has a row of zeros.

SOLUTION Partition A into rows, and suppose the ith row is zero:

$$A = \begin{pmatrix} A_1 \\ A_2 \\ \vdots \\ A_i \\ \vdots \\ A_m \end{pmatrix} = \begin{pmatrix} A_1 \\ A_2 \\ \vdots \\ 0 \\ \vdots \\ A_m \end{pmatrix}$$

Then

$$
AB = \begin{pmatrix} A_1 B \\ A_2 B \\ \vdots \\ 0B \\ \vdots \\ A_m B \end{pmatrix} = \begin{pmatrix} A_1 B \\ A_2 B \\ \vdots \\ 0 \\ \vdots \\ A_m B \end{pmatrix},
$$

so the ith row of AB is zero. $\square$

Exercises B.0

1. Let $A = \begin{pmatrix} 4 & 3 & -2 \\ 2 & -5 & 6 \end{pmatrix}$ and let $B = \begin{pmatrix} 0 & -1 & 3 \\ 2 & -1 & 6 \\ 5 & 2 & 1 \end{pmatrix}$.

   (a) Find AB (from the definition) as a 2×3 matrix.

   (b) Partition A as $\begin{pmatrix} A_{11} & | & A_{12} \end{pmatrix}$ and B as $\left(\begin{array}{c|c} B_{11} & B_{12} \\ \hline B_{21} & B_{22} \end{array} \right)$, where both A_{11} and B_{11} are 2×2 matrices, that is, say what each of $A_{11}, A_{12}, \cdots, B_{22}$ are.

   (c) Determine each of the relevant products from (b) above and find AB as a partitioned matrix.

$\Diamond$ 2.   Explore how your software handles block matrices.

   (a) Enter the matrices

   $$
   A = \begin{pmatrix} 1 & 2 & -1 \\ 0 & -1 & 4 \end{pmatrix}, \quad B = \begin{pmatrix} 1 & -1 \\ 3 & 1 \end{pmatrix}
   $$

   $$
   C = \begin{pmatrix} 4 & 2.6 & 0 \\ 3 & -.3 & 8 \end{pmatrix}, \quad D = \begin{pmatrix} 3 & -2 \\ 1.5 & 4 \end{pmatrix}
   $$

   (b) Make a 4×5 matrix E from the matrices $A, B, C,$ and D to get

   $$
   E = \left(\begin{array}{ccc|cc} 1 & 2 & -1 & 1 & -1 \\ 0 & -1 & 4 & 3 & 1 \\ \hline 4 & 2.6 & 0 & 3 & -2 \\ 3 & -.3 & 8 & 1.5 & 4 \end{array} \right)
   $$

   You probably do not need to retype all the entries! Note that $E = \begin{pmatrix} A & B \\ C & D \end{pmatrix}$

   (c) Make a 4×4 matrix F from E by deleting the last column of E.

3. Show that if A is an $m \times n$ matrix and B is an $n \times p$ matrix whose k^{th} column is zero, then the kth column of AB is zero.

4. Prove the assertion of Theorem B.4 that the rows of AB are the rows of A times B.

5. Prove: If a matrix is multiplied on the right by a diagonal matrix, the jth column of the product is the jth diagonal entry times the the jth column of the original matrix. (Compare with Example B.5 and Exercise 1.3.6.)

6. Suppose A is a square matrix partitioned as

$$A = \left(\begin{array}{c|c} X & Y \\ \hline 0 & Z \end{array} \right)$$

where X and Z are square invertible matrices and 0 is a zero matrix.

  (a) Find formulas for $P, Q, R,$ and S so that the block matrix

$$\left(\begin{array}{c|c} P & Q \\ \hline R & S \end{array} \right)$$

is A^{-1}. (*Caution:* matrix multiplication is not commutative!) If you are successful, you will have shown that matrices with the given block form are invertible.

  (b) Use your formula to find A^{-1} when $X = (\, -1 \,), Y = (\, 1 \quad -1 \,)$, and $Z = \begin{pmatrix} 1 & 1 \\ 2 & 3 \end{pmatrix}$

(Note that $Z^{-1} = \begin{pmatrix} 3 & -1 \\ -2 & 1 \end{pmatrix}$)

7.(a) Suppose A is a square matrix partitioned as $A = \left(\begin{array}{c|c} X & Y \\ \hline 0 & Z \end{array} \right)$, where X and Z are square and 0 is a zero matrix. Consider using row operations to evaluate the determinant of A, then explain why

$$\det A = (\det X)(\det Z)$$

  (b) Use part (a) and Problem 2.6.7. to find the determinant of

$$A = \begin{pmatrix} -1 & 3 & -4 & 2 & 1 & -1 \\ 0 & -1 & 1 & 3 & -1 & 1 \\ 3 & 0 & -2 & -1 & 4 & -3 \\ 2 & -1 & 4 & 3 & -2 & 0 \\ 0 & 0 & 0 & 0 & 3 & -2 \\ 0 & 0 & 0 & 0 & 5 & 1 \end{pmatrix}$$

  (c) Give an example to show that if A is a square matrix partitioned as $A = \left(\begin{array}{c|c} P & Q \\ \hline R & S \end{array} \right)$, then it is *not always true* that

$$\det A = (\det P)(\det S) - (\det Q)(\det R)$$

Chapter 1

Section 1.2, page 17

1.2.1. (a) $\begin{pmatrix} 4 \\ -4 \\ 12 \end{pmatrix}$ (c) $\begin{pmatrix} 4 \\ 0 \\ 4 \end{pmatrix}$ (e) $\begin{pmatrix} 11 \\ -7 \\ 13 \end{pmatrix}$

1.2.2. (a) $\begin{pmatrix} 4 \\ 3 \\ -6 \\ -7 \end{pmatrix}$ (d) $\begin{pmatrix} 2\alpha + \beta + 2\gamma \\ \alpha \\ 3\beta + 6\gamma \\ -3\alpha - \beta - 2\gamma \end{pmatrix}$

1.2.3. (a) $\alpha = 2, \beta = -3$ (b) Not possible to write $(3, -1, 2, 0)$ in this way.
(c) $\alpha = -2.5, \beta = 5$ (d) Many possibilities: $\alpha = -1, \beta = 0$; or $\alpha = 0, \beta = -1/2$; or $\alpha =$

1.2.4. (a) $\begin{pmatrix} 0 & 6 & -3 \\ 9 & 3 & 3 \end{pmatrix}$ (c) $\begin{pmatrix} -1 & 2 & 3 \\ 4 & 2 & -1 \end{pmatrix}$ (e) $\begin{pmatrix} 0 & 3 \\ 2 & 1 \\ -1 & 1 \end{pmatrix}$ (h) $\begin{pmatrix} 2 & 7 \\ 6 & 1 \\ -11 & 7 \end{pmatrix}$

Section 1.3, page 30

1.3.2. (a) $\begin{pmatrix} 5 & 1 \\ 1 & 11 \end{pmatrix}$ (b) $\begin{pmatrix} 9 & 3 & 3 \\ 3 & 5 & -1 \\ 3 & -1 & 2 \end{pmatrix}$

1.3.3. (a) $\begin{pmatrix} 3 & -1 & -2 \\ 2 & 1 & -7 \end{pmatrix}$ (c) $\begin{pmatrix} 0 & 0 \\ 0 & 0 \end{pmatrix}$ (e) $\begin{pmatrix} 1 & -2 \\ 1 & -2 \end{pmatrix}$ (f) undefined

1.3.4. (a) $\begin{pmatrix} 11 \\ 35 \\ 36 \end{pmatrix}$ (d) $\begin{pmatrix} 5 \\ 12 \\ 10 \end{pmatrix}$

1.3.5. (b) $S = S'$, so S is Hermitian. (d) $S^3 = \begin{pmatrix} -6 & 21 & 0 \\ 21 & 33 & 57 \\ 0 & 57 & 45 \end{pmatrix}$

1.3.9. $E^{-1} = \begin{pmatrix} \frac{1}{3} & \frac{1}{3} \\ \frac{2}{3} & -\frac{1}{3} \end{pmatrix}$

Chapter 2

Section 2.2, page 45

2.2.1. $w = 5, x = -8, y = 2,$ and $z = -13$

2.2.4. $w = -3z/2, x = 0, y = z,$ and z is arbitrary

2.2.5. (a)
$$\begin{pmatrix} 1 & 0 & 0 & 0 \\ 2 & 1 & 0 & 0 \\ 1 & 1 & 1 & 0 \\ 1 & -1 & 2 & 1 \end{pmatrix} \begin{pmatrix} w \\ x \\ y \\ z \end{pmatrix} = \begin{pmatrix} 5 \\ 2 \\ -1 \\ 4 \end{pmatrix}$$

2.2.6. $x = -8, y = 5.5$

2.2.11. (a) Any multiple of $(x_1 - x_2)$, for example, $5(x_1 - x_2) = (5, 5, -5, 0, 10)$

2.2.12. (b) $X = 2Y$ solves $AX = 2b$

2.2.13. (a) $X = (1, 0, 0, 0)$ (b) $X = (0, 1, 0, 0)$ (e) $X = (1, 2, -1, 1)$

Section 2.3, page 73

2.3.1. $a = 2$ and $b = -1$ is the unique solution.

2.3.5. There are infinitely many solutions: $x = \frac{2}{3}y + \frac{7}{3}$ and y is arbitrary. For example, $x = \frac{7}{3}, y = 0$; $x = 3, y = 1$; and $x = 1, y = -2$ are solutions.

2.3.6. $x = 3$ and $y = -2$ is the unique solution.

2.3.7. There are no solutions to this system.

2.3.8. $x = 3, y = 2$ and $z = 1$ is the unique solution.

2.3.12. There are no solutions to this system.

2.3.13. The solution of the system is $x = 4z - 2$, $y = -3z + 1$, and z is arbitrary.

2.3.14. The only solution of the system is $x = 0$, $y = 0$, and $z = 0$.

2.3.20. $w = 1$, $x = -1$, $y = 5$, and $z = -1$ and it is unique.

2.3.27. $X = (-0.5649, -0.5322, 0.5063, -0.9967, 0.9654, 0.0300)$

Section 2.4, page 81

2.4.2. $\begin{pmatrix} 1 & 0 & 0 \\ 1 & 1 & 0 \\ -1 & 1 & 1 \end{pmatrix}$

2.4.8. The given matrix is not invertible.

2.4.13. $\begin{pmatrix} 36 & -3 & -10 & -12 \\ 29 & -3 & -8 & -10 \\ 11 & -1 & -3 & -4 \\ -6 & 0 & 2 & 2 \end{pmatrix}$

2.4.14. $\begin{pmatrix} \frac{3}{2} & -2 & -\frac{1}{2} & -1 \\ -\frac{3}{2} & -3 & \frac{1}{2} & -1 \\ 1 & 3 & 0 & 1 \\ -3 & 4 & 1 & -1 \end{pmatrix}$

2.4.16. (a) A does not have a right inverse.

$$\begin{pmatrix} 2 & -1 & 0 \\ 1 & -1 & 0 \end{pmatrix} \text{ and } \begin{pmatrix} 0 & -3 & 2 \\ 0 & -2 & 1 \end{pmatrix}$$

are left inverses of A as well as many other matrices.

(b) E does not have a left inverse.

Section 2.5, page 90

2.5.4. $\begin{pmatrix} 1 & 2 \\ 3 & 4 \end{pmatrix} = \begin{pmatrix} 1 & 0 \\ 3 & 1 \end{pmatrix} \begin{pmatrix} 1 & 2 \\ 0 & -2 \end{pmatrix}$

2.5.8. $\begin{pmatrix} -2 & 0 & -1 & 0 \\ 2 & 1 & 1 & 0 \\ 6 & 2 & 6 & 1 \\ 4 & 1 & 3 & 0 \end{pmatrix} \begin{pmatrix} 1 & 0 & -\frac{3}{2} & 0 & \frac{5}{2} \\ 0 & 1 & 2 & 0 & -1 \\ 0 & 0 & 0 & 1 & -2 \\ 0 & 0 & 0 & 0 & 0 \end{pmatrix}$

Section 2.6, page 103

2.6.2. -6

2.6.4. -12

2.6.6. -80

2.6.7. -142

2.6.9. 114

2.6.11. $\dfrac{1}{25}\begin{pmatrix} 7 & 1 \\ -4 & 3 \end{pmatrix}$

2.6.17. $x = 2,\, y = 3,\, z = -1$

Section 2.7, page 122

2.7.7.   kitchen: \$105,570, novelty: \$69,680, toy: \$154,740

2.7.8.   Yolanda: \$43.60, Zeke: \$38.40

2.7.10. pill: \$565,300

2.7.12. $1,715,884,494,940$

2.7.13. Acme: 600; Better Biscuit: 300; Income: \$1980

Chapter 3

Section 3.2, page 135

3.2.2.   not closed under addition

Section 3.3, page 145

3.3.2.   yes, that is the meaning of linear combination: we can take $k = 2,\, c_1 = 2,$
$v_1 = u,\, c_2 = -3,$ and $v_2 = v$ in the definition on page 136.

3.3.4. $w = x + 2y + 0z$

3.3.5. $y = 2w + 5x$

3.3.6. $(1, -1, 5, -5) = -3(1, 1, -1, 1) + 2(2, 1, 1, -1)$

3.3.7. $(1, -1, 1, 2)$ is not a linear combination of $(1, 1, -1, 1)$ and $(2, 1, 1, -1)$.

3.3.8. $3(1, 0, 1, 1) - 2(0, 1, 1, -1) - (2, 1, 1, 3) = (1, -3, 0, 2)$

3.3.10. For example, $u = (-3, 2, 0, 1)$ and $v = (5, -3, 0, 1)$

3.3.13. Yes, they agree. $v_1 = u_1 + u_2$, $v_2 = 2u_1 + u_2$, and $v_3 = u_1 - u_2$ so every vector that can be written as a linear combination of the v's can also be written as a linear combination of the u's. Conversely, $u_1 = -v_1 + v_2$ and $u_2 = 2v_1 - v_2$, so every vector that can be written as a linear combination of the u's can also be written as a linear combination of the v's (actually of v_1 and v_2).

Section 3.4, page 155

3.4.2. This calculation does not show that the vectors are independent nor does it show they are dependent: $0u + 0v + 0w = 0$ is true for every three vectors u, v, and w, whether they are dependent or independent.

3.4.4. dependent: $(2, 3, 1) + 2(-1, 1, 2) - 5(0, 1, 1) = (0, 0, 0)$

3.4.5. independent

3.4.7. dependent: $-2(1, 1, -1, 1) + (2, 1, -1, 2) + (1, -1, 1, 1) - (1, -2, 2, 1) = (0, 0, 0, 0)$

3.4.8. independent

3.4.12. independent

3.4.14. dependent

3.4.16. $v_4 = 2v_1 + 3v_2 - v_3$

3.4.17. $u = v_1 - v_2 + 2v_3 - 2(2v_1 + 3v_2 - v_3) = -3v_1 - 7v_2 + 4v_3$

Section 3.5, page 165

3.5.3. There are many bases; one is $\{(1, 0, 3), (0, 1, 2)\}$

3.5.5. (a) $(1 \ -1 \ 1 \ 0 \ -2)$ and $(2 \ 0 \ 1 \ 1 \ -1)$ form a basis for the row space
        (b) $(1, 2, 1)$ and $(-1, 0, -3)$ form a basis for the column space.

3.5.7. $-3(1, 0, 0) + 2(2, 1, 0) + (1, -2, 1) = (2, 0, 1)$

3.5.9. (b) coordinates: $(1, -2, -1.5)$, that is, $(0, 0, 1) = 1(1, 2, 1) - 2(-1, 1, 0) - 1.5(2, 0, 0)$

3.5.11. $u = 2w - 3v$, so $z = -7v + 5w$ works

3.5.14. $e_1 = .5u + .5v$ so $w = e_1$ will not work; but $w = e_2$ or $w = e_3$ will work.

Section 3.6, page 182

3.6.1. do not span: $(7, -3, 1)$ is not in the span of these vectors

3.6.2. span

3.6.5. dimension is 2

3.6.6. dimension is 3

3.6.9. $(1, 2)$ and $(-1, 1)$ form a basis for the range, so its dimension is 2. The nullspace is just $\{0\}$, so it has no basis and so its dimension is zero.

3.6.10. $(-1, 0, 2)$ is a basis for the range, so its dimension is 1. $(1, 2)$ is a basis for the nullspace, so its dimension is 1.

3.6.13. $(1, 2, 1, 0)$, $(-1, 1, 1, -1)$, and $(1, 0, 3, 4)$ form a basis for the range, so its dimension is 3. $(3, 4, 1, -10)$ is a basis for the nullspace, so its dimension is 1.

3.6.15. $(1, 2, -1, 1)$ is a basis for the range, so its dimension is 1. $(1, 0, 0, -2)$, $(0, 1, 0, 1)$, and $(0, 0, 1, 0)$ form a basis for the nullspace, so its dimension is 3.

3.6.21. (a) (i) (b) (iv)

Section 3.7, page 194

3.7.1. A basis for the column space is $(-2, 4)$, a basis for the nullspace is $(1, 2)$, a basis for the column space of the adjoint is $(-2, 1)$, and a basis for the nullspace of the adjoint is $(2, 1)$. The dimension of each of these subspaces is 1.

3.7.3. dimensions of column space and row space are 2, dimension of nullspace is 1, dimension of nullspace of adjoint is 0.

3.7.5. dimensions of column space and row space are 2, dimension of nullspace is 2, dimension of nullspace of adjoint is 1.

3.7.7. A basis for the nullspace is $(4, -1, 5, 2)$. Dimensions of the column space and row space are 3, dimensions of the nullspace of the matrix and its adjoint are 1.

3.7.9. The dimension of the range of A is $9 - 4 = 5$, so the dimension of the range of A' is 5, and the dimension of nullspace A' is $5 - 5 = 0$.

3.7.13. John and Mary agree and the dimension of the subspace is 2

Section 3.8, page 203

3.8.1. (a) (α) ii, iv (β) i, iv (γ) v

3.8.4. (a) $\dim \mathcal{N}(C) = 2$, $\dim(\mathcal{R}(C)) = 2$, $\dim(\mathcal{R}(C')) = 2$, $\dim(\mathcal{N}(C')) = 1$

3.8.6. (a) $3t^2 + 2t - 1$

3.8.8. (a) true (c) sometimes true, sometimes false (d) true.

Chapter 4

Section 4.2, page 219

4.2.1. $\|v\| = \sqrt{6}$; $\langle v, w \rangle = -2$

4.2.5. $\|v\| = \sqrt{7}$; $\langle v, w \rangle = 5 - 8i$

4.2.6. $\|v\| = \sqrt{\pi}$; $\langle v, w \rangle = 0$

4.2.7. $\theta = .9303$ (radians)

4.2.12. $\theta = \pi/2$ (radians); the vectors are orthogonal

4.2.23. (b) $v = \frac{1}{2}w_1 + 2w_2 + \frac{1}{2}w_3$

4.2.24. (b) $\sqrt{170}$

4.2.28. (b) $w = v_1 + \frac{1}{2}v_2 - \frac{1}{2}v_3 + v_4$

4.2.29. (b) By Theorem 4.9, w is not in $\text{span}(\{v_1, v_2\})$
(c) $z = (1, -4, -5)$ or any multiple of z

Section 4.3, page 231

4.3.4.

$$u_1 = \frac{1}{\sqrt{3}}\begin{pmatrix} 1 \\ 1 \\ -1 \end{pmatrix},\ u_2 = \frac{1}{\sqrt{78}}\begin{pmatrix} 5 \\ 2 \\ 7 \end{pmatrix},\ u_3 = \frac{1}{\sqrt{26}}\begin{pmatrix} 3 \\ -4 \\ -1 \end{pmatrix}$$

4.3.6.

$$u_1 = \frac{1}{\sqrt{2}}\begin{pmatrix} 1 \\ 0 \\ -1 \end{pmatrix},\ u_2 = \frac{1}{\sqrt{6}}\begin{pmatrix} 1 \\ 2 \\ 1 \end{pmatrix}$$

4.3.11.   (a) $u_1 = \dfrac{1}{\sqrt{2}}(1,1,0,0),\ u_2 = \dfrac{1}{\sqrt{6}}(-1,1,2,0),\ u_3 = \dfrac{1}{\sqrt{12}}(1,-1,1,3)$

Section 4.4, page 243

4.4.3. $w = (7/3, 4/3, 1/3, -1)$ and $u = (2/3, -1/3, -1/3, 1)$

4.4.4.   (b) $w = (-4/5, -1/5, -3/5, 3/5)$ and $u = (-1/5, 1/5, 8/5, 7/5)$

4.4.8.   (a) $(1, 0, 2)$ and $(0, 1, -1)$. In fact, any pair of columns of A will form a basis for the column space.
(b) $(1, 0, 3, -1)$ and $(0, 1, 1, 2)$
(c) v_1 and v_2
(d) $(-2, 1, 1)$

4.4.9.   (a) 6 (b) 6 (c) 2 (d) 5

Section 4.5, page 255

4.5.6. $w = (0.2222, 0.3889, 0.0000, -0.1667, 0.2222)$
and $u = (0.7778, -0.3889, 0.0000, 1.1667, 0.7778)$

4.5.7. $w = (2, 2/3, -2/3, -4/3, 0),\ u = (1, 1/3, 2/3, 4/3, -1)$

4.5.12. $s_a = (1.6296, -1.3704, 0.5185, 0.7778)$
and $s_n = (-0.6296, -0.6296, -1.5185, 1.2222)$

Section 4.6, page 264

4.6.2.

$$Q = \begin{pmatrix} \dfrac{1}{\sqrt{2}} & -\dfrac{1}{\sqrt{6}} & \dfrac{1}{\sqrt{12}} \\[2mm] \dfrac{1}{\sqrt{2}} & \dfrac{1}{\sqrt{6}} & -\dfrac{1}{\sqrt{12}} \\[2mm] 0 & \dfrac{2}{\sqrt{6}} & \dfrac{1}{\sqrt{12}} \\[2mm] 0 & 0 & \dfrac{3}{\sqrt{12}} \end{pmatrix} \quad \text{and} \quad R = \begin{pmatrix} \sqrt{2} & \dfrac{\sqrt{2}}{2} & 0 & \dfrac{\sqrt{2}}{2} \\[2mm] 0 & \dfrac{\sqrt{6}}{2} & \dfrac{\sqrt{6}}{3} & -\dfrac{\sqrt{6}}{6} \\[2mm] 0 & 0 & \dfrac{\sqrt{12}}{3} & \dfrac{\sqrt{12}}{3} \end{pmatrix}$$

4.6.4. $u_1 = (-0.2294, -0.6882, -0.2294, -0.4588, -0.4588)$,
$u_2 = (0.4652, 0.0145, 0.7415, -0.4507, -0.1745)$,
$u_3 = (0.0712, 0.6574, -0.3400, -0.2202, -0.6314)$,
and $u_4 = (-0.6898, 0.3066, 0.1788, -0.5365, 0.3321)$ are a basis for the range.
$v = (0.2954, 0.1074, -0.6713, -0.6444, 0.1880)$ is a basis for the nullspace.

Section 4.7, page 270

4.7.1. $(-2, 1, -3, 3)$ is a basis for $M \cap N$ and $\dim(M + N) = 3$

Chapter 5

Section 5.2, page 280

5.2.1. (a) $(3, -1, 1)$ is not in the range of the coefficient matrix.
(b) $x = -1, y = 1$

    (c) $b_0 = (1, 1, 0)$ is the closest point of the range to $(3, -1, 1)$ and $X_0 = (-1, 1)$ is the unique solution of $AX = b_0$.

5.2.7. $a = 0.8571$, $b = -0.2857$, $c = 0.3571$

5.2.9. $x = 5$, $y = -3$; consistent

Section 5.3, page 287

5.3.1. $T = .2376v - 8.9421$

5.3.3. $\alpha = 0.3235$ and $\beta = 1.7616$

5.3.4. $y = 69/26 - 21/(13x)$

5.3.7.   (b) $R = 206.77$

5.3.9.   (c) $w = 3q_1 - q_2 + 2q_3$ and distance is $\|b - w\| = \sqrt{5}$
    (d) $X = (3, -1, 2)$

Section 5.4, page 303

5.4.1.   Closest point is $w = (2/3, -1/3, 1/3, 1)$; distance is $1/\sqrt{3}$.

5.4.2.   (c) $w = (0, 1, 1)$ and $u = (1, 1, -1)$

5.4.6.   "dawn"

5.4.8.   (a)

$$Q = \begin{pmatrix} \frac{1}{\sqrt{2}} & -\frac{3}{\sqrt{22}} & -\frac{4}{\sqrt{297}} \\ \frac{1}{\sqrt{2}} & \frac{3}{\sqrt{22}} & \frac{4}{\sqrt{297}} \\ 0 & -\frac{2}{\sqrt{22}} & \frac{12}{\sqrt{297}} \\ 0 & 0 & -\frac{11}{\sqrt{297}} \end{pmatrix} \quad \text{and} \quad R = \begin{pmatrix} \sqrt{2} & \frac{1}{\sqrt{2}} & \frac{1}{\sqrt{2}} \\ 0 & \frac{\sqrt{22}}{2} & \frac{1}{\sqrt{22}} \\ 0 & 0 & \frac{\sqrt{297}}{11} \end{pmatrix}$$

    (b) $x = -0.4074$, $y = 0.5185$, $z = 1.2963$

Chapter 6

Section 6.2, page 323

6.2.1. u is an eigenvector for the eigenvalue $\lambda = 4$; v is not an eigenvector.

6.2.3. $(4, 1)$ is an eigenvector for the eigenvalue 3.

6.2.5. $(2, -1, 1)$ is a basis for the eigenspace $\lambda = -1$
$(-2, 1, 0)$ and $(-1, 0, 1)$ form a basis for the eigenspace $\lambda = 2$.

6.2.7. $(3, 1)$ is a basis for the eigenspace $\lambda = 3$
$(1, 1)$ is a basis for the eigenspace $\lambda = 5$

6.2.10. $(2, 1, -2)$ is a basis for the eigenspace $\lambda = 5$
$(1, -2, 0)$ and $(0, 2, 1)$ form a basis for the eigenspace $\lambda = -4$.

6.2.11. $\lambda = 1$ is the only eigenvalue and $(1, 0, 1)$ is a basis for that eigenspace.

6.2.16. $\lambda = 3$ is an eigenvalue and $(2, 0, 1)$ is a basis for that eigenspace
$\lambda = 0$ is an eigenvalue and $(1, -1, 1)$ is a basis for that eigenspace
$\lambda = -2$ is an eigenvalue and $(1, 2, -1)$ is a basis for that eigenspace

Section 6.3, page 337

6.3.1. (c) eigenvalue -3; $Y(t) = e^{-3t}(2, -3)$,
that is, $y_1(t) = 2e^{-3t}$ and $y_2(t) = -3e^{-3t}$
(d) $y_1(t) = -2e^{2t} + 2e^{-3t}$ and $y_2(t) = e^{2t} - 3e^{-3t}$.

6.3.3. (a) $w_1 = (1, -2)$ is a basis for the eigenspace $\lambda = 2$
and $w_2 = (1, 1)$ is a basis for the eigenspace $\lambda = -1$.
(b) $u(t) = -(8/3)e^{2t} - (1/3)e^{-t}$ and $v(t) = (16/3)e^{2t} - (1/3)e^{-t}$.

6.3.5. The eigenvalues of the coefficient matrix are -2 and -5.
$y_1(t) = 18e^{-2t} - 17e^{-5t}$ and $y_2(t) = -12e^{-2t} + 17e^{-5t}$ is the solution of
the initial value problem.

6.3.7. The eigenvalues of the coefficient matrix are 0 and 1.
$u(t) = 5 - 2e^t$, $v(t) = -7 + 3e^t$, and $w(t) = 3 - e^t$ is the solution of the
initial value problem.

6.3.10. $y(t) = (9/2)e^{-3t} - (7/2)e^{-5t}$

Section 6.4, page 351

6.4.1. $\mathrm{Re}(y(t)) = e^{-2t}\cos(3t)$ and $\mathrm{Im}(y(t)) = e^{-2t}\sin(3t)$

6.4.3. $y_1(t) = -e^{2t}\cos(3t) + 4e^{2t}\sin(3t)$ and $y_2(t) = e^{2t}\sin(3t) + 4e^{2t}\cos(3t)$

6.4.8. $y(t) = e^{-t}(2\cos(2t) + .5\sin(2t))$

Section 6.5, page 362

6.5.1. (a) $-4u$ (b) $11v$ (d) $-12u - 22v + 44w$

6.5.4. $(-1/3)w$

6.5.6. P has eigenvalues 0 and/or 1.

6.5.8.

$$S = \begin{pmatrix} 1 & 2 & 1 \\ -1 & -3 & -2 \\ 1 & 2 & 0 \end{pmatrix} \quad \text{and} \quad D = \begin{pmatrix} -1 & 0 & 0 \\ 0 & 1 & 0 \\ 0 & 0 & 2 \end{pmatrix}$$

Section 6.6, page 369

6.6.2. $y_1(t) = 2 + 3t - \frac{1}{2}t^2 + \frac{1}{6}t^3$

6.6.3. $e^{tA} = I + tA + \frac{t}{2}A^2$

6.6.4.

$$\begin{pmatrix} .8e^{3t} + .2e^{-2t} & .8e^{3t} - .8e^{-2t} \\ .2e^{3t} - .2e^{-2t} & .2e^{3t} + .8e^{-2t} \end{pmatrix}$$

Section 6.7, page 373

6.7.1. The first two components of the solution are $y_1(t) = e^{3t}(1 + t + \frac{t^2}{2})$ and $y_2(t) = e^{3t}(-2t - t^2)$

Chapter 7

Section 7.1, page 387

7.1.7. One solution is

$$U = \begin{pmatrix} \frac{1}{\sqrt{5}} & \frac{4}{\sqrt{45}} & \frac{2}{3} \\ 0 & \frac{5}{\sqrt{45}} & -\frac{2}{3} \\ -\frac{2}{\sqrt{5}} & \frac{2}{\sqrt{45}} & \frac{1}{3} \end{pmatrix} \quad \text{and} \quad \Lambda = \begin{pmatrix} 5 & 0 & 0 \\ 0 & 5 & 0 \\ 0 & 0 & -4 \end{pmatrix}$$

7.1.8. (b) One solution is

$$U = \begin{pmatrix} \frac{1}{\sqrt{2}} & 0 & -\frac{1}{2} & \frac{1}{2} \\ 0 & \frac{1}{\sqrt{2}} & -\frac{1}{2} & -\frac{1}{2} \\ \frac{1}{\sqrt{2}} & 0 & \frac{1}{2} & -\frac{1}{2} \\ 0 & \frac{1}{\sqrt{2}} & \frac{1}{2} & -\frac{1}{2} \end{pmatrix}$$

7.1.12. (a)

$$U = \begin{pmatrix} \frac{1}{\sqrt{2}} & \frac{1}{2} \\ \frac{i}{\sqrt{2}} & -\frac{i}{\sqrt{2}} \end{pmatrix} \quad \text{and} \quad D = \begin{pmatrix} 2+i & 0 \\ 0 & 2-i \end{pmatrix}$$

(b) For example, if $p(x) = 2 + ix$, and

$$A = \begin{pmatrix} 0 & -i \\ i & 0 \end{pmatrix}$$

then $p(A) = G$

Appendix A

page 398

A.0.1. (a) 4   (b) -5 *(not $-5i$!)*   (c) $\sqrt{41}$ (d) $4 + 5i$.

A.0.2.   (a) $7 - i$ (b) $4 - 58i$ (c) $-5 + 2i$ (d) $-5 - 2i$ [*i.e. the conjugate of (c)*]
         (e) $\frac{3}{25} - \frac{4}{25}i$ (f) $1 - 3i$ (g) $.26 + 1.18i$ (h) $\frac{|1+3i|}{|2-i|} = \frac{\sqrt{10}}{\sqrt{5}} = \sqrt{2}$

A.0.3.   The roots are 1, $1 + 2i$, and $1 - 2i$.

Appendix B

page 408

B.0.6.   (b) $\left(\begin{array}{c|cc} -1 & 5 & -2 \\ \hline 0 & 3 & -1 \\ 0 & -2 & 1 \end{array} \right)$

B.0.7.   (b) -1846